THE DARWINIAN HERITAGE
AND SOCIOBIOLOGY

THE DARWINIAN HERITAGE AND SOCIOBIOLOGY

Edited by
Johan M.G. van der Dennen,
David Smillie, and Daniel R. Wilson

Human Evolution, Behavior, and Intelligence
Seymour W. Itzkoff, Series Editor

PRAEGER

Westport, Connecticut
London

Library of Congress Cataloging-in-Publication Data

The Darwinian heritage and sociobiology / edited by Johan M.G. van der
    Dennen, David Smillie, and Daniel R. Wilson
        p. cm.—(Human evolution, behavior, and intelligence, ISSN
    1063–2158)
    "These papers were presented at a conference of the European
    Sociobiological Society which held its 18th annual meeting at Christ
    College, Cambridge in 1995"—Introd.
    Includes bibliographical references (p.) and index.
    ISBN 0–275–96436–1 (alk. paper)
    1. Sociobiology—Congresses. 2. Social Darwinism—Congresses.
I. Dennen, J. van der. II. Smillie, David, 1925– . III. Wilson, Daniel R.,
1956– . IV. Series.
HM106.D24   1999
304.5—dc21   99–011537

British Library Cataloguing in Publication Data is available.

Library of Congress Catalog Card Number: 99–011537
ISBN: 0–275–96436–1
ISSN: 1063–2158

First published in 1999

Praeger Publishers, 88 Post Road West, Westport, CT 06881
An imprint of Greenwood Publishing Group, Inc.
www.praeger.com

Printed in the United States of America

∞™

The paper used in this book complies with the
Permanent Paper Standard issued by the National
Information Standards Organization (Z39.48–1984).

10 9 8 7 6 5 4 3 2

We dedicate this volume to Jan Wind, the founding father and subsequently the secretary of the European Sociobiological Society, who died October 30, 1995. The 1995 Cambridge meeting was his last ESS conference.

Contents

Contents

Introduction

Charles Darwin's epic account of evolution *On the Origin of Species* appeared almost 150 years ago, a long time in the light of the dramatic changes that have been taking place in biology in recent years. Darwin was both a wide-ranging empiricist and a remarkable theorist who radically changed the way we think about ourselves and the natural world. In addition, he explored a wide variety of intellectual topics outside the sciences, touching on philosophy and the humanities in addition to his theories of nature. His work and his life continue to be inspirational today both within biology and, most recently, within the social sciences and even, to some extent, the humanities.

The present volume reflects this inspirational quality of Darwin's work, covering, as it does, many disciplines and different perspectives. These chapters originated as papers presented at the eighteenth annual meeting of the European Sociobiological Society at Christ's College, Cambridge, in 1995. This meeting place was particularly appropriate since it was here that Darwin received his degree, shortly before he departed on his trip around the world on the *Beagle* as a naturalist in 1831. It was on this trip that Darwin began to collect the information that ultimately culminated in his seminal work on evolutionary theory.

With respect to the conference, just a few words about the new field of sociobiology are in order. In 1975 E. O. Wilson produced a book titled *Sociobiology: The New Synthesis* that examined the nature of social organization in a wide range of animals from colonial microorganisms to humans. This work was stimulated by his earlier study of the ecology and evolution of social insects. In *Sociobiology* Wilson not only surveyed the field of animal sociality, but also spelled out relevant evolutionary principles to account for the evolution of these taxa and even predicted that such themes would come to encompass the disciplines in the social sciences, gradually replacing traditional nonbiological disciplines.

When his book appeared, it initiated a good deal of controversy in the United States since it seemed to some to imply genetic determinism applied to human behavior. It is also no longer clear whether the term ''sociobiology'' covers Wilson's broad field of cross-species investigation or should be confined to the application of evolutionary principles to humans. Such terms as ''behavioral ecology'' and ''comparative psychology'' have come to be preferred to cover Wilson's broad field of cross-species comparisons. Elsewhere in the world, ''sociobiology'' is still widely utilized to cover such work where it is directed toward understanding human nature in its biological and social dimensions. In the present volume the term seems quite appropriate given the sponsorship of the European Sociobiological Society. We are at present in a good position to gain an idea of the extent to which Wilson's prediction that a new biological, evolutionary perspective will encompass former social science disciplines is being accomplished.

In the present volume we find contributions from sociobiologists in anthropology, sociology, political science, psychology, and the humanities. These chapters are interwoven in an interesting fashion, given that they have come from a variety of different cultural contexts in Europe, America, and the Middle East. While these efforts represent the work of scholars pursuing their projects quite independently, they can be related to one another in a meaningful way, reflected by the order in which they are presented. Not only do these chapters bring new perspectives from the field of Darwinian evolution, they also suggest the possibilities of integration not previously found in traditional social sciences. Biology has emerged as a foundation that can help us see how the different fields of the social sciences can be understood within the broad context of biological evolution. Humans can be seen as a part of the natural world rather than as separate from nature. This is in the spirit of Wilson and certainly is consistent with Darwin's own vision from the previous century.

This volume offers a variety of speculative themes about how Darwinian thought is finally being applied to our understanding of ourselves. It covers a broad range of topics and presents possibilities of that integration foreseen by Wilson in his highly important book on sociobiology. While a number of the themes are still somewhat speculative, they are also bold in showing how different paths may lead to a common solution even though we have not yet arrived at definite conclusions. That such speculations are interspersed with empirical studies assures us such an effort is open to the sobering effect of findings that have the potential for disproof. We offer, then, a move toward real scientific integration that may end in a different place from which it started. But this is the way of science. It seems quite unlikely that such an approach can do anything but progress. We are pleased to bring these chapters together and hope that they both will provide an introduction to the field of human sociobiology and inspire further work in this growing discipline.

We would like to express our sincere thanks to Vincent Falger and Robin Allott for the initiation and planning of the conference at Cambridge and especially to Robin, who managed the details of the conference in a gracious manner that made it a memorable occasion for all of us.

PART I

BIOLOGICAL FOUNDATIONS

This volume opens with three chapters that explore the underlying explanatory framework of sociobiology, the evolutionary path by which Darwinian natural selection might have brought into being social entities consisting of both conflicting and integrated elements. Michael J. C. Waller contrasts Richard Dawkins's idea of selfish genes as the basic units of selection with Elliott Sober's account of group selection. Waller, however, raises doubts about whether selfish genes really translate directly into organismic selection, as Dawkins proposes, or into the converse view of group selection proposed by Sober. Instead, Waller holds that selfish genes are the sole basis of a Darwinian solution to cooperative entities. David Smillie finds that Darwin himself took two different views toward natural selection, the Malthusian perspective that focused on competitive organisms and, in contrast, a concern with the generation of variation leading in a quite different direction. The latter point of view has some real advantages over the former. Peter A. Corning sees the need to encompass both competition and cooperation between various entities at different hierarchical levels, and he shows how such interactions can result in larger synergistic unities. This is a perspective that he finds applicable to human societies as well as to the rest of the biological world. In all three chapters there is a search for viewpoints on selection that will reveal how social cooperation can arise along with competition between parts.

1

Group Selection and the Selfish Gene: The Units-of-Selection Problem Revisited

Michael J. C. Waller

THE DAWKINS/SOBER DEBATE

Elliott Sober has taken Richard Dawkins to task as a primary contributor to a "gaggle of sloppy arguments against group selection" (Sober, 1993, p. 106). Although he does not entirely dismiss Dawkins's theory of the selfish gene, he suggests that it is of limited value because it cannot deal with traits arising from organismic adaptation and group adaptation. If this proposition is true, it presents a profound challenge to the thesis upon which Dawkins has built his professional reputation. As Sober points out, Dawkins is aware of these two potential sources of difficulty and has sought to dispose of them separately. The obvious reality of organismic adaptation cannot be ignored. Instead, Dawkins claims that it can readily be incorporated within the genic pale: "There are two ways of looking at natural selection, the gene's angle and that of the individual. If properly understood they are equivalent" (Dawkins, 1989, pp. viii–ix). With group adaptation Dawkins takes a very different approach. In Sober's colorful words, "Much of the effect of Williams's (1966) book and Dawkins's (1976) popularisation has been to cast the concept of group adaption into the outer darkness" (1993, p. 106). Given that in the 1989 edition of *The Selfish Gene* Dawkins describes V. C. Wynne-Edwards as an "academic heretic" who is "wrong in an unequivocal way" (p. 297) for no greater offense than sticking tenaciously to a group-selectionist viewpoint, Sober's use of language does not seem unfair.

Although Sober finds fault with both of these arguments, he is prepared to accept that at one level, genic selectionism does meet the all-encompassing claims Dawkins and Williams make for it. He acknowledges the special property of evolutionary longevity that genes uniquely enjoy. Dawkins puts the point particularly clearly: "Natural selection in its most general form means the differential survival of entities. Some entities live and others die but, in order for this selective death to have any impact on the world, an additional condition must be met. Each entity must exist in the form of lots of copies, and at least some of the entities must be potentially capable of surviving—in the form of copies—for a significant period of evolutionary time. Small genetic units have these properties: individuals, groups and species do not" (1989, p. 33). But to Sober this observation is little more than a truism. The main thrust of his argument is to reinstate natural selection as a multilevel phenomenon. He therefore dismisses the longevity issue as either a trivial restatement of Mendelism or a deeply flawed attempt to dispose of group selectionism. There is considerable advantage to Sober in thus marginalizing the main thrust of Dawkins's argument. It enables him to concentrate his fire on the weakest part of Dawkins's case, the assertion that there is a fundamental equivalence between the evolutionary interests of genes and those of individual organisms. However, the purpose of this chapter is to argue that while Sober is correct in suggesting that Dawkins's account deals inadequately with some examples of what seems to be group selection at work, Sober himself makes a much more egregious error in describing as trivial the selfish-gene thesis at the core of Dawkins's argument. Shorn of the constraints Dawkins puts on it by broadly equating the interests of genes with the interests of individuals, the selfish-gene approach has even greater potential than has yet been realized to revolutionize our understanding of the process of natural selection, at least as it affects sentient species. The implications for the social sciences, as well as the life sciences, are immense.

Certainly, Sober has no difficulty in providing unequivocal examples of clashes of evolutionary interests between some genes and the individuals that carry them. He points to driving genes that, in theory at least, can sweep to fixation solely on the basis of what might be called superior copying productivity. Success in intergenetic competitions of this type can be secured without conferring any adaptive advantage on the carrying organism. Worse, a driving gene with a deleterious effect upon its carriers may work continuously to preclude the fixation of a rival gene with positive adaptive value. At the population level, the result may be a stable polymorphism that, by disadvantaging some individuals, shows that the interests of genes and individuals are most certainly not invariably "equivalent." Sober develops this argument beyond the theoretical level by citing Lewontin and Dunn's (1960) study of a driving gene that renders homozygous male house mice sterile. The initial research hypothesis assumed a frequency for the gene reflecting a gene/individual conflict of interest of the type just sketched. In fact, the frequency was significantly lower, reflecting the occasional loss of complete breeding units when all males chanced to be carriers of the gene. Thus in this case, the gene's (suicidal) "interest" in achieving fixation is in perpetual conflict with selective pressures operating at the organismic and group levels.

Such examples demonstrate that it is simply not true that there is an invariable equivalence between the evolutionary interests of genes and the individuals that carry them. But other aspects of Sober's critique are far more tendentious, albeit, no doubt, unintentionally so. In at least two cases he seems to redefine Dawkins's terms in ways that preclude his doing full justice to the latter's argument. First, he uses the term "genic selection" as coterminous with the idea of the "selfish gene" and then, by implication, reduces the scope of the latter by treating the effects of natural selection on genes as but one of a triad of processes of which organismic and group adaptation are the other equal partners: "My own view is that the idea of the 'selfish gene' is a good description of some traits, even though it is not a good description of all. In this section, I'll describe a few biological examples that help establish the scope and limits of the three possibilities we have considered. The goal is to clarify how the concepts of genic adaptation, organismic adaptation, and group adaptation are related" (1993, p. 106). This has the effect of reducing the majestic sweep of Dawkins's all-encompassing argument to the far narrower business of genes winning or losing because of properties relevant at the level of gamete formation, which Sober specifically distinguishes from the organismic and group levels. This is not what Dawkins is talking about. He builds on the longevity argument quoted earlier by asserting that the only entities that, over evolutionary timescales, natural selection can favor are genes. This means that examples of what Sober calls organismic adaptation or group adaptation can either be explained in terms of the persistence of the genes that set up the characters in question, or they cannot be explained at all. In short, genic selection is not just one of three processes of natural selection; it is *the* process of natural selection at its highest and purest level of abstraction.

As I have already made clear, Sober does not deny this but considers it little more than a truism: "If all they mean by their position is that Mendelism is true or that Weismannism is true, or that evolution can be described in terms of what happens to genes, then their position is trivial" (1993, p. 103). That the neo-Darwinist thesis is infinitely more than this is well illustrated in Dawkins's attack upon Wynne-Edwards (1986). Wynne-Edwards's offense arises from decades spent studying the behavior of the red grouse. He has noted that the number of individual breeding territories within a given area appears to be determined by overall optimization of resources, not the number of males in competition for them. Once all the available, optimally sized territories have been occupied, the unsuccessful males seem quietly resigned to a probable death by slow starvation. Realizing the enhanced survival prospects for the group as a whole this pattern of behavior gives, Wynne-Edwards concludes that this is the reason it has been favored by natural selection. To Dawkins this is nonsense. He suggests that a male who sits and starves is a product of countless evolutionary experiments that have shown that his best strategy is "to wait in the hope that somebody [i.e., a successful male] will die, rather than squander what little energy he has in futile fighting" (1989, p. 118). It is not that Dawkins thinks that

Wynne-Edwards is wrong in suggesting that there is an overall advantage to the group. His attack is based on the assumption that this cannot be the reason why this behavior has been favored by natural selection. As groups, as opposed to their individual members, lack corporality, they must also lack genes. Group benefits cannot, therefore, be selected for, and Wynne-Edwards's position is logically untenable. In Dawkins's view the explanation for self-sacrificial behavior has to be kin selection, reciprocal altruism, or some other subtle form of self-interest built into the genes of individual organisms. As those who benefit from the deaths of the unsuccessful males—their successful rivals—are, in the main, genetic strangers, kin selection cannot be a significant factor. As the probable outcome for the starving males is death, there can be no element of reciprocity in the straightforward sense of forgoing immediate advantage in the expectation of a future payback. Ergo, making the best of a bad job by sitting, waiting, and hoping has to be the underlying rationale.

Dawkins seems to be offering an explanation couched in terms of organismic adaptation in place of Wynne-Edwards's group-adaptation proposal, but, at root, the whole debate is about genes. Dawkins is saying that any gene that codes for the behavior "forgo your own reproductive interests in favor of those of the better-fitted strangers" buys for itself nothing more than a one-way ticket out of the gene pool. In contrast, a gene that codes for "conserve energy and hang on in" is likely to achieve fixation. The bedrock strength of Dawkins's argument arises from the harsh but irrefutable fact that it is at this level, and only at this level, that the debate has to be conducted. If the group-benefit element of Wynne-Edwards's argument is to be vindicated, and I believe that it can be, the vindication will have to take the form of a "selfish" genetic strategy that favors the fixation of the gene (or genes) that codes for the apparently altruistic behavior Wynne-Edwards has observed. Attempting to resolve this kind of issue by granting organismic and group selection equivalent and autonomous status in relation to genic selection simply will not do.

The other area in which I believe that Sober weakens Dawkins's case by unintentionally misrepresenting it directly concerns altruism. Dawkins's own definition of this is clear: "An entity . . . is said to be altruistic if it behaves in such a way as to increase another such entity's welfare at the expense of its own" (1989, p. 4). Sober offers something significantly different: "A defining characteristic of evolutionary altruism is that altruists are less fit than selfish individuals within the same group" (1993, p. 98). Perhaps to labor the obvious, the distinction can be made clear by using an analogy from athletics. To Dawkins the only participants in a race who can properly be called altruists are those who damage their own chances of winning in order to help another competitor. A good example would be pacemakers, participants who initially run at a speed too fast to be optimal in terms of their own success in order to help maximize the performance of another runner. However, Sober's definition also embraces those who lose despite doing their best to win. He takes this line in order to explain how, in apparent defiance of the selfish gene, altruism can persist over evolutionary time and grow in frequency. He cites work carried out by Lewontin (1970) in connection with the supplanting in Australian rabbits of virulent strains of the myxoma virus by much less virulent ones. The key factor in this context is that myxoma is transmitted by a fly that only bites live rabbits. The virulent virus kills its hosts so quickly that there is very little time for the fly to transfer some of its infected blood to another rabbit. The weaker virus either does not kill or kills much more slowly, and this gives the fly much greater scope in which to operate. As a result, the mild strain grew and prospered, while the virulent strain nose-dived into near-terminal decline. Sober quotes approvingly Lewontin's claim that this process "involves group selection" (1993, p. 109) and then goes on to confirm his own view that "if low-virulence viruses replicate more slowly than high-virulence strains, lower virulence is a form of altruism" (1993, p. 110).

There is no doubt that this is an interesting finding, but it is most inappropriate to use it in support of either the proaltruism or group-selection argument. In the United Kingdom, in the 1960s, a film was produced based on a book by Bill Naughton. Both were called *Alfie* and concerned the amoral life of a typical 1960s antihero. One of the pieces of advice Alfie offered his audience was summarized in the phrase "little, but often." What he meant by this was that if a cashier working for a company steals £1,000 in one week, it is sure to be missed; but if he or she takes £10 per week for 100 weeks, nobody will notice. This, surely, is the "strategy" adopted by genes responsible for the less virulent virus. As nobody would ever have accused Alfie of being altruistic, it seems inappropriate to so designate the genes in question. In terms of the standard genetic metaphor, what they are actually displaying is intelligent self-interest, that is, smart selfishness. The fact that less virulent groups prosper and virulent ones do not can be quite satisfactorily explained in terms of successful and unsuccessful selfish genetic strategies. It offers no evidence whatsoever of the long-term viability of altruism or of the existence of discrete group-selection processes.

Again Sober seems greatly to underrate what Dawkins is talking about. In what I take to be a dismissive comment, he speaks of "defin[ing] altruism as what cannot evolve" (p. 110). Yet, as the Wynne-Edwards debate makes clear, this is precisely what Dawkins is interested in. The unsuccessful red grouse do not replicate more slowly than the successful ones; they do not replicate at all. This is the real challenge Sober fails to grapple with. Wynne-Edwards tells us that the effects at group level are clearly beneficial. Yet, inevitably, the genes that specify the behavior perish with their carriers as a direct consequence of its enactment. Fudging the issue by redefining altruism will not do. The only way this issue can be satisfactorily resolved is by proposing a strategy at gene level in which the elimination of the losers is part of a wider pattern beneficial to the long-term survival of the genes that specify it. This is best achieved by reexamining Dawkins's selfish gene in the light of Sober's overwhelming evidence that the evolutionary interests of genes and the interests of the organisms that carry them are not identical.

SEXUAL-REPRODUCTION GENES AS AN EXEMPLAR

Such an approach entails taking an entirely gene-centric view and not allowing the interests of individuals to intrude in any way. That this has not been the case in the past is well illustrated by a *New Scientist* article by Ridley entitled "Is Sex Good for Anything?" The following paragraph reveals the individual-centricity of most neo-Darwinist thinking:

This idea, that individual selection and not group selection explains animal selection, crystallised in 1966 with the publication of "Adaption and Natural Selection" by George Williams. That book did for biology what Adam Smith had done for economics: it explained how collective effects could flow from the actions of self-interested individuals. But Williams realised that there was one troubling exception to his thesis: sex. The traditional explanation for sex was based on group selection. It demanded that an individual altruistically share its genes with those of another individual when making young, because if it did not the species would not innovate and would, a few hundred thousand years later, be "outcompeted" by other species that did. Sexual species, it said, were better off than asexual species. But were sexual individuals better off than asexual ones? If not, sex could not be explained by the Williams "selfish" school of thought. Therefore, either there was something wrong with those theories and true altruism could indeed emerge, or the traditional explanation of sex was wrong. And the more Williams and his allies looked the less sense sex seemed to make for the individual, as opposed to the species. (Ridley, 1993, p. 37)

The reason Ridley gives for sex disadvantaging individuals is that any individual that gave it up in favor of asexual reproduction "would contribute twice as much [genetic material] to the next generation and would soon be left in sole possession of the genetic patrimony of the species" (p. 37). The detail of Ridley's article is highly instructive, particularly the account he gives of a very elegant field study by Lively and Vrijenhoek demonstrating the value of sex, in a suitable environment, as a means of continually countering the invasive successes of parasites. However, the issue I want to draw attention to is the disastrous red-herring effect the idea of the selfish individual has had on exploration of this topic. If the longevity argument had been fully comprehended, individuals should have been ignored from the outset in favor of genes. There is only one question that needs to be answered if we are to understand the evolutionary persistence of sex: Under what circumstances might genes that code for the suite of behavioral, physiological, and psychological phenomena that constitute sexual reproduction be expected to achieve fixation?

Where sexual reproduction involves mate choice, the key to answering this question lies in the hoary old joke about the brain being the body's largest sexual organ. Although we generally view natural selection as a blind mechanical process, the emergence of brains capable of making discriminating choices introduced a new dimension. Their involvement in activities such as the determination of which of a group of prey is the most vulnerable, deciding whether or not to challenge a rival, and of particular importance here, mate choice, means that, in effect, Dawkins's concept of the blind watchmaker (1988) acquired some very useful, sighted little helpers. Indeed, with sexually selecting life forms there are obvious parallels with the activities of human stockbreeders. In both cases, intelligence is at work in selecting breeding partners. The stockbreeder has commercial considerations in mind but in the natural world what might be called ''evolvability'' is of all-consuming interest. (This is not to say that characteristics with a poor evolvability quotient have never acted as sexual attractants, but simply that, by definition, those who make bad evolutionary choices are soon flushed out of the gene pool.) As Ridley makes clear, the advantage of sexual reproduction is that it provides a rapid means of generalizing useful genes without reducing genetic diversity. The more intelligent the process of selection is, the greater will be the advantage. This is why most animals with the cerebral equipment to do so are very choosy about the quality of their sexual partners. But to make evolutionary sense, this type of explanation has to be reformulated in terms of genes. Translated into the metaphor of the selfish gene, sexual reproduction can be seen for what it is: a strategy whereby elements of the brains and bodies of their bearers are defined by sexual-reproduction genes (SRGs) in ways that ensure that (a) the SRGs are always carried by the evolutionarily most successful members of any breeding line in which they are found; and (b) genetic diversity is maintained in order to give the SRGs the best possible prospects for the future. Proportionate transfers of parental genetic material have no relevance whatsoever. The fundamental evolutionary effect of sexual reproduction is the perpetuation of SRGs. Individuals are puppets, not puppeteers. Think, for example, of the ancient species of small fish that progressively mated itself into becoming a reptilian species, then a land mammal, only to return to the sea eventually to become porpoises. The individuals at each stage are unrecognizably different from their forebears, reflecting the fact that the great majority of the defining genes have been dropped in favor of diverse others, better suited to new conditions. But the SRGs, which drove the whole process, survive, both preserved and in considerable measure unaffected by these multiple metamorphoses.

Looking at sexual reproduction in this way enables four key features to be drawn out, all with direct relevance to the rehabilitation of Wynne-Edwards. First, in evolutionary terms the overall beneficiaries of sexual reproduction are not individual organisms, but the small subcoalition of genes that actually define the structures, processes, and behaviors that make sexual reproduction possible. Second, where mate choice is involved, genes that specify those aspects of the organism's brain that enable it to make discriminating choices between sexual partners are vital members of this subcoalition. Third, because the SRGs are general (i.e., they are carried by all members of a species), it is a matter of complete indifference to the genes if some individuals carrying them suffer in consequence of their strategy. Their interest lies in continued association with the abler, more attractive individuals who secure a partner at the dance: wall-flowers denied their chance to breed are no more than insurance policies that, in the event, did not have to be drawn upon. Fourth, one of the advantages of sexual reproduction is the speed with which it facilitates adjustment to changing circumstances. This is at its greatest where brains are involved. Asexual repro-duction can only adapt by mutation; sexually reproducing plants pollinated by the wind, or other species that simply release their seed, must rely upon chance. Not so with animals who pick their mates. The effectiveness and speed of the blind processes of natural selection are enhanced immeasurably by overlaying a random process with an intelligent one.

COMPARATOR GENES

This is the type of analysis that will now be brought to bear on Wynne-Edwards's red grouse. Imagine another subcoalition of genes that, in this case, specifies what might be termed a comparator mechanism. With the same tran-scending circularity, the primary evolutionary effect of the comparator mecha-nism is to perpetuate the comparator genes that specify it. As with sexual selection, the comparator mechanism is built into the bearer's brain. Its function is to make its bearers continually assess themselves in relation to their peers. The next step is a linkage between the results of these self-assessments and emotional state. High self-assessments engender a sense of emotional well-being, and low self-assessments cause conditions such as depression and chronic anx-iety. The final step is for emotional state to become a determinant of physical well-being. On the negative side, low self-esteem leads to physiological and behavioral changes that result in elimination from the gene pool. Although there is not scope here for elaboration, there is overwhelming evidence from psycho-somatic medicine and animal studies that such a linkage actually exists. On the positive side, comparatively successful performers experience a sense of well-being that increases their contribution to the gene pool by stimulating their libidos and improving their health. In the notes to the second edition of *The Selfish Gene* Dawkins calls actual examples of this "the Duke of Marlborough Effect" (1989, p. 286).

The composite effect of mechanism is to act as a multiplier of the effects of natural selections. Genes carried by the comparatively successful are likely to be reproduced in even higher numbers in the following generation, while those of the comparatively unsuccessful will be even further reduced or eliminated. Most important, such a mechanism would operate even if, objectively, the mar-gin of advantage was small and the attributes of the less successful were more than adequate to meet the practical requirements of self-replication. Although the actual beneficiaries would be the genes that defined the process, the internal drive to do better or get out would act for the continual refinement of the species

Figure 1.1
Interaction between Comparative Fitness and Ability to Assess Own Worth and That of Others

		Comparative fitness	
		High	Low
Judgmental Skills	High	The favored group. Individuals scoring highly on both counts disproportionately selected for.	Elimination accelerated by self-awareness of inferiority.
	Low	Poor judgment of own worth and that of others leads to elimination despite high potential.	Low comparative fitness leads to elimination regardless of own ideas.

as a whole. It might be at its strongest when environmental pressures were at their most aggressive; but even when external pressures were comparatively weak, the operation of the comparator mechanism would still be strongly felt.

There is no requirement for a shared criterion of perfection. As with mate choice, the process operates solely on the basis of individual and entirely subjective assessments of relative performance. Inevitably, successful individuals will have been eliminated because they underestimated their own abilities or overestimated those of others. However, again as with mate choice, from an evolutionary standpoint this is no more than the standard process by which inefficient traits give way to more effective ones. Nor does self-delusion pay. The whole process is kept on track by natural selection. Any individual equipped with a comparator mechanism can be looked at from two standpoints. First, how does it rate in terms of comparative fitness? Second, how good are its judgmental skills in assessing both its own fitness and that of prospective mates and rivals? The interaction between these two factors is summarized in figure 1.1.

One implication of this matrix is that it does not matter how far comparator genes carried by individual animals are separated in terms both of time and biological relatedness. Natural selection continually ensures that those carried by the successful representatives of any given species remain concerned with virtually identical fitness-related performance criteria. In effect, comparator genes are the most selfish genes of all. They can only engage with environmental pressures at the level of the organism, but in terms of the standard metaphor their real concern is the construction of successful genetic coalitions that will carry them into future generations. To an even greater degree than sexual-reproduction genes, they have secured for themselves a virtually unassailable position. By coding for a multiplier on the effects of natural selection, they ruthlessly select or reject their fellow genes to ensure that they are always on the side that is winning. Like boxing managers, they do not enter the ring themselves. They just sit there in the front row working out whom to back and whom to dump. It all seems so clever that it is hard to remember that in attributing such a strategy to them, we speak only metaphorically. Like all other genes, those responsible for the comparator mechanism neither have nor are capable of having the slightest understanding of what they are about. They are, after all, just mindless specks of DNA.

It is the quality of contingency that marks comparator-gene theory out as something new. In the past, neo-Darwinists have been able to reject group-selection theories on the grounds that they seek to treat individuals as "pawn(s) in the game, to be sacrificed when the greater interest of the species as a whole requires it" (Dawkins, 1989, p. 7). Dawkins reasons that because altruistic individuals will by definition reproduce less frequently than their selfish peers, "after several generations of this natural selection, the 'altruistic group' will be overrun by selfish individuals, and will be indistinguishable from the selfish group" (1989, p. 8). This type of argument is typical of a debate that seems to have been conducted exclusively in terms of an altruism/selfishness dichotomy; that is, individuals are presumed to be genetically programmed permanently as either altruistic or selfish. Yet ethology abounds in examples of contingent behavior. An animal at the top of the pecking order adopts appropriately dominant behavior; transferred to a new social group in which it cannot push its way to the top, it behaves in an entirely different way. Nor does contingency stop at behavior. In the face plates of dominant orangutan males we have an example of contingent physiological structures. While all males carry genes that create the potential for face plates, only those who achieve alpha status actually produce them. This is the basis upon which comparator-gene theory avoids the usual pitfalls. It differs from classical group-selection theories in two key areas. First, the beneficiaries of altruistic behavior are not species as a generality, but the set of genes responsible for the comparator mechanism. Second, it does not argue that some individuals are born altruistic and others selfish. Each individual carries within itself a multioption program, with possibilities ranging from extreme self-interest through more moderate shades of selfishness and then into increasing degrees of altruism that finally culminate in self-elimination. It operates in accordance with what artificial-intelligence researchers call condition/action rules: if you can assess yourself as relatively successful, then go forth and multiply; but if you assess yourself as a comparative failure, then get out of the gene pool. Inevitably a substantial proportion of each generation is greatly disadvantaged by the mechanism. However, there are consequential benefits to the winners, and it is they who put the genes that define the mechanism into the next generation.

There are still two problems. First, there is a specific form of genetic selfishness to which the comparator might be vulnerable. This would be encoded by a rival coalition of genes whose strategy would differ in only the following respect from that outlined above. Rather than self-eliminating, the least well-adapted carriers of "mock-comparator" genes would be kept in play to give a crucial, albeit highly marginal, reproductive advantage to their coalition. After all, even a maladapted organism will have some potential for reproducing. Seemingly there would be no compensatory advantage for the comparators. With their rivals gaining just as much as they from the group beneficial consequences of the comparators' sacrifices, it seems axiomatic that the reproductive successes of all the other carriers of the opposing coalitions would balance out—hence the apparently crucial importance of that tiny advantage enjoyed by the mock-comparators as a result of keeping their poorest performers alive. It therefore looks as if the comparators—along with self-destructive behaviors—would be steadily out-evolved. However, in a paper currently in preparation, I draw on kin selection theory to argue that among species that practice mate-selection, the comparative fittedness of a prospective mate's close kin is unlikely to have been ignored in the evolution of mate-choice criteria. This should have been particularly so with regard to the identification of deleterious recessives. I have termed this "stigma theory," its main corollary being that the mere presence of comparatively maladapted kin is highly likely to act against the reproductive interests of close relatives. If so, there will have been significant kin-selection advantages to the evolution of maladaption-related, self-destructive behaviors, something that would strongly tip the balance against mock-comparators.

The second problem lies in explaining how such a devastating mechanism could have arisen and become fixed in the first place. One strong possibility is that it followed in the wake of our evolved mechanisms for pain and pleasure. These may be assumed to have arisen for one primary reason: pleasure encourages sentient organisms to do things that enhance evolutionary potential; pain discourages them from doing those things that do not. Initially, they are likely to have been experienced as random events, unrelated to the requirements of survival and replication. As such, they will have been maladaptive, and most of the genes responsible would soon have been eliminated from the gene pool. However, by chance, a few embryonic systems must have become crudely aligned to activities favorable or unfavorable to the reproductive success of their bearers. The normal processes of natural selection would then have done the rest. The heightened evolvability of genes that make their bearer feel good and want to repeat the experience whenever it does something to the gene's benefit is obvious. It only requires one or two 10,000:1 mutations to have feelings of pleasure stimulate physical responses that make the individual better able to repeat the performance. The same is true of pain. The old proverb "He who fights and runs away, lives to fight another day" makes explicit the advantage to genes of also being able to discourage their bearer from doing something threatening to the genes' prospects of survival. One way of achieving this, at least in the context of disputes between two members of the same species, is to have an emotion such as fear trigger physiological reactions that make submissive behavior inevitable, an idea first mooted (at the organismic level) by Price (1967). Because continually to enter contests and then collapse into submission is not a very fruitful form of existence, natural selection can be relied upon to favor individuals who have the capacity to learn from such experiences.

Developed in strict accordance with the principles of neo-Darwinism, this brief outline provides a hypothetical explanation for the emergence of individuals with three key characteristics. First, behavior advantageous to the "sentient genes" that define it is reinforced emotionally and physically; second, behavior that threatens the genes is discouraged emotionally and physically; and third, those individuals have an inbuilt predisposition to repeat activities that give pleasure and avoid those that give pain. Up to this point the Dawkins precept holds true: there is no difference between the interests of the sentient genes and those of the individuals that carry them. What is good for the genes is good for the organism. But once this pattern was well established, mutation could with advantage produce some individuals in whom these characteristics were more generalized. We already accept that this has happened. Kin selection, which must have arisen by mutation, is a process in which genes specify some behaviors that, although they work to the genes' advantage, impose great costs on those individuals in whom the behavior is expressed. The emergence of comparator genes calls for a similar development, also by mutation. Individuals bearing a mutant version of the sentient genes would suffer not only when they were under direct threat, but also when they were in danger of "letting their genes down" by not performing reproduction-critical activities as well as their peers.

Such a development may have arisen simultaneously within several individuals within a family group. But even if only one bearer emerged, the odds are that the gene responsible would become widespread very rapidly. The exception would have been cases where the individual affected was unable to raise its performance above the average. In all other cases such an arrangement would invariably work to the evolutionary advantage of the genes that specified it. If the bearer was of average potential, living alongside others who were indifferent to their relative performance, the drive for comparative success would act as a powerful spur, and any improvement that had positive implications for reproductive success would serve to spread the genes. The best prospect of all would have been an individual with comparator genes plus some other variation conferring advantage, no matter how marginal. Such a combination would have led to maximum exploitation of the marginal reproductive advantage, whatever its nature. This would have resulted in a disproportionate rate of transfer both for the comparator genes and the advantageous genes with which they were associated. Regardless of the gender in which the comparator mechanism first appeared, sexual selection would have ensured that in the next generation it appeared in both. Thereafter, since a disproportionate number of those bearing the gene would be high achievers, inclined to seek out high achievers as partners, it would become universal with extraordinary rapidity.

While comparator genes were spreading through a breeding group, their effects may have been restricted by natural selection to making their abler bearers

more fecund and eliminating those who proved incapable of raising their performance above the comparatively mediocre. But once comparator genes are
fixed within the population, an infinity of possibilities opens up. Provided the
sacrifice has good evolvability potential for the ubiquitous comparator genes, it
does not matter who sacrifices his or her reproductive interests in favor of whom,
or how distantly they are related. The possibility arises of complex social relationships in which some individuals who do not themselves reproduce derive
the sense of success on which their lives depend by being relatively better
supporting actors than other aspirants to the same roles. Another option for the
comparatively unsuccessful would be to seek to exploit different food sources
or new territories. With every member of the group under pressure to succeed
in relative as well as absolute terms, the imperative to compete, complement,
diverge, or die would act as a continual stimulus to the comprehensive exploitation of any advantage, no matter how marginal, the continual refinement of
social behavior, and the rapid adoption of novel patterns of behavior culminating
in speciation. But for those who continue to assess themselves as relative failures
because none of these avenues offers an escape, death seems to be the only way
out. The instrument might be a predator, a parasite, a disease, or the sudden,
catastrophic failure of a physical process vital to life. At the level of the individual this can be an unmitigated disaster, but to the comparator genes all this
signifies is that another suboptimal gene coalition has been taken out of the gene
pool.

For the comparator genes, one ever-present danger remains. Even after they
have become fixed within the population, it is always possible that mutation
could produce individual organisms in which they are not represented. What if
one of these proved better adapted to the prevailing circumstances than the rest
of the breeding group? Sexual selection offers the first line of defense. The
operant principle is this: if it is better, mate with it. This would immediately act
to bring the most useful new genes into the ambit of the comparator mechanism.
But let us imagine that the threat comes from a different species altogether, a
life form that is simply better at converting the available energy supplies into
reproductive success. The strategy here is that which generals are supposed to
follow as a precaution against the emergence of a new, unexpected threat, the
constant refinement and improvement of their own troops. This does not guarantee success against an unexpected enemy with a devastating new weapon, but
it does mean that the enemy must attain a much higher standard to win. The
prospects for the comparator genes are probably better. New entrants would be
extremely unlikely to reach the required standard unless they too were the products of a ruthless internal selection procedure that operated even when environmental pressures were low.

This is the basis on which Wynne-Edwards's observations can be reconciled
with the core of Dawkins's argument. Remember the losing red grouses, sitting,
starving in the Scottish heather rather than acting to increase their own chances
of survival by attempting to enforce a reduction in the size of breeding territo-

ries? Comparator-gene theory suggests that they were doing exactly what Wynne-Edwards has always said they were doing, quietly accepting death by starvation rather than spoil the chances of better-fitted members of the flock. The whole process is orchestrated by comparator genes, copies of which are carried by every individual. Fine-tuned to perfection by natural selection, they use the grouses' own brains, in a process identical to sexual selection, to identify those individuals whose preservation offers them the best possible prospects of survival. Again as with sexual selection, the criteria are so clear that, subject to a few squabbles at the margin, both winners and losers very rapidly achieve a common understanding as to who is in and who is out. Then the brain of each loser simply goes into a state of self-abnegation that we call depression, only reverting to purposeful self-interest if misfortune overtakes one of the winners. To Wynne-Edwards and to Sober it looks like group selection, but the reality is that, underneath it all, quintessentially selfish genes are working their tickets to immortality.

NOTE

An article based on this chapter and entitled "Darwinism and the Enemy Within" was published in the *Journal of Social and Evolutionary Systems*, 18(3), 1995.

REFERENCES

Dawkins, R. (1988). *The blind watchmaker*. London: Penguin Books.

———. (1989). *The selfish gene* (2nd ed.). Oxford: Oxford University Press.

Lewontin, R. (1970). The units of selection. *Annual Review of Ecology and Systematics*, 1, pp 1–14.

Lewontin, R., & Dunn, L. (1960). The evolutionary dynamics of a polymorphism in the house mouse. *Genetics, 45*, pp 705–722.

Price, J. S. (1967). The dominance hierarchy and the evolution of mental illness. *Lancet*, No. 2, pp 243–246.

Ridley, M. (1993, December 4). Is sex good for anything? *New Scientist*, pp. 36–40.

Sober, E. (1993). *Philosophy of biology*. Oxford University Press.

Williams, G. (1966). *Adaptation and natural selection*. Princeton, NJ: Princeton University Press.

Wynne-Edwards, V. C. (1986). *Evolution through group selection*. Oxford: Blackwell Scientific Publications.

2

The Implications of Darwin's Variational Paradigm

David Smillie

In this chapter I want to challenge a position that has become, over the past thirty years, something of a basic view of natural selection. I take as a recent statement of this view a passage from Steven Jay Gould (1995): "Natural selection is a theory of ultimate individualism. Darwin's mechanism works through the differential reproductive success of individuals who, by fortuitous possession of features rendering them more successful in changing local environments, leave more surviving offspring. Benefits accrue thereby to species in the same paradoxical and indirect sense that Adam Smith's economic theory of *laissez faire* may lead to an ordered economy by freeing individuals to struggle for personal profit alone" (p. 7). Rather than dispute Gould's quite conventional claim, I shall go back to Darwin and examine his approaches to natural selection.

Darwin had two different ways of describing the process of natural selection—two ways that seemed quite consistent to him and to many of those who have followed him. On the one hand, he viewed natural selection as analogous to the selective processes utilized by an animal or plant breeder; just as the breeder selects for breeding among different animals and plants those that carry some approximation of desired traits or qualities, so also over the long run do natural conditions select among extant populations of organisms those that carry traits leading to survival and procreation. This theme is thoroughly explicated in *The Variation of Animals and Plants under Domestication* (Darwin, 1868, 1875). Here I speak of this analogy as Darwin's variational paradigm since it focuses on the selection of variants making for organismic adaptation to surrounding circumstances.

On the other hand, Darwin, following Malthus, also saw that natural process of reproduction, operating to bring about a geometric increase in the numbers of a species, would lead after a time to the species outstripping the available resources, which could only increase arithmetically. The production of more offspring than can survive inevitably introduces competition for scarce resources, both within and between species. The economy of nature is thus characterized by a continual struggle for existence. Those organisms that are successful in this struggle will always leave more offspring than their rivals, and this will bring about an increase in those qualities adapted to natural conditions. I call this Darwin's Malthusian paradigm.

For Darwin, the survival of organisms and the survival of variations were more or less equivalent since he had no clear understanding of hereditary processes and what we might today call "genetic survival," or the persistence of alleles in a sexually breeding species. Competition between organisms and survival of those qualities that make for success seemed to him, and has since seemed to many others, to constitute a unified view of the primary mechanism of evolutionary change and stability.

In the past 150 years major changes have taken place such that we now have a quite adequate account of heredity, especially that of sexually breeding organisms. The collection of fossil evidence that Darwin presumed would always be insufficient to test his theory is now much improved. Thus we should do our best continually to apply what knowledge we have so as to reconsider Darwin's beliefs and to modify his theory where this is required.

I argue in this chapter that Darwin's two paradigms are not consistent with one another. The first, the variational paradigm emphasizing the selection of traits or genes, provides deeper insights into the mechanisms governing evolution than the second, the Malthusian paradigm, which gives Darwin's theory its "ultimate individualism" (as Gould claims). True, however, it is the latter view that predominates in most discussions of natural selection as applied to evolutionary theory. I will briefly review each of these perspectives in a contemporary light and show that the variational perspective is quite consistent with developments in modern genetics, while the Malthusian perspective offers only limited insights and is misleading when it is applied to all natural selection.

DARWIN'S VARIATIONAL PARADIGM

For Darwin, as for contemporary evolutionary biologists, the presence of variation is crucial to the operation of natural selection. Where do such variations come from? As Darwin indicates in chapters 1, 2, and 5 of On the Origin of Species, this is a crucial question, but one that continued to puzzle him throughout his life. Indeed, it was not until forty years after the publication of Origin in 1859 and eighteen years after Darwin's death that Mendel's answer was rediscovered: variations are the embodiment of what Mendel called "traits," "characters," or "elements" (Mayr, 1982, p. 736). Today, of course, we speak of genes and alleles, but it is useful to keep in mind the notion of inherited traits or characters as well. Sexual reproduction allows for the interplay of a wide number of characters within an interbreeding population and thus for the possibility of the selection of characters well adapted to natural conditions, which Darwin understood as being the entities being subject to selection. From his variational perspective, then, it is quite reasonable to think of nature as a selective agent or cause, and the entities being selected as genes or traits that persist across generations. This is a view espoused by Williams (1966) and by Dawkins (1976, 1982), namely, genic selection. But both of them make the unwarranted assumption that in the vast majority of cases, this amounts to the same thing as the selection of individuals. If we consider sexually breeding organisms, it seems reasonable to say that individuals, or genomes, could not possibly be what are selected since they are ephemeral beings that survive only momentarily, to be continually replaced by others having novel genetic combinations, new or even unique patterns of inherited traits. Indeed, sexuality can be understood as a process of continually mixing genes so as to produce novel combinations. Because of the mixing process, it is traits that are being tested in each generation and traits that constitute the subject matter of natural selection as a process, as Bateson (1982), for one, pointed out some time ago. This conclusion is quite consistent with Darwin's analogy between natural and artificial selection, since it is traits that are also the focus of the breeder. It takes someone who can pick out the traits of animals or plants leading toward a desired end to be successful as a breeder.

DARWIN'S MALTHUSIAN PARADIGM

Influenced by his reading of Malthus, Darwin came to believe that competition between organisms constituted the motivating force behind natural selection. Once organisms had filled the available living spaces in the world, a condition Lyell and others spoke of as "plenitude" (Stanley, 1981), the only way a new species or individual could make its way into the living space was to displace those already there. Darwin (1859) said, "The face of nature may be compared to a yielding surface, with ten thousand sharp wedges packed close together and driven inwards by incessant blows, sometimes one wedge being struck, and then another with greater force" (p. 67).

This role for competition in the evolutionary scenario represents a focus on individual organisms, some gaining access to scarce resources, others not surviving due to the lack of these same resources. Being a naturalist fascinated by organisms and not having a knowledge of Mendelian genetics, Darwin inevitably understood the story of evolution as constituted by the success and failure of living beings. Indeed, he said, "All that we can do, is to keep in mind that each organic being is striving to increase at a geometrical ratio" (1859, pp. 78–79), as though sexual organisms were reproducing themselves in their offspring. But closer consideration shows that sexual organisms do not have offspring; only pairs of individuals have offspring, which are always different from either one of their parents. The mixing of genes and traits in sexual reproduction means that many of the trait combinations, successful in any one individual, are being destroyed in the next generation. Thus if each organic being is striving to increase, each is doomed to failure.

When we consider that it is traits that survive generation after generation, the Malthusian paradigm no longer applies as a necessary characteristic of selection. It is only by a great stretch of a metaphor that we can say that traits are struggling with one another over scarce resources. Anatol Rapoport (1960) distinguished between fights and games. In fights there is an attempt by one party to eliminate the other, as in Malthusian competition over scarce resources. But in games competition takes place to find out which of two parties is superior in a particular activity. Both players survive to try again. Scarce resources really do not play a part in games at all. Traits, appearing over and over again in different combinations as they arise in new organic phenotypes, might be said to be involved in a contest to see which is superior; they are playing a game rather than having a fight. Multiple traits (or alleles) may well survive, participating in a population as polymorphisms, as D. S. Wilson (1994) has emphasized. It is really only through the ambiguity in the definition of competition that the Malthusian view of natural selection has survived.

A CONTEMPORARY PERSPECTIVE ON THE TWO PARADIGMS

From a contemporary perspective, then, we can get a better grasp of why these two paradigms do not constitute a neat blend, with one supporting the other. Let us retain as a frame of reference, as Darwin and most subsequent evolutionists have done, evolution within sexually reproducing species. It may seem at a superficial first glance that the animal breeder's selection of some animals to breed is equivalent to the competitive survival of some organisms over others. However, on closer examination the animal breeder is selecting, not simply the offspring of mating pairs, but rather those animals carrying desired traits. In nature, outcrossing means that such traits are appearing over and over again throughout the entire population and ultimately throughout the species. This means that the species will be constituted by traits tested repeatedly, traits appearing in organismic contexts where reproduction has been adequate. In natural selection the environments, instrumental in making some organisms successful at breeding and other organisms not, are very diverse and very complex, while in artificial selection the animal breeder is apt to have had one or only several traits in mind in the selection process.

In support of a Malthusian perspective one might argue, as Gould does, that even though selection focuses on traits, those particular traits highlighted by competitive interactions between organisms struggling for scarce resources are the ones being chosen. In some instances this is certainly true. What Darwin called sexual selection, contrasting it, by the way, with natural selection, emphasizes male-male competition over access to females. This, in conjunction with female choice, may well select for male traits such as size and aggressiveness, and display features under conditions in which females represent a scarce resource. But under these same conditions females may well be selected for their capacities to identify features in males that will make for the probable success of their mutual offspring, and such selection is not a matter of competition in Darwin's sense. Selection in this case does not operate under conditions of scarce resources since males are in plentiful supply.

Gould argues that sexual selection in males supports the contention that Darwin's theory is fundamentally individualistic, since such competitive activity could not possibly exist for the good of the species. But a thoroughgoing individualism such as he proposes leads to a much simpler solution to the burdensome costs of sex: simply get rid of males altogether. With one female parent, the costs of sex are dramatically eliminated. Yet species continue to evolve in the direction of sexuality. If individuals are the sole objects of selection, how are we to understand this phenomenon?

More broadly, we may say, new traits are selected not only where ecological pressures and social conditions create scarce resources but also where new conditions open up opportunities for some novel traits in comparison with others. For example, when a habitat changes for a species of plant, with the appearance of extended dry spells creating desertlike conditions, those organisms that carry genes for water-retention mechanisms will be favored, and there will be selection for a new set of traits quite apart from competitive interactions between individuals. I am here drawing my example from Darwin, who commented: "I use the term Struggle for Existence in a large and metaphorical sense. . . . [A] plant on the edge of a desert is said to struggle for life against the drought, though more properly it should be said to be dependent on moisture" (1859, p. 62). We may speak of this as "opportunity selection" to emphasize the fact that selection operates through the capacity of traits to exploit new opportunities in contexts quite apart from competition with conspecifics over scarce resources.

SELECTIVE ENVIRONMENTS

One advantage in retaining Darwin's variational paradigm is its somewhat unconventional emphasis on environments as causally active over evolutionary time spans. As breeders are active causal agents in selecting some traits over others, so natural environmental conditions can be understood to be causal agents in the context of natural selection. Such a phraseology, in which environments adopt the role of causal agents and organisms are simply effects, runs against our usual way of seeing the world. Even Ernst Mayr, the originator of the important distinction between ultimate and proximate causality, when writing for a lay audience, says: "There is no particular selective force in nature, nor a definite selecting agent. . . . It is not the environment that selects, but the organism that copes with the environment more or less successfully. There is no external selection force'' (1991, pp. 86–87). In contrast to this commonsense view, I have emphasized, as Mayr had done earlier, that causally active organisms represent a framework distinct from that of neo-Darwinian selection, that of proximate causality. I argue here that natural selection is basically a theory about changes in persisting organismic traits rather than an account of ephemeral individuals and thus it utilizes the framework of ultimate causality.

If this viewpoint had been taken seriously during the past hundred years, we would now have a science of environments corresponding to the science of genetics, and we would not see the divisions that currently exist between ecologists and evolutionists. From an evolutionary perspective we need to understand and classify the wide number of environmental influences that operate in natural selection, environments involving the ecology of species and their interactions, environments constructed internally to organisms by developmental processes, and, important in the context of sociobiology, the many aspects of social environments and their consequent selective effects, as next described.

SOCIAL ENVIRONMENTS

I have already adverted to the selective effects arising from social environments where one sex competes for access to another, what Darwin called ''sexual selection.'' I have indicated that this competition is a social selective factor operative only under quite particular conditions. Of very real interest are the social selective opportunities available in social environments. Such opportunities may involve an increase in food resources, the enhancement of mating opportunities, protection from predators, or multiple other opportunities that have often been ignored or underemphasized in the sociobiological literature. One may think of these opportunities as equivalent to social synergisms (Corning, 1983, and Chapter 3, of the present volume.)

When organisms aggregate for any reason, there is a persistent tendency for the coadaptation of individuals, or as we might better say from the variational perspective, there are advantages arising from traits that make for increased adaptation to the group. Given the emergence of some pattern in which individuals utilize the available social opportunities for enhancement of both self and others, genes begin to accumulate that further a social strategy. Such a process leads to the increase of traits able to exploit novel social opportunities. Once this process is started, there will then be a tendency toward the increase of those inherited traits that serve to consolidate a persisting social strategy, including defenses against those mutations that might undermine existing patterns.

Putting this in the language of Darwin's variational paradigm, social environments tend to select for an increase in social traits, consistent with the analogy of breeders selecting for traits that enhance qualities already present to some extent in the breeding population. Sociality in organisms may then evolve, providing ''benefits'' for the allelic patterns that constitute them; genes function so as to constitute organisms that will utilize the opportunities available in a social life, and this ultimately serves their own long-term evolutionary survival.

CONCLUSIONS

I have argued in this chapter that the somewhat conventional view of Darwinian selection, articulated in the present instance by Gould but held widely by many evolutionists, represents a limited view of Darwin's contribution to evolutionary theory. By my lights, Darwin had two different arguments, one focusing on traits, or variations, the other on individuals. The individualistic argument of Gould and many others fits a variety of presuppositions present in the Western European worldview and has come to be seen as at the heart of the theory of natural selection. However, a consideration of developments that took place after Darwin's death has subsequently emphasized the evolutionary role of traits and genes, this being consistent with Darwin's variational paradigm. It is time for those of us in sociobiology to review once more the selective role of inherited traits and the genes that mediate them. Particularly in the case of sexually reproducing organisms we can see that the one story told in the language of genes and the other told in the language of competing organisms move off in quite different directions. It is my view that Darwin's variational paradigm contains implications that have been largely overlooked. I believe that it opens up new possibilities for the future of sociobiology.

REFERENCES

Bateson, P. P. G. (1982). Behavioural development and evolutionary processes. In King's College Sociobiology Group (Ed.), *Current problems in sociobiology* (pp. 133–151). Cambridge: Cambridge University Press.

Corning, P. (1983). *The synergism hypothesis.* New York: McGraw-Hill.

Darwin C. (1859). *On the origin of species.* London: John Murray.

———. (1868, 2nd ed., 1875). *The variation of animals and plants under domestication.* 2 vols. London: Murray.

Dawkins, R. (1976). *The selfish gene.* Oxford: Oxford University Press.

———. (1982). *The extended phenotype.* Oxford: W. H. Freeman.

Gould, S. J. (1995). Spin doctoring Darwin. *Natural History, 104,* 6–70 (passim).

Mayr, E. (1982). *The growth of biological thought: Diversity, evolution, and inheritance.* Cambridge, MA: Belknap Press of Harvard University Press.

———. (1991). *One long argument.* Cambridge, MA: Harvard University Press.

Rapoport, A. (1960). *Fights, games, and debates.* Ann Arbor: University of Michigan Press.

Stanley, S. (1981). *The new evolutionary timetable.* New York: Basic Books.

Williams, G. C. (1996). *Adaptation and natural selection.* Princeton, NJ: Princeton University Press.

Wilson, D. S. (1994). Adaptive genetic variation and human evolutionary psychology. *Ethology and Sociobiology, 15,* 219–235.

3

Cooperative Genes: Synergy and the Bioeconomics of Evolution

Peter A. Corning

Metaphors can exert a powerful influence over our perceptions, for better or worse. Adam Smith's "invisible hand" is one such potent image. Another is Richard Dawkins's "selfish gene." But, as Anatol Rapoport warned in an article on "Ideological Commitments in Evolutionary Theories" (1991), the imagery we use to characterize nature or human societies may also become a distorting lens that biases our interpretation of a more complex reality; we oversimplify at our peril.

In a new preface accompanying the revised edition of *The Selfish Gene*, Dawkins acknowledged that his famous metaphor is, after all, a heuristic device. Responding to critics who had accused him of extremism, Dawkins deployed another metaphor, the Necker cube (a two-dimensional drawing that can be perceived in different ways), to illustrate the intent behind his inspired bit of anthropomorphism. "My point was that there are two ways of looking at natural selection, the gene's angle and that of the individual. . . . It is a different way of seeing, not a different theory" (1989 [1976], pp. ix–x).

Actually, there are more than two ways of looking at natural selection, and what I propose to do here is to rotate the Necker cube in such a way that our focus shifts to the functional interactions that occur within and among various "units" of selection. To borrow Dawkins's terminology, we will focus on the role played by the "vehicles" rather than the "replicators." In recent years, various theorists have advocated a multilevel, hierarchical (some prefer "holarchical") model of the evolutionary process. A somewhat different approach will be taken here. We will focus specifically on the "economics" of cooperation at various levels of biological organization, with particular emphasis on the causal role of synergy (the combined effects produced by two or more elements, parts, or individuals), along with cybernetic (communications and control) processes. I will suggest that the vehicles of selection at various levels not only exhibit a ''shared fate'' but also produce synergies—interdependent functional effects that have served as the proximate causal mechanisms (the payoffs) underlying the emergence and persistence of more inclusive levels of organization. (Of necessity, this discussion will also be very abbreviated; more extensive treatments can be found in Corning, 1983, 1995a, 1995b, 1996a, 1996b, 1997.)

COOPERATION IN EVOLUTION

It is important to begin with a definition of cooperation. The reason is that even such a commonplace and intuitively obvious term has been used in a variety of ways in the scientific literature, sometimes with unfortunate consequences. As the term will be utilized here, cooperation refers to a relationship—a condition in which two or more elements, parts, or individuals ''operate together''; it connotes a functional interaction. Cooperation in this sense may or may not also be considered selfish or altruistic, mutualistic or exploitative, or positive or negative; such attributes involve additional, post hoc judgments about the consequences of a cooperative relationship with respect to some separately specified goal or value. By the same token, a cooperative relationship may or may not be voluntary. Slavery, in nature and human societies alike, involves a form of involuntary cooperation, and so does the host's role in a parasitic relationship. Indeed, there are innumerable cases in nature in which, for better or worse, a cooperative relationship is mutually obligatory, whether the cooperators like it or not. For instance, honey bees and the flowers that are pollinated by their activities are completely dependent on one another. One complication with this definition involves cases of virulent parasitism, or cooperation that leads to the death of the host. Perhaps this can be finessed by redefining virulent parasitism as predation in slow motion.

Once a clear distinction is drawn between mutualism, altruism, parasitism, and cooperation, a significant source of confusion and misinterpretation in the literature, especially in sociobiology, dissolves. This can be illustrated by an example from symbiosis, by definition a sustained functional interaction (cooperation) between members of two or more species. The so-called VAM (vesicular-arbuscular mycorrhizal) fungi are generally considered to be models of mutualism with many species of plants. Careful studies have shown that VAM fungi do in fact enhance plant growth in low-phosphorus soils, but in high-phosphorus soils or in low-sunlight conditions (when photosynthetic activity is reduced), they may actually become parasitic and reduce plant growth. In a similar vein, avian brood parasitism, such as the cowbirds' practice of infiltrating their eggs into other species' nests, is ordinarily harmful to the ''hosts'' who end up nurturing somebody else's nestlings. But cowbird chicks also eat botfly larvae, which can infect a host's nestlings and lower their fitness. Whenever there are heavy botfly infestations, the parasitic cowbirds may actually enhance their hosts' reproductive success.

Thus symbiotic interactions are sometimes mutualistic and sometimes parasitic, but in both cases they involve "cooperative" effects. In one case, we call it mutualistic because both partners benefit; in the other case, we call it parasitic because the costs and benefits, while equally real and measurable, are asymmetrical. But in both cases, the interactions that occur between the two species produce interdependent functional consequences; they are cooperative effects. Moreover, in this case as in others that we will consider later, the functional consequences may be affected by changes in the context—that is, in various ancillary factors that are external to the focal relationships per se. Often these "ecological" factors may involve three-way cooperative relationships in which parasites are assisted by mutualists, or vice versa. An example is the bark beetle of the genus *Dendroctonus*, which utilizes a symbiotic fungus (genus *Ceratocystus*) to suppress tree-tissue resistance to invasion by the beetle, so it can breed in healthy trees while, at the same time, spreading the fungus to new trees.

Conversely, many mutualistic relationships in nature appear to have evolved as a means of counteracting parasites—for example, the fifty-odd species of cleaner fish that remove parasites from their larger (often predatory) hosts, or the numerous bird species that remove parasites from large ruminant animals. By the same token, "mutual grooming" is a widespread practice in social animals, the benefits of which have nothing to do with "grooming" as we know it but rather with parasite removal (and perhaps some social functions as well).

A second point about cooperation as a functional concept is that it is found at every level of living systems. Beginning with the very origins of life, it is the common denominator in all of the various formal hypotheses about the earliest steps in the evolutionary process (see the detailed review in Maynard Smith and Szathmáry, 1995.) At the level of the genome, genes do not act alone, even when major single-gene effects are involved. As Dawkins observed in one of the less frequently quoted but more important passages of *The Selfish Gene*, the genes are not really free and independent agents: "They collaborate and interact in inextricably complex ways, both with each other and with their external environment. . . . Building a leg is a multi-gene co-operative enterprise" (1989 [1976], p. 37). To underscore this point, Dawkins employed a metaphor from rowing: "One oarsman on his own cannot win the Oxford and Cambridge boat race. He needs eight colleagues. . . . Rowing the boat is a co-operative venture" (p. 38). Furthermore, Dawkins noted: "One of the qualities of a good oarsman is teamwork, the ability to fit in and co-operate with the rest of the crew" (p. 39).

By the same token, the evidence that "intragenomic conflicts" do in fact exist does not negate the overall conclusion that, by and large, such conflicts are exceptions rather than the rule. Cosmides and Tooby (1981), in their original paper on the subject, pointed out that "there can be little doubt that the great proportion of cytoplasmic gene expression does act symbiotically with nuclear genes. . . . No part of the genome can act to any great extent against the interests of the rest, or it will be rendered inert by the rest of the genome through balancing selective processes, constituting a sort of 'parliament of the genes' [after Egbert Leigh, 1971]'' (pp. 86, 88).

The origin of chromosomes likewise may have involved a cooperative symbiotic process. This hypothesis, suggested by various theorists, has been developed as a formal model by Maynard Smith and Szathmáry (1993). The model suggests that if two (theoretically separate) genes interact cooperatively so that the two in combination are "synergistic" (functionally superior to each one alone), then there is clearly an advantage to becoming linked so that they will replicate and segregate together.

Sexual reproduction, one of the major outstanding puzzles in evolutionary theory, is also a cooperative phenomenon as the term is used here. Although there is still great uncertainty about the precise nature of the benefits—which are presumed to be very substantial given the (presumed) genetic costs—it can safely be assumed that sexual reproduction is, by and large, a mutually beneficial joint venture.

Eukaryotic cells, it is now generally agreed, can also be characterized as a cooperative venture—an obligate federation that may have originated as a symbiotic union (parasitic, predatory, or perhaps mutualistic) between ancient prokaryote hosts and what have now become cytoplasmic organelles, particularly the mitochondria, the chloroplasts, and, possibly, eukaryotic undulipodia (cilia) that may have evolved from structurally similar spirochete ancestors. The functional consequence, which we will discuss in greater depth later, was a union of specialists and a melding of complementary functions, both a "division of labor" and a cooperative "combination of labor."

The phenomenon of symbiosis, by definition a class of cooperative relationships in nature, provides yet another example. Symbiosis has traditionally been treated by mainstream biologists as an evolutionary sideshow. However, in the past decade or so it has emerged from the shadows. Not only has the darker side of symbiosis—parasitism—gained new prominence as a source of evolutionary change, but more benign commensalistic and mutualistic forms of symbiosis are also more widely appreciated. The case for "symbiogenesis" as a significant factor in evolution was documented by participants at a 1989 conference on the subject and in a subsequent volume edited by Margulis and Fester (1991). Among the supporting evidence:

- Mutualistic or commensalistic associations (not to mention parasitism) exist in all five "kingdoms" of organisms.

- Documentation that twenty-seven of seventy-five phyla in the four eukaryotic kingdoms (or 37 percent) exhibit symbiotic relationships.

- Over 90 percent of all modern land plants establish mycorrhizal associations.

- Land plants represent a joint venture between fungi and green algae.
- Approximately one-third of all known fungi are involved in mutualistic symbioses (e.g., lichens).
- Virtually all species of ruminants, including some 2,000 termites, 10,000 wood-boring beetles, and 200 Artiodactyla (deer, camels, antelope, and the like) are dependent upon endoparasitic bacteria, protoctists, or fungi for the breakdown of plant cellulose into usable cellulases.

Sociobiology is also, by definition, concerned with cooperative relationships. The social interactions in nature among members of the same species may be perturbed by free riders, "defectors," exploiters, or conspecific "parasites," and yet the fact remains that within-species cooperative behaviors are fairly common and encompass a broad array of survival-related functions, including (1) hunting and foraging collaboratively, which may serve to increase capture efficiency, the size of the prey that can be pursued, or the likelihood of finding food patches; (2) joint detection, avoidance of, and defense against predators, the forms of which range from mobbing and other kinds of coordinated attacks to flocking, herding, communal nesting, and synchronized reproduction; (3) shared protection of jointly acquired food caches, notably among many insects and some birds; (4) cooperative movement and migration, including the use of formations that increase aerodynamic or hydrodynamic efficiency and reduce individual energy costs and/or facilitate navigation; (5) cooperation in reproduction, which can include joint nest building, joint feeding, and joint protection of the young; and (6) shared environmental conditioning and thermoregulation.

One caveat should be mentioned at this point about where the boundary line should be drawn in using the term "cooperation." Although the definition proposed earlier could be stretched to include all of the nearly infinite number of functional interactions that occur in nature, including interactions with the physical environment, our usage is confined to those relationships and interactions that involve biological (and biosocial) processes—the interactions that occur within and between organisms. (One borderline category includes the "tools" and other artifacts—nests, hives, dams, galleries, spider webs, and the like—that are the products of biological processes.)

COOPERATION VERSUS COMPETITION

Is nature in fact red in tooth and claw? Is it a gladiatorial arena that is characterized by implacable competition? Charles Darwin's references to competition and the "struggle for existence," as many Darwin scholars have noted, were derived from the Malthusian assumption of a relentless increase in numbers and the presumption of inevitable ecological scarcity. However, many ecologists and paleontologists question (or at least qualify) the Malthusian assumption. Ecologists observe many patterns of interaction in nature that have the effect of mitigating direct ecological competition, such as outmigration, "habitat tracking,'' specialization and niche partitioning, and a variety of feedback-driven population-adjustment processes. Indeed, there are also niche-creating processes that may serve to reduce competition.

A number of other prominent biologists have called for a shift in the focus of the theoretical Necker cube. For instance, Edward O. Wilson asserts that ''the key remaining questions of evolutionary biology are more ecological than genetic in nature'' (1987, p. 1). A shift of focus away from a preoccupation with the differential reproductive success of genes to the process of adaptation—the causal dynamics of natural selection in the ''economy of nature''—has a number of implications, as we shall see. One implication that should be mentioned at this point is that cooperative interactions of various kinds (including those that are parasitical as well as those that are commensalistic or mutualistic) will assume much greater significance relative to direct ecological and reproductive competition. It may not be an exaggeration to say that competition and cooperation are coequal phenomena in terms of the lifetime process of earning a living and reproducing. In fact, there is very often in nature a complex interplay between competition and cooperation; many animals use cooperation as a means for more effective competition. Also, there are many cases in which the two modes of interaction coexist in a multifaceted relationship that combines competitive costs and offsetting cooperative benefits, as was pointed out many years ago by Alexander (1974). To cite a few specific examples:

• Eusocial insect species can generally occupy a broader spectrum of habitats and are often able to dominate and even exclude potential competitors among solitary and primitively social species. Nevertheless, they are not the harmonious communities that we once supposed.

• Prairie-dog colonies experience greater interpersonal aggression and a higher incidence of parasite infestations (conditions that can reduce fitness), but these disadvantages are more than offset by the joint increase in fitness due to greater protection from predators.

• Members of African lion prides cooperate and complete with one another in a variety of ways: Females typically hunt large prey in groups, share food, and may even share in guarding cubs and defending the pride. However, there is also much intracoalition competition for mating privileges among males.

• Most primates live in groups. But, as Walters and Seyfarth note in a chapter on conflict and cooperation in primate societies (in Smuts et al., 1987, p. 1306), ''life within a primate group is . . . delicately balanced between competition and co-operation.'' Aggressive interactions are commonplace. Nevertheless, the fitness benefits manifestly outweigh the costs.

Traditional neo-Darwinian theory, as purified by the selfish-gene perspective, attributes evolutionary change to competition among the ''replicators,'' the ultimate units of selection. In the neo-Darwinian model, cooperation plays a decidedly subsidiary role. But if we rotate the Necker cube and view evolution as an ecological/economic process—a survival enterprise in which living systems and their replicators are embedded—then differential reproductive success may be viewed as the result of a complex interplay of competitive and cooperative interactions (along with a variety of other factors), both within and among functionally interdependent "units" of ecological interaction. Our focus shifts to the activities of the "vehicles" (in Dawkins's terminology) or the "interactors" (in the terminology of the well-known philosopher of science, David Hull). It might be appropriate to label this approach "bioeconomics."

BIOECONOMICS

The term "bioeconomics" was first used, as far as I can determine, by an obscure turn-of-the-century theorist, Hermann Reinheimer, one of whose provocatively titled books was *Evolution by Co-operation: A Study in Bioeconomics* (1913). Reinheimer was intent on showing how cooperation produced benefits of various kinds in nature and human societies. Today, the term is utilized in a variety of ways, ranging from Georgescu-Roegen's (1977) thermodynamic approach to economics to the analyses of human-environment interactions. There is also a rapidly developing interdisciplinary bioeconomics movement, centered in economics, that has a variety of theoretical foci.

A somewhat different approach will be taken here. A Darwinian approach to the "economy of nature" focuses on "adaptation"—the means that are employed to earn a living in the environment. In other words, the basic problem is survival and reproduction. In a rash moment, George Williams (1966) called adaptation an "onerous concept." Presumably he was frustrated by the practical problem of defining it in such a way that it could be translated cleanly into selection coefficients in population-genetic models. However, bioeconomics (in this incarnation) is focused on a different problem, namely, how a given organism pursues its personal "survival enterprise" in a given environment. The focus is not on explaining how a trait may have evolved but on the functional interactions between an organism and its environment, how an animal uses its capabilities and resources to meet its survival and reproductive needs.

Some features of a bioeconomics paradigm are as follows. First, an implicit "core" assumption is that organisms are not oriented to evolutionary competition; they are oriented to real-time biological/ecological problems. In this view, direct reproductive competition is ecologically driven. Thus bioeconomics is not primarily concerned with reproductive success but with how an organism "earns its living"—how it meets its specific biological needs in a specific environment. Because the problem is always context specific, so must be the responses. Bioeconomics is focused not on "optimizing" but on Nobelist Herbert Simon's concept of "satisficing"—how efficiently an organism uses its resources and capabilities to meet its needs.

Second, the basic calculus involves the relationship between costs and benefits in connection with biological needs satisfaction. The "level of analysis" in this framework is not fixed. It involves a nested hierarchy/holarchy of biological units based on the degree to which there are functional interdependencies in connection with various aspects of the overall problem of needs satisfaction. In any given situation, the unit of analysis could be an individual organism, a symbiotic relationship, a mating pair, a socially organized group, a relationship between groups, or an ecosystem. Indeed, various aspects of any given organism's survival enterprise could be organized simultaneously at several levels. Finally, there is no reason in principle why a bioeconomics framework could not be applied evenhandedly to all living organisms, inclusive of humans.

There are numerous propositions and predictions that might be derived from this perspective. A few "hopeful monsters"—after Richard Goldschmidt's (1948, 1952) much-derided but maybe not entirely hopeless theory of macromutations—are as follows: (1) The calculus of direct bioeconomic costs and benefits will provide a better predictor of cooperative behaviors than will the degree of biological relatedness (inclusive fitness). The latter consideration may bias "all-other-things-being-equal" situations, but biological kinship is neither necessary nor sufficient as a condition for cooperation. (2) Organisms do not as a rule maximize the number of offspring. They can be expected to strike a balance between viability and reproduction and will often trade off some potential fecundity for an increased likelihood of survival. (3) Direct (zero-sum) reproductive competition is more likely to be a result of direct bioeconomic/ecological competition than vice versa.

In an overview and analysis of cooperative behaviors, Jerram Brown (1983, p. 29) noted: "Natural selection is an ecological process and cannot be understood solely from genetic considerations. Relatedness to nondescendants does not determine the direction or product of natural selection; it only supplies an additional cost or benefit." Moreover, Jon Seger (1991), echoing Darwin's proposed explanation for human evolution in *The Descent of Man*, points out that the various hypothesized explanations for social life (inclusive fitness, parental manipulation, group selection, and mutualism) are not mutually exclusive and in many cases might reinforce one another. The implication of these statements is that we should expect to find a significant degree of decoupling between the degrees of genetic relatedness and cooperative behaviors.

There are four sources of evidence for this proposition. First, there is the entire domain of symbioses. Here we have a wide range of cooperative relationships that can only be accounted for in bioeconomic, cost-benefit terms. Kinship is largely irrelevant. Second, there is supporting evidence in the various game-theoretic models of cooperation between unrelated individuals, along with the substantial research literature that these models have inspired. (We will discuss these further later.) Third, there is the entire category of outbreeding reproduction, a class of cooperative behaviors that, by definition, falls outside of the inclusive-fitness model. Finally, over the past decade or so there have been many field and laboratory studies of cooperation among conspecifics that are inconsistent with inclusive-fitness theory and/or suggest that the particular behaviors in question are more satisfactorily explained in bioeconomic terms.

Following are a few examples from the more extensive reviews of the literature that can be found in Corning (1995a, 1996a, 1997): In birds, Ligon and Ligon (1982) analyzed the communal nesting and extensive helping behaviors among green woodhoopoes (*Phoeniculus purpureus*), both among closely related and unrelated birds. They found that this behavior pattern markedly increased the woodhoopoes' likelihood of survival and reproductive success in an East African environment characterized by a severe shortage of suitable nest sites. Sherman et al. (1988) hypothesized that genetic diversity within social hymenoptera may have a previously unrecognized group-level advantage as a buffer against parasites and pathogens. In social carnivores, Packer and Pusey (1982) found that breeding coalitions of African male lions included nonrelatives much more commonly than kin-selection theory would predict. Scheel and Packer (1991) found a similar pattern in the hunting and cub-guarding behaviors of female lions.

Cooperative hunting is also a phenomenon that is found in a diverse range of species, from canids to felids, birds, and even some spiders. In a systematic reanalysis of the data from twenty-eight studies encompassing sixty species, Packer and Ruttan (1988) concluded that cooperative hunting is a behavioral pattern that is likely to occur only when there are "synergies"—per capita food acquisition that is greater than each participant could obtain alone. For example, they cite Kruuk's descriptions of how a solitary hyena would be unable to separate a wildebeest calf from its mother. But when hyenas hunt in pairs, there is a division of labor; one engages the mother while the other catches the calf. Packer and Ruttan also found that the propensity to hunt collaboratively varies greatly across species. They noted that the critical factor is not relatedness, but the size, abundance, and character of the prey and the degree of preexisting gregariousness among the hunters, which might be due to other ecological benefits, particularly protection against predation.

The extensive cost-benefit studies in vampire bats (*Desmodus rotundus*) by Wilkinson (1984, 1988, 1990) provide a particularly significant example of the capacity of functional/bioeconomic factors to transcend the influence of genetic relatedness in shaping cooperative behaviors. Wilkinson concludes: "Reciprocity is likely to be more beneficial than kin selection—provided that cheaters can be detected and excluded from the system" (1990, p. 82).

SYNERGY AND THE BIOECONOMICS OF COOPERATION

The common denominator in all of these diverse examples of cooperation is synergy—the combined effects produced by two or more elements, parts, or individuals. Derived from the Greek word *synergos* ("working together"), the term "synergy" is used here with reference to the functional effects that are produced by things that "operate together" (cooperation). Although the term is frequently associated with the slogan "the whole is greater than the sum of its parts," or "2 + 2 = 5," actually this is a caricature of a much more subtle and multifaceted concept. I prefer to say that the effects produced by wholes are different from what the parts can produce alone. In any case, a key to understanding synergy is that it involves effects that are jointly produced and interdependent. Take away a major part and the synergy will attenuate or dissolve. Here are few examples:

- Hemoglobin is a tetrameric protein, and each of its four monomers binds oxygen. However, hemoglobin also displays the remarkable property of positive cooperativity; binding activity by one monomer increases the binding affinity of the others.

- The observed error rate in normal cellular DNA replication is remarkably low (about 10^{-10} to 10^{-8} per base pair) compared with the theoretical potential, given the ambient sources of decay, damage, and copying errors of about 10^{-2}. The reason for this discrepancy is that it is the combined result of a complex set of mechanisms that "work together," including proofreading by DNA polymerases, methylation-instructed mismatch correction, enzymatic systems that repair or bypass potentially lethal or mutagenic DNA damage, processes that neutralize or detoxify mutagenic molecules, the regulation of nucleotide precursor pools, and, of course, the redundancy achieved by double-stranded genetic material (Haynes, 1991).

- Emperor penguins (*Aptenodytes forsteri*) are able to buffer themselves against the intense cold of the antarctic winter by huddling together in dense heat-sharing colonies numbering in the tens of thousands. Experiments have shown that, in so doing, the penguins are able to reduce their energy expenditures by as much as 50 percent (Le Maho, 1977). Similarly, honey bees, through joint heat production or fanning activities, as the need arises, are able to maintain the "core" temperature of their lives within a narrow range (Gould and-Gould, 1995).

- The African honey-guide (*Indicator indicator*) is an unusual bird species that utilizes beeswax as a food source. However, in order to obtain beeswax, these birds must rely on a coordinated search-and-destroy effort with a symbiont, such as the African badger (*Mellivora capensis*), which has the capacity to dismember the hive (and to consume only the honey, leaving the beeswax behind). However, this unusual example of symbiotic predation depends on a third symbiont, a gut bacterium associated with the honey-guide that produces an enzyme that can break down wax molecules (Bonner, 1988).

- Rowing seems to be a popular sport with biologists. In addition to Dawkins's example of the Oxford-Cambridge boat race, Maynard Smith and Szathmáry (1995) use as a metaphor specifically for synergy the image of two men in a rowboat, each with one oar. If only one oarsman is rowing, the boat will go in circles. A quantitative example can also be added to these nautical metaphors: A world-class "varsity eight" (plus a coxswain) can cover 2,000 meters over the water in about 5.5 minutes. However, a single sculler can at best row the same distance in about 7 minutes. The difference is a synergistic effect.

As these examples suggest, there are many different kinds of synergistic effects. Some involve threshold effects, "phase transitions," or density-dependent phenomena. Although each of the parts in such cases may have an additive relationship to any other part, the parts in combination may produce nonlinear effects. For instance, the players in a classic tug-of-war might be evenly divided, so that there is a stalemate. But if you add one more player to either side, the war may soon be over.

A second type of synergy involves "emergent phenomena." Although biologists frequently use the term broadly, it is restricted here to situations in which two or more "parts" merge in such a way that a new "whole" arises with distinctive new chemical, physical, and/or functional properties. Indeed, emergent phenomena are ubiquitous in the biological realm, ranging from protein molecules to the products of diploid reproduction to the elaborately constructed hives of honey-bee colonies.

Another kind of synergy involves what might be called a "combination of labor," sort of a mirror image of the division of labor. Many symbiotic partnerships fall into this category; the symbionts typically provide complementary functional capabilities or resources for one another. Lichen is an obvious case in point; the fungal partner provides surface-gripping and water-retention capabilities, while the cyanobacterium or green alga brings photosynthesizing capabilities to the relationship. Moreover, a comparative study by Raven (1992) documented that lichen symbioses typically result in an enhancement of nutrient and energy uptake compared with the performance of their asymbiotic "ancestors." Among members of the same species, communal nesting, joint foraging, reproductive "coalitions," collective migration, heat sharing, mutual defense and mobbing behaviors, pleometrosis, and cooperative hunting could all be considered combinations of labor with synergistic effects.

What is generally referred to as a "division of labor" is, of course, a particularly important generator of synergies in nature and human societies alike. As defined here, a division of labor differs from a combination of labor only in the sense that in the latter case, the focus is on breaking up a global task into parts in such a way that "differentiation" and specialization can be used to achieve efficiencies. (Numerous examples, from cyanobacteria to algae, insects, birds, and mammals of various kinds, are cited in Corning, 1996a.)

It should be emphasized that synergistic effects are measurable and quantifiable even when we may not completely understand the underlying material causes. There are numerous measuring rods: economics of scale, reduced energy expenditures, higher yields, lower mortality rates, a larger number of viable offspring, and so on. However, it should also be stressed that synergy is always context specific and contingent: for example, African lions can do better by hunting small, isolated prey alone; if wild dogs were ruminants, sociality would certainly not have any nutritional benefits; and collective defense by hamadryas baboons is relevant only because of the presence of large predators.

A related point is that synergy is a phenomenon that can have both positive and negative consequences from the point of view of the various participants. This is obviously the case in zero-sum types of parasitism, where the parasite's gain is at the expense of the host. It is also true in the many instances in which cooperative behaviors involve tradeoffs. For instance, the benefits of joint protection against predation may be offset by increased susceptibility to parasites, or increased competition for mating privileges, or perhaps even increased conspicuousness to potential predators. Often, in fact, the balance between positive and negative synergies may be critical in determining the net benefits, if any, and the ultimate likelihood that cooperation will occur. Also, there are many threshold phenomena in which more of a good thing may become bad.

It should also be noted that use of the synergy concept by evolutionary biologists has increased considerably in the past decade or so, although it is still most frequently employed in a relatively narrow spectrum of research specialties. (A computer search of a biological sciences database for the year 1993 identified over 10,000 references, but most were associated with biochemistry, endocrinology, pharmacology, and related disciplines.) Some explicit uses of the synergy concept in evolutionary biology and sociobiology include Kondrashov's (1982, 1988) hypothesis regarding the basis of sexual reproduction, which relies on synergistic linkages between deleterious mutations. Similarly, Maynard Smith and Szathmáry's (1993) theory of the origin of chromosomes postulates a synergistic relationship among primordial genes. Szathmáry (1993) also utilizes the concept in a model derived from metabolic control theory which suggests that under some conditions, two mutations affecting a metabolic pathway could act synergistically. Rosenberg (1991) postulates a necessary role for "synergistic selection" in the evolution of warning coloration (aposematism) in marine gastropods. The synergy arises when a potential predator has multiple "distasteful" encounters with the same morph, which enhances the joint selective value for each bearer. Hurst (1990) suggests that parasite diversity in a given cell or organism may be more burdensome than a similar quantity of uniform types because various synergistic interactions among different parasites may enhance their mutual effects. Hurst proposes that diploidy, multicellularity, and anisogamy may be antiparasite mechanisms; they might serve to reduce parasite diversity.

Synergy has also been deployed in some recent sociobiological studies. Santillán-Doherty and his colleagues (1991), in a study of stump-tailed macaques (*Macaca arctoides*), found nonlinear synergistic effects among three variables—kinship, sex, and rank—in shaping the interactions among the animals in their study population. Packer and Ruttan (1988) also explicitly recognized the role of synergy in cooperative hunting. They observed that when individual hunting success is already high, there is little to be gained by cooperating. Cooperation depends on synergy—an increase in the average individual feeding efficiency through joint efforts. "An increase in hunting success with group size therefore indicates synergism from co-operation, whereas a decrease indicates some form of interference [negative synergy]" (1988, p. 183).

Maynard Smith's use of the synergy concept deserves special note. Known for his introduction of game-theory models into evolutionary theory (among other contributions), Maynard Smith (1982) coined the term "synergistic selection'' more or less as a synonym for D. S. Wilson's (1975, 1980) concept of "trait group selection'' and a similar formulation by Matessi and Jayakar (1976), both of which sought to account for the evolution of altruism without the need for inclusive-fitness theory. The general approach involved temporary (functional) interactions among nonrelatives in nonreproductive groups. The key feature of the synergistic-selection model, according to Maynard Smith, was a fitness gain to interacting altruists that was greater than the gain to an altruist and a nonaltruist. (At this point, Maynard Smith, like many other theorists, was conflating altruism and cooperation.)

Maynard Smith and Szathmáry also make liberal use of the synergy concept in their volume on the evolution of complexity, *The Major Transitions in Evolution* (1995). However, in their treatment, synergy moves on and off stage at frequent intervals. At times, they abjure the term "synergy'' altogether in favor of more conventional synonyms (or else in favor of a more process-oriented term). Thus "autocatalysis'' is featured in relation to the origins of life but not its synergistic functional results. Similarly, hypercycles are featured but not their synergies, although synergy is mentioned in Szathmáry's "stochastic-corrector'' model. (Maynard Smith, in a personal communication, acknowledges that the "universal'' significance of synergy only became apparent to them after their volume was completed.)

THE EVOLUTION OF COOPERATION

In *The Synergism Hypothesis* (Corning, 1983; also see Corning, 1996a), I proposed that synergistic phenomena of various kinds have played a key causal role in the evolution of cooperation generally and in the evolution of complex systems in particular; I argued that a common functional principle has been associated with the various steps in this important directional trend. The reasoning behind this hypothesis can be briefly summarized. The cardinal point is this: It is the functional (bioeconomic) effects or consequences of various organism-environment pattern changes, insofar as they may impact on differential survival, that constitute the "causes'' of natural selection. Another way of putting it is that causation in evolution runs backwards from our conventional view of things. It is the "proximate'' functional effects that may result from any change in the organism-environment relationship that are the causes of the "ultimate'' (transgenerational) selective changes.

This is where synergistic phenomena fit into the process. Cooperative interactions in nature that produce positive functional consequences, however they may arise, can become "units'' of selection that differentially favor the survival and reproduction of the "parts'' (and their genes). In other words, it is the proximate advantages (the payoffs) associated with various synergistic interactions (in relation to the particular organism's needs) that constitute the underlying cause of the evolution of cooperative relationships and complex organization in nature.

This theory of cooperation/complexification in evolution is particularly relevant to symbiosis and sociobiology. The hypothesis is that the synergies that may result from cooperative behaviors are the very cause of their systematic evolution over time via their impacts on differential survival and reproduction. Moreover, many of these evolutionary changes originate (and are initially adopted) at the behavioral level. For, as Ernst Mayr (1960) observed, behavioral innovations are often the "pacemaker" of evolutionary change. In C. H. Waddington's words, "It is the animal's behavior which to a considerable extent determines the nature of the environment to which it will submit itself and the character of the selective forces with which it will consent to wrestle" (1975, p. 170).

One way of bringing this theoretical perspective into better focus might be to revisit the plethora of formal models of inclusive fitness and cooperative behavior that have appeared in recent years. A key assumption that burdened the birth pangs of sociobiology was the notion that cooperation and altruism are synonymous. However, this assumption collapses once genetic altruism (and kinship) is removed as a precondition, and game theory by its very nature challenged the inclusive-fitness assumption. Game theory suggested that cooperative behaviors could evolve independently of kinship, given an appropriate set of strategic circumstances and behaviors. Although the focus has always been on the behavioral context and the strategies of the "players," if one looks closely at the various game-theory formalizations, they tacitly depend on an interaction between the behavior of the players and the structure of the payoff matrix. If one looks closely at the payoff matrices in some of the "classic" formulations, like tit-for-tat, the cooperative strategies in turn depend on synergy.

Many other game-theory models have flourished since the pioneering work of Maynard Smith, Axelrod and Hamilton, and others, but to my knowledge, none have utilized anything other than a payoff matrix that posits higher joint rewards for cooperation. In retrospect, the reason is obvious. In a general analysis of these models, Peck (1993, p. 195) concluded that "the position of [stable] equilibria (and hence the frequency of co-operators) depends on the size of the various payoffs that define the Prisoner's Dilemma game." (As Maynard Smith has pointed out, if a cooperating group of n individuals can produce $3n$ offspring, it pays to cooperate.) In addition, many theorists have observed that cooperation can best be maintained among individuals who have a continuing relationship, regardless of kinship.

How can "cheating" or "defection" (the prisoner's dilemma) be prevented? Maynard Smith and Szathmáry's response is rendered in terms of game theory. They posit two different kinds of game situations. The first is a "sculling" model in which two oarsmen each have two oars and row in tandem. In this model it is easy for one oarsman to slack off and let the other one do the heavy work. This corresponds to the prisoner's dilemma game. However, in a two-person "rowing" model, each oarsman has only one opposing oar. Now their relationship to the performance of the boat is interdependent. If either oarsman slacks off, the boat will go in circles. In this case, mutual cooperation becomes an evolutionarily stable strategy (ESS). Maynard Smith and Szathmáry conclude that the rowing model is a better representation of how cooperation evolves in nature: "The intellectual fascination of the Prisoner's Dilemma game may have led us to overestimate its evolutionary importance" (1995, p. 261).

Some implications associated with the application of a synergy perspective in evolutionary biology and sociobiology include the following:

- The interdependency of parts and wholes implies that one cannot focus only on one level in attempting to understand either the causal dynamics or the functional properties of a "system." Interactions between levels may be critically important to understanding the dynamics at a given level.

- Group selection at the level of the whole may or may not be disadvantageous to the parts; the theoretical card deck has been unnecessarily "loaded" by the assumption that sociality is equivalent to altruism. As Darwin noted long ago, various levels of selection may reinforce one another.

- One can test for the presence of a functional/trait group (synergistic selection) with the following type of experiment (or "thought experiment"): take away any major part and observe the consequences for the other parts.

- An important corollary is the "paradox of dependency." Cooperative interactions may produce fitness-enhancing synergies, but a tradeoff may be that the "parts" become dependent on the "whole." Wholes may become "obligatory" survival units, one consequence of which may be that a breakdown of the whole results in the demise of the parts.

Earlier, it was asserted that the bioeconomic synergies produced by the parts in combination are the underlying cause of the evolution of wholes; in this dynamic, wholes become the "units" of selection (i.e., group selection). Among the supporting evidence for this proposition, reviewed in Corning (1995a, 1996a, 1997), some examples cited by Maynard Smith (1982) are representative: In orb-web spiders (*Metabus gravidus*), groups of fifteen to twenty females may cooperate in building a joint web to span a stream where prey are abundant; tropical wasps (*Metapolybia aztecoides*) often establish joint nests for their mutual benefit; and coalitions of lion males cooperate in taking over and holding a pride, thereby achieving collective effects that would not otherwise be possible.

Does this shift of focus in the Necker cube from individual interactions to group-level benefits make any difference? Yes, because we are then focused on the causal level—the functional synergies and their influence on natural selection. A "functional-group-selection" framework also reinvigorates a concept that Williams may have thought he had banished from evolutionary biology: the "good of the group" or "downward causation." The term "downward causation" was coined by psychobiologist Roger Sperry (1969, 1991, and elsewhere) in connection with the functional organization and operation of the human brain.

(Psychologist Donald Campbell, 1974, evidently developed the concept independently; he used it with reference to control processes in hierarchically organized biological systems.) Sperry was fond of using as an illustration the
metaphor of a wheel rolling downhill; its rim, all of its spokes, and, indeed, all
of its atoms are compelled to go along for the ride.

We will use the term here in a slightly different sense. "Downward causation" in this context refers to the selective influences that have shaped the
evolution of complexity. Why do selfish genes cooperate in ways that lead to
their becoming interdependent? How did morphological castes and a division
of labor evolve in naked mole rats? How do reproductive controls evolve in
mutualistic symbioses in which, as Margulis (1993) points out, there must of
necessity be reproductive synchronization if the relationship is to remain stable?
Downward causation in an evolutionary context refers to the fact that the functional (synergistic) properties of the "whole" become a selective "screen"—a
significant influence on the differential survival and reproduction of the parts.
Sometimes the parts might be disadvantaged (e.g., nonreproductive workers),
and kin selection may help us to understand how such "sacrifices" for the
common good may occur. But, as the evidence cited earlier indicates, kinship
is not a sine qua non. The whole may be sustained by fitness tradeoffs; the costs
may be offset by commensurate benefits. For instance, an animal that is at risk
from predators might suffer a reduction in its relative reproductive fitness in a
social group setting, but it may also enjoy greatly enhanced odds of survival
and absolute fitness. (This may well help to explain why "defeated" contenders
for breeding privileges sometimes stay on in the group and may even serve as
helpers.)

What about the problem of cheating? From a functional (synergy) perspective,
if cooperation offers sufficient benefits, it may be in the interest of some individuals to coerce cooperation from others (see the evidence reviewed in Boyd
& Richerson, 1992; Clutton-Brock & Parker, 1995). Inclusive fitness provides
one possible explanation for punishment as a successful strategy in social
groups. Another might be the sort of individual fitness tradeoffs referred to
earlier. But Leigh (1991) has proposed that group selection could also provide
a "mechanism." The enforcement of cooperation in "the common interest"
might have significant fitness-enhancing value for the members of groups that
are in competition with other groups or other species.

SOME IMPLICATIONS

Among the many implications of a synergic perspective, I will briefly mention
five. First, neither the "selfish-gene" nor the "holistic" perspective is sufficient
to encompass the causal dynamics of the evolutionary process. The genes are
in fact encased in a hierarchy/holarchy of functionally interdependent "systems"; nature exhibits a complex structure of parts-wholes relationships. Furthermore, each level of wholeness represents an emergent unit of causation. The

appropriate focus for the Necker cube should be on functional relationships and their bioeconomic consequences. Consider the textbook example of Lamarck's giraffes. Was it genetic variability among *Giraffidae* that "caused" the evolution of their distinctive necks (as emphasized in the classical Darwinian view), or was it the adoption by ancestral giraffes of the distinctive "habit" of eating the leaves in the tops of acacia trees (as Lamarck proposed)? The correct answer, of course, is both. A behavioral innovation by the whole organism exerted "downward causation" that selectively favored certain available gene combinations.

A second implication is that the segregation of research and theory between symbiosis/parasitology, in one camp, and sociobiology, in the other, is neither warranted nor productive. A focus on the bioeconomics of cooperation puts genetic relatedness in its proper place as one factor in the equation, but by no means is it the driving force behind cooperation in nature. It is not inclusive-fitness considerations but the functional consequences of cooperative interactions that account for their evolution, an explanatory paradigm that applies equally to symbiosis and sociobiology.

Third, as suggested earlier, this approach also brings an interactional perspective to game theory. The strategies of the players may play a role, but the numbers in the payoff matrix and the real-world synergies they represent are also crucial. More important, this perspective highlights a major limitation of game theory. Game theory has served admirably to shed light on cooperative phenomena from the point of view of the individual "players" and their behavior. Many valuable insights have been gained from the use of this important theoretical tool. The problem is that game theory views all social interactions as if they were analogous to dyadic transactions in an economic "market." (Indeed, game theory was imported from microeconomics.) The players are presumed to have only limited "exchange" relationships with one another, and no overriding common or joint interests. Even Adam Smith knew better than that. While many cooperative relationships in nature may involve marketlike transactions, many more resemble the more complex relationships that are often found in human societies—alliances for the production of "collective goods," "cooperatives," "mutual aid societies" (which we now call insurance companies), "joint-venture partnerships," and even contractual corporations ("fictitious individuals" in legalese).

Accordingly, a fourth implication is that symbiology and sociobiology could benefit from the application of a "systems" perspective. One example is what might be called a "corporate-goods" model, as distinct from the "collective-goods" models that have become fixtures in economic theory (Olson, 1965). Some of the distinctive characteristics of a corporate-goods model (which, not coincidentally, resembles human organizations) are the following: (*a*) there is an overarching goal (implicit or explicit), which can often be broken down into a number of component subgoals, some or all of which may be pursued simultaneously; (*b*) individual "stakeholders" in the corporation may at one and the same time have exclusive personal "interests" and varying degrees of overlapping joint stakes in the viability of the corporation; (c) some of the outputs of the corporation may take the form of indivisible, jointly shared "collective goods," while others ("corporate goods") may be divisible and may be distributed more or less symmetrically; (d) likewise, the "costs" associated with producing the goods may be borne "equitably" or may be highly skewed; in other words, the ratio of costs and benefits to various participants may vary considerably; (e) role specialization is common but is not a necessary condition; (f) corporate organizations can range from being facultative to being completely obligatory; and (g) "government" (cybernetic control processes) is a necessary concomitant of any social group, and its importance intensifies as the degree of behavioral coordination and integration increases. (Indeed, communication is important in any cooperative activity. However, cybernetic controls may be "centralized" or distributed, or both, and the mechanisms for achieving cybernetic control range from chemical "signals" to sophisticated visual and auditory "languages.")

The fifth and final point returns us to our leading question, "Why are selfish genes so often cooperative?" The answer, in a nutshell, is not that they are surreptitiously altruistic but that cooperation represents an often advantageous survival strategy and a way to compete more effectively. The paradox, however, is that by cooperating in the pursuit of their own interests they also advance the interests of others. Moreover, invention has become the mother of necessity; selfish genes have come to be dependent on one another. Thus a complex organism, or superorganism, also represents a "collective survival enterprise," a functional unit of survival and reproduction in which the corporate interest of the "whole" becomes a filter or screen that differentially affects the survival and reproduction of the parts. It is time to focus more intently on this aspect of the evolutionary process.

REFERENCES

Alexander, R. D. (1974). The evolution of social behavior. *Annual Review of Ecology and Systematics, 5*, 325–383.

Bonner, J. T. (1988). *The evolution of complexity by means of natural selection.* Princeton, NJ: Princeton University Press.

Boyd, R., and Richerson, P. J. (1992). Punishment allows the evolution of cooperation (or anything else) in sizable groups. *Ethology and Sociobiology, 13*, 171–195.

Brown, J. L. (1983). Cooperation—A biologist's dilemma. *Advances in the Study of Behavior, 13*, 1–37.

Campbell, D. T. (1974). Downward causation in hierarchically organized biological systems. In T. Dobzhansky & F. J. Ayala (Eds.), *Studies in the philosophy of biology* (pp. 85–90). Berkeley: University of California Press.

Clutton-Brock, T. H., and Parker, G. A. (1995). Punishment in animal societies. *Nature, 373*, 209–216.

Corning, P. A. (1983). *The synergism hypothesis: A theory of progressive evolution*. New York: McGraw-Hill.

———. (1995a, August 2–6). *The co-operative gene: Synergy and the bioeconomics of evolution*. Paper prepared for the Annual Meeting of the European Sociobiological Society, Cambridge, United Kingdom.

———. (1995b). Synergy and self-organization in the evolution of complex systems. *Systems Research*, *12*(2), 89–121.

———. (1996a). The co-operative gene: On the role of synergy in evolution. *Evolutionary Theory*, *11*, 183–207.

———. (1996b). Synergy, cybernetics, and the evolution of politics. *International Political Science Review*, *17*(1), 91–119.

———. (1997). Holistic Darwinism: 'Synergistic selection' and the evolutionary process. *Journal of Social and Evolutionary Systems*, *20*(4), 363–400.

Cosmides, L., & Tooby, J. (1981). Cytoplasmic inheritance and intragenomic conflict. *Journal of Theoretical Biology*, *89*, 83–129.

Dawkins, R. (1989 1976). *The selfish gene* (new ed.). Oxford: Oxford University Press.

Georgescu-Roegen, N. (1977a). A bioeconomic viewpoint. *Review of Social Economy*, *35*, 361–375.

———. (1977b). Bioeconomics: A new look at the nature of economic activity. In L. Junker (Ed.), *The political economy of food and energy* (pp. 105–133). Ann Arbor: University of Michigan Press.

Goldschmidt, R. B. (1948). Egotype, ecospecies and macroevolution. *Experientia, 4*, 465–472.

———. (1952). Evolution as viewed by one geneticist. *American Scientist*, *40*, 84–98.

Gould, J. L., and Gould, G. C. (1995). *The honey bee*. New York: Scientific American Library.

Haynes, R. H. (1991). Modes of mutation and repair in evolutionary rhythms. In L. Margulis & R. Fester (Eds.), *Symbiosis as a source of evolutionary innovation*. Cambridge, MA: MIT Press.

Hurst, L. D. (1990). Parasite diversity and the evolution of diploidy, multicellularity, and anisogamy. *Journal of Theoretical Biology, 144*, 429–443.

Kondrashov, A. S. (1982). Selection against harmful mutations in large sexual and asexual populations. *Genetical Research, 40*, 325–332.

———. (1988). Deleterious mutations and the evolution of sexual reproduction. *Nature, 336*, 435–440.

Le Maho, Y. (1977). The emperor penguin: A strategy to live and breed in the cold. *American Scientist, 65*, 680–693.

Leigh, E. (1971). *Adaptation and diversity*. San Francisco: Freeman, Cooper & Co.

———. (1991). Genes, bees and ecosystems: The evolution of a common interest among individuals. *Trends in Ecology and Evolution (Tree), 6*: 257–262.

Ligon, J. D., & Ligon, S. H. (1982). The cooperative breeding behavior of the green woodhoopoe. *Scientific American, 247*(1), 126–134.

Margulis, L. (1993). *Symbiosis in cell evolution* (2nd ed.). New York: Freeman.

Margulis, L., & Fester, R. (Eds. 1991). *Symbiosis as a source of evolutionary innovation: Speciation and morphogenesis*. Cambridge, MA: MIT Press.

Matessi, C., & Jayakar, S. D. (1976). Conditions for the evolution of altruism under Darwinian selection. *Theoretical Population Biology, 9*, 360–387.

Maynard Smith, J. (1982). The evolution of social behavior: A classification of models.

In King's College Sociobiology Group (Ed.), *Current problems in sociobiology* (pp. 28–44). Cambridge: Cambridge University Press.

Maynard Smith, J., & Szathmáry, E. (1993). The origin of chromosomes: I. Selection for linkage. *Journal of Theoretical Biology, 164*, 437–446.

———. (1995). *The major transitions in evolution*. Oxford: Freeman Press.

Mayr, E. (1960). The emergence of evolutionary novelties. In S. Tax (Ed.), *Evolution after Darwin*, Vol I. Chicago: University of Chicago Press.

Olson, M. (1965). *The logic of collective action: Public goods and the theory of groups*. Cambridge, MA: Harvard University Press.

Packer, C., & Pusey, A. E. (1982). Cooperation and competition within coalitions of male lions: Kin selection or game theory? *Nature, 296*, 740–742.

Packer, C., & Ruttan, L. (1988). The evolution of cooperative hunting. *American Naturalist, 132* (2), 159–198.

Peck, J. R. (1993). Friendship and the evolution of co-operation. *Journal of Theoretical Biology, 162*, 195–228.

Rapoport, A. (1991). Ideological commitments in evolutionary theories. *Journal of Social Issues, 47*(3), 83–99.

Raven, J. A. (1992). Energy and nutrient acquisition by autotrophic symbioses and their asymbiotic ancestors. *Symbiosis, 14*, 33–60.

Reinheimer, H. (1913). *Evolution by co-operation: A study in bio-economics*. London: Kegan Paul, Trench, Trubner.

Rosenberg, G. (1991). Aposematism and synergistic selection in marine gastropods. *Evolution, 45*, 451–454.

Santillán-Doherty, A. M., et al. (1991). Synergistic effects of kinship, sex, and rank in the behavioural interactions of captive stump-tailed macaques. *Folia Primatologica, 56*, 177–189.

Scheel, D., & Packer, C. (1991). Group hunting behaviour of lions: A search for cooperation. *Animal Behaviour, 41*(4), 697–710.

Seger, J. (1991). Cooperation and conflict in social insects. In J. R. Krebs and N. B. Davies (Eds.), *Behavioral ecology: An evolutionary approach* (2nd ed.), (pp. 338–373). Oxford: Blackwell Scientific Publications.

Sherman, P. W., et al. (1988). Parasites, pathogens, and polyandry in social hymenoptera. *American Naturalist, 131*(4), 602–610.

Smuts, B., et al. (Eds.). (1987). *Primate societies*. Chicago: University of Chicago Press.

Sperry, R. W. (1969). A modified concept of consciousness. *Psychological Review, 76*, 532–536.

———. (1991). In defense of mentalism and emergent interaction. *Journal of Mind and Behavior, 12*(2), 221–246.

Szathmáry, E. (1993). Do deleterious mutations act synergistically? Metabolic control theory provides a partial answer. *Genetics, 133*, 127–132.

Waddington, C. H. (1975). *The evolution of an evolutionist*. Ithaca, NY: Cornell University Press.

Weismann, A. (1889). *Essays upon heredity and kindred biological problems*. Oxford: Clarendon Press.

Wilkinson, G. S. (1984). Reciprocal food sharing in the vampire bat. *Nature, 308*, 181–184.

———. (1988). Reciprocal altruism in bats and other mammals. *Ethology and Sociobiology, 9*, 85–100.

———. (1990). Food sharing in vampire bats. *Scientific American, 262*(2), 76–82.

Williams, G. C. (1966). *Adaptation and natural selection: A critique of some current evolutionary thought*. Princeton, NJ: Princeton University Press.

Wilson, D. S. (1975). A general theory of group selection. *Proceedings of the National Academy of Sciences, 72,* 143–146.

———. (1980). *The natural selection of populations and communities*. Menlo Park, CA: Benjamin/Cummings.

Wilson, E. O. (1987). Causes of ecological success: The case of the ants. *Journal of Animal Ecology, 56,* 1–9.

PART II

SOCIOBIOLOGY AND CULTURE

The chapters in this part deal with the evolution of culture, certainly a fundamental theme in understanding the human species with its many cultures around the world. Peter Meyer sees human social systems as "autopoietic," designed around reciprocal patterns of interaction and quite consistent with a large number of biological social systems. In addition, and as we will see expressed in later chapters, he holds that human social systems are capable of utilizing the integrative forces coming from kin selection even though the biological relationships in human culture are much more diffuse than those utilizing a strictly individualistic view of natural selection. Thus there is no reason to assume that human culture operates in a frame apart from that of the evolution of animal societies. Robin Allott takes a related view of culture that focuses on the key influence of communication in the form of human language as a general capacity but also considers the significance of a multiplicity of human languages, each operating in a particular culture. Tracing language back to its origins, he finds this rooted in motor signals leading to the motor aspects of speech. This perspective of Allott's may be supported by a view of language comprehension recently emphasized in the linking of Wernicke's area in humans and chimpanzees.

Dennis Werner adds a philosophical dimension to these evolutionary perspectives by showing that mind and language can be brought together in an evolutionary frame quite consistent with studies of epistemology. Linguistics, cultural anthropology, and the philosophy of knowledge are only different facets of a larger whole. John Constable adds an interesting perspective in taking the view that science, reflected in biological science, can provide an understanding of the products of culture and mind found in the humanities. Rather than a humanistic rejection of scientific reductionism, it is better that we seek a scientific integration of the humanities along with our study of language, culture, and mind, thus allowing for a single unified perspective.

4

The Sociobiology of Human Cooperation: The Interplay of Ultimate and Proximate Causes

Peter Meyer

In the history of the social sciences, the causes underlying human cooperation have been a major problem. In fact, any social theory has to deal with the problem of whether cooperation and other forms of social behavior arise from the self-interest of individuals, or whether the foundations for cooperation are laid by society and therefore depend on a sort of cultural invention. This chapter suggests that sociobiology with its emphasis on genetic interests sheds some fresh light on the causes underlying conflicts of interest, as well as on those underlying cooperation. Before we embark upon sociobiology's contributions to an improved understanding of this problem, it may be worthwhile to look at some of the theoretical concepts put forward in the history of social theory.

One of the most influential social theorists to address the present issue was Thomas Hobbes. According to his view, individual human beings are engaged in continuous competition, their behavior driven mainly by their interest in self-preservation, leading to an endless striving for power over their competitors. Hobbes's solution to the nightmare of endless struggle was the establishment of a "commonwealth," an institution designed to protect the most cherished interests, mainly the lives and property, of individual citizens by seizing part of their powers, stripping them of some of the means for the continuation of strife. Regarding the origin of cooperation, Hobbes's theory suggests that in order to avoid endless struggle and continuous "fear of death," people will be compelled by "some coercive Power . . . to the performance of their Covenants, by the terrour of some punishment, greater than the benefit they expect by the breach of their Covenant" (Hobbes, 1968, p. 202). Adam Smith, another eminent social theorist, censured Hobbes's neglect of "sympathy," a natural source of human cooperation, as envisaged by Smith (Raphael, 1991, p. 14; Smith, 1986). This chapter suggests that Smith's criticism is plausible and is in fact in keeping with findings of evolutionary theory. Both theorists, while disagreeing on the origin of human cooperation, felt that cooperation must somehow be related to individuals' interests and the more or less rational ways in which they pursue them.

Unlike these theorists, some modern sociologists would not conceive of the individual as the ultimate source of human cooperation, but would rather point to the functional necessities of social systems as the causes for the evolution of cooperation. According to Niklas Luhmann, an influential modern social theorist, sociology should not be concerned with individuals at all, but rather specialize in the study of the nature of social systems. Further, social systems are to be regarded as completely independent of individual behavior. In fact, individuals are merely environments of these systems, with no sizable impact upon their operations (Luhmann, 1985, p. 358). This approach suggests that cooperation is not caused by individual considerations or interests, but is an outcome of the functional requirements of social systems.

Viewed from a metatheoretical perspective, these traditions of social theory proceed from two antagonistic positions, the former beginning at the "bottom," that is, the individual human being, the latter departing from the "top," that is, the social system and its functional requirements. Consequently, the former tradition may be termed "individualistic," the latter "collectivistic" or "systemic." Obviously, there is also disagreement as to the nature and origin of human cooperation: Whereas individualistic social theorists, on the one hand, emphasize the impact of "natural" individual tendencies and volitions, collectivistic theorists, on the other hand, typically tend to diminish such influences and put emphasis on systemic needs as the primary source of cooperation. With respect to the problem of the origin of human cooperation, it should be noted that the collectivistic approach, particularly Luhmann's systemic theory, fails to identify causes for the evolution of cooperation between people. Richard Münch has demonstrated that modern systems theorists, with their emphasis on the causes underlying systemic integration of modern society, must be accused of a total neglect of the causes underlying social integration. It should be added that Münch, in following Durkheim's views, puts emphasis on "bonds of solidarity" as the foundations of social integration, whereas systemic integration applies "money, political power or law" (Münch, 1995, p. 11). Leaving aside the causes of systemic integration, it is interesting to note that Münch, while pointing to solidarity as the basis of social integration, fails to identify the causes of solidarity. According to Münch, the neglect of the forces underlying social integration is primarily responsible for modern social theorists' failure to account for the recent return to nationalism, as well as for various forms of ethnic strife in the modern world (Münch, 1995, p. 16). Following Münch's view, I suggest that an improved understanding of the causes underlying human cooperation may well shed some fresh light on the causes underlying solidarity and social integration or its breakdown in some parts of the modern world.

According to the individualistic view, "egotism" and/or "sympathy" originate from the natural endowment of human individuals, whereas the collectivistic perspective maintains that these as well as other individual inclinations emanate from the functional requirements of social systems. Regarding the origin of cooperation, there are also opposing views, the individualistic perspective emphasizing conflicts of interest, and the collectivistic perspective stressing the "functional" effects of cooperation for the system as a whole. Unlike these collectivistic and systemic approaches, this chapter proceeds from an individualistic starting point, suggesting that sociobiology can account for some of the causes underlying the preference for cooperation over purely selfish behavior.

It may be obvious from sociobiology's emphasis on "genes" as the unit of selection (Dawkins, 1978) that it is a thoroughly individualistic approach. Despite some plausible criticisms of the notion of genes as units of selection (Smillie, 1995, p. 248), the "selfish-gene" concept will nevertheless be utilized in the present context, mostly for the sake of brevity. Unlike some of the individualistic theorists mentioned before, sociobiologists have "kin-selection" theory at their disposal, enabling them to account for egoistic behavior and conflicts of interest as well as for altruistic behavior within the boundaries of one theoretical framework. From a sociobiological point of view, conflicts of interest are likely to arise whenever two individuals do not "carry identical sets of genes" (Haig, 1993, p. 496), and, according to Haig, this is even the case in a mother's relation with her fetus. There is a conflict of interest during pregnancy when the fetus attempts "to remain in the womb until the nutritional benefits of remaining inside are not worth the increasing risks of delivery" (Haig, 1993, p. 516).

With regard to this type of conflicts of interest, it should be stressed that since there is no conscious decision making on the part of either mother or fetus, the underlying conflict must be clearly distinguished from other types of conflicts of interest in which individuals consciously pursue their interests. Therefore Smillie's warning that the concept of conflicts of interest must not be applied to situations similar to the mother-fetus relation other than in a metaphorical manner should be seriously considered (Smillie, 1995, p. 239). Unlike conflicts of interest in which individuals more or less rationally pursue their interests, the fetus cannot deliberately influence its mother's behavior. On the contrary, its behavior and its mother's behavior, as well as the conflict of interest between them, are products of natural selection.

It should be emphasized once again that the majority of conflicts studied by kin-selection theory are of an impersonal and involuntary nature. It should be expected that altruism, that is, the investment of precious time and energy by one individual in another individual, should be selected against. However, according to kin-selection theory, it is in fact favored by natural selection. It was a major breakthrough in social theory when Hamilton's kin-selection theory (Hamilton, 1978) presented a resolution to the problem of "why natural selection does not select against all types of altruistic behavior" (Voland, 1996,

p. 96). According to this theory, altruistic acts will be the more likely the closer donor and recipient are genetically related, because by his investment the donor may enhance his reproductive success by supporting his relative's offspring. According to kin-selection theory, an altruistic act is favored by natural selection because although the act is not necessarily directly reciprocated by the recipient, due to their genetic relatedness, the investment is likely to enhance both the donor's and the recipient's fitness and hence serves common interests. Taking Haig's findings into consideration, it should be added that despite close genetic relatedness, conflicts of interest are likely to arise in certain situations, for instance, because "offspring may be selected to take more from a parent than the parent is selected to give" (Haig, 1993, p. 497).

Although there is "no direct evidence, regarding the degree of reciprocal altruism practiced during human evolution nor its genetic basis today ... it is reasonable to assume that it has been an important factor in recent human evolution and that the underlying emotional dispositions affecting altruistic behavior have important genetic components" (Trivers, 1971, p. 48). The fruitfulness of the concept of reciprocal altruism for an understanding of human behavior has been further enhanced by Alexander's work on "indirect reciprocity." According to Alexander, "indirect reciprocity" occurs when the first investor is not repaid by the original debtor, but by a third person who had not been involved in the primary exchange (Alexander, 1986, p. 107). The occurrence of "indirect reciprocity" presupposes the mental capacities for observing the original transaction on the part of a third person and, above all, a competence for taking over the "moral" obligations from this transaction.

With regard to the behavioral mechanisms underlying reciprocity, there may be no objection to regarding "indirect reciprocity" as a foundation of peculiarly human patterns of cooperation, as suggested by Alexander. It should, however, be emphasized that reciprocal altruism is a part of this foundation as well (Badcock, 1991, p. 259). According to Miller, reciprocal altruism "is time-delayed reciprocity. . . . Because reciprocity occurs, this kind of altruism is consistent with inclusive fitness" (Miller, 1993, p. 237). In further following Miller's view, it may be suggested that the behavioral basis for this simple type of reciprocity may be "some instinctual precursor of emotional sympathy" (Miller, 1993, p. 238) that is active in lower animals as well as in man. In the mammals, particularly in the human species, emotional processes replace most of these instincts. Emotional processes "can correlate the motivations of individuals" in such a way that by sheer physical proximity, "the biorhythms . . . of mammals, including humans, become synchronized" (Miller, 1993, p. 238). This synchronization, in turn, compels individuals to create social bonds "that motivate individuals to constrain mutually destructive, self-interested pursuits of power" (Miller, 1993, p. 238) or of pure egotism. With regard to Alexander's indirect reciprocity, it may be gathered from Miller's considerations that although the existence of mental capacities is a precondition for the occurrence of this type of reciprocity, due consideration of emotional processes is even more important for an understanding of the evolutionary underpinnings of human reciprocity. A later section of this chapter will turn to some relevant aspects of emotional processes. Regarding human behavior in general, it may be summarized that "over the lengthy course of hominid evolution, a predisposition to co-operate with other members of one's community enhanced the individual's ability to maximize inclusive fitness" (Crippen, 1994, p. 326), and this is the ultimate cause why cooperation was favored as compared to pure egotism.

Various levels of complexity of the behavioral mechanisms underlying reciprocity must be taken into account. As Miller observes, even individual reptiles manage to imitate other individuals' behaviors at times, thereby synchronizing individual behaviors. Imitation plays a role in human behavior too, but other causes, such as emotions and mental operations, overlap with it. In human evolution, the combination of these distinct causes led to very complex patterns of reciprocity, including the long-term "storage of social obligations" (Wiessner, 1982, p. 65), which are major foundations of human social life. According to Wiessner, the !King San people create relationships of mutual reciprocity and use to store such social obligations "until the situation of have and have not is reversed" (Wiessner, 1982, p. 67). Although these partnerships "are not economic contracts, but rather grounds of mutual help" (Wiessner, 1982, p. 68), obligations may be retained in terms of food if partners agree this is a fair deal. In summarizing, it may be suggested that natural selection should favor the evolution of various types of reciprocity, provided that "by creating coalitions of reciprocators" (Brown, 1991, p. 107) the benefits exceed individual investments. According to Corning, reciprocity is likely to be "more beneficial than kin selection—provided that cheaters can be detected and excluded from the system" (Corning, 1998, p. 23).

Maynard Smith and Szathmáry have further endorsed the view that reciprocal altruism among kin is probably only one of the causes underlying the evolution of cooperation. According to them, there "is some evidence that . . . rowing may be a better model than sculling of the situations in which cooperation evolves. Once it is common, cooperation is evolutionarily stable. The problem is how it becomes common in the first place, because defection is also stable. A possible answer is that cooperation starts between relatives, and later, once it is common, spreads to non-relatives" (1995, pp. 261–262). As opposed to the rowing game, sculling is "identical to the familiar Prisoner's Dilemma" (Maynard Smith & Szathmáry, 1995, p. 261). Similarly, Boyd and Richerson have pointed out that "cooperation can arise via reciprocity when pairs of individuals interact repeatedly" (1992, p. 172). If for some reason punishment is introduced into this game, "then everyone is best off conforming to the norm" (Boyd & Richerson, 1992, p. 173). The reason for this preference for conformity, according to these authors, is that "although nobody lives forever, social groups often persist much longer than individuals. When they do, individuals can expect to be punished up until their own last act . . . and in many societies one's family is also subject to retribution" (Boyd & Richerson, 1992, p. 185).

Against this background, it may be suggested that the sociobiological notion of "altruism," which is meant to designate the reproductive consequences of a given behavior, applies only to a limited scope of cooperative behaviors. As is pointed out by Maynard Smith and Szathmáry, once rowing has been introduced, defection no longer pays off, and cooperation becomes evolutionarily stable. Regarding the ultimate cause underlying human cooperation, it may be inferred from these considerations that natural selection has designed human beings in such a way that they can detect cheaters and free riders, at least to some extent. These selective forces are further supported by introducing punishment: As pointed out by Boyd and Richerson, this will stabilize cooperation because relatives may be made liable for retribution. It thus becomes evident why parents should instruct their children on social rules and on the ways breaches can be detected, as well as on the need to comply with the rules. Therefore, it may be suggested that both the detectors provided by natural selection and the social rules securing reciprocity are established on a common level of ultimate causality. According to Daly and Wilson, the term "ultimate" addresses the question why "certain goals have come to control behavior at all" (1988, p. 7), whereas the term "proximate" refers to motives and goals preceding overt behaviors.

For instance, hunting with primitive weaponry requires a high degree of cooperation because these weapons are neither as far-reaching nor as accurate as modern weapons are. Therefore, successful hunting with primitive weapons generally requires large numbers of hunters who carefully coordinate their behaviors in order to substitute for lack of accuracy, as well as for the limited range of their weapons. In both cases, however, larger per capita food acquisition is to be expected than each participant can obtain alone (Corning, 1983, p. 274). Regarding the origin of cooperation, it may be gathered from this example that by coordinating individual behaviors, each individual is likely to gain in terms of important resources, irrespective of his genetic relationship.

With regard to the distinction between ultimate and proximate causes, the greater success of food acquisition from cooperative hunting seems to suggest that cooperation must have been favored as a general goal by natural selection, because net gains per individual in most environments are likely to be larger in cooperative hunting than could be expected from individual hunting expeditions. Despite these distinct advantages of cooperation, there can be no strictly genetic determination of human behaviors in general, and of cooperation in particular, because human populations live in vastly different environments (Boehm, 1989, p. 925). According to Boehm, one would be ill advised to expect more from explanations in terms of ultimate causality than an account for the general causes underlying the preference for cooperation in human populations. Therefore, a full understanding of any particular cooperative act presupposes due consideration of proximate causes, such as environmental conditions, as well as other factors directly preceding actual behavior, because human behavior, in addition to the ultimate causality underlying the general preference for certain goals, entails on-the-spot decision making that is designed to ensure adaptation to specific environments.

While the fruitfulness of analyses in terms of ultimate causes underlying the evolution of any particular human behavior cannot be denied, one would certainly be ill advised to explain away the proximate causes that are the product of environmental adaptation, as well as direct causal antecedents of any human behavior. According to Boehm, Netsilik Eskimos had to adapt their hunting tactics to environmental conditions. Sometimes hunters "had to disperse and hunt alone"; by contrast, in other environments, "food resources sufficiently clumped . . . that hunting could be done collectively" (Boehm, 1989, p. 925). Take as another example the conflicts of interest between male and female reproductive strategies in the human species: While sociobiologists succeeded in disclosing the ultimate cause underlying these strategies, namely, sexual selection, nothing could be gained from failing to notice the impact of historically variable socioeconomic conditions on human reproduction. Due to the environmental adaptability of human populations, there is considerable variance in human reproductive behavior, typically conferring reproductive privileges on high-ranking males.

Whereas many tribal societies relate an individual male's social status to his success in warfare (Chagnon, 1979), modern society establishes social status mainly on individual economic success or educational level, as well as on some other factors. However, high social status is a guarantee for reproductive rights of males in modern society as well as in tribal society, irrespective of major differences between these types of society (van den Berghe & Whitmeyer, 1990, p. 38). Due to the availability of advanced reproductive technology, as well as due to social changes, according to Cliquet, there is a universal trend in European society toward "below-replacement fertility" (1996). In the present context this fact may further endorse the view that whereas sexual desires prevail under the most heterogeneous social conditions, their consequences in terms of reproduction are dependent upon sets of proximate causes that result from the availability of reproductive technologies, as well as from social change.

With respect to the proximate causes underlying human cooperation, the impact of inventions and of other social novelties that occurred several thousands of years ago should be given due consideration. For instance, the domestication of plants and animals, as well as the invention of new organizational patterns, exerted a major impact upon patterns of cooperation. According to Phillips, "Between 5000 and 2000 B.C., an insignificant or accidental 'kick' resulted in 'experiments' . . . with maize that produced genetic changes in that cereal. . . . Starting with what may have been (initially) accidental deviation in the system, a positive feedback network was established which eventually made maize cultivation the most profitable single activity in Mesoamerica" (Phillips, 1987, p. 235). It may be assumed that cultural inventions disclosed new levels of human cooperation in agricultural society, as well as in warfare, levels of cooperation unprecedented in overall social complexity in human history. Since

fertile land as well as water supplies became scarce resources, competition between adjacent societies was elevated to a new level (Meyer, 1995). Hence
warlike values exercised a growing influence upon these societies. While the
causes for these major transitions in the evolution of human cooperation cannot
be dealt with in this context, it should be stressed that human cooperation has
in fact transcended pristine levels. In view of this fact, while I emphasize the
biological underpinnings of human cooperation, I do not deny the feasibility of
the evolution of ever more complex patterns of cooperation in human history.
Due to the trend toward higher levels of social complexity, new patterns of
systemic integration were needed and in fact were introduced, putting a strain,
however, on the individual motives underlying social integration.

Turning to the main issue of this chapter, I suggest that due to evolutionary
processes, cooperation is favored over egoistic behavior in the human species.
More specifically, it may be proposed that the human individual has been provided with a set of emotional cues that indicate to him or her the reproductive
value of a given behavior. Given that cooperation is in fact selectively favored
over purely selfish behavior, it should be expected that human emotions are
designed in such a way that individuals should in fact prefer cooperation over
egoistic behaviors. Some findings endorsing this view will be presented after a
description of some relevant aspects of the human emotional system.

Unlike in some lower species, the lifetime of individual organisms is limited
in the mammals, including the primates and *Homo sapiens*; that is, individual
organisms are born and they die. According to evolutionary theory, this truism
holds some major implications for an understanding of the internal structure of
species-specific behavioral patterns. Since birth and death are the limiting events
for any individual and his or her total time budget, it should be expected that
his or her behavior patterns have been designed by selective forces in such a
way that he or she tends to make an "economic" use of this budget. Since any
behavior also represents an investment of energy, and the physiological and
motor aspects of behavior require continuous energetic inputs, "economic" solutions to the individual energy budget should also be selectively favored. According to evolutionary theory, the scarcity of time and energy seems to be the
focal point for an understanding of social behavior, particularly of reproductive
behavior. Essentially, scarcity implies a limitation of reproductive chances. It
should therefore be expected that selection has designed species-specific behavioral appetites and drives in such a fashion that they tend to serve the individual
organism's ultimate reproductive goal.

This species-specific pattern of appetites and drives puts all members of a
species into a "participatory universe" (Prigogine & Stengers, 1981, p. 267):
Whatever their genetic differences and conflicts of interest, they share appetites
or drives, as well as a basic temporal direction of behaviors. It is this similarity
in temporal orientation that allows conspecifics to establish more complex social
patterns. Due to this basic similarity of their participatory universe, conspecifics

may engage in conflicts of interest as well as in cooperative behavior. Despite this parallelism of behaviors and aspirations of conspecifics, it is important to note that due to the general scarcity of resources, there necessarily is a competitive aspect in their life as well. As pointed out by mainstream Darwinism, scarcity of certain resources gives the clue to an understanding of the functional integration of the most variegated types of behavior of a species. According to this line of thought, agonistic behaviors, the rearing of offspring, and dominance hierarchies must somehow contribute to reproduction; otherwise, they would not exist. Therefore, it is taken for granted that differential reproduction may still be considered the mainstay of evolutionary reasoning without denying Smillie's argument against an oversimplified concept of scarcity (Smillie, 1995, p. 248).

I suggest that the human emotional system is part of a more embracing system of physiological processes employing inbuilt perceptory thresholds, namely, drives, as well as temporal structures. These processes are in turn firmly established upon genetic information. Therefore, there is some reason to assume that there is a universal basis of human emotionality. Due to this common genetic heritage, "we classify, learn, and respond to those things that have had high survival value" in the past (Shaw & Wong, 1989, p. 82). For instance, we respond emotionally to kin categories "because survival value is involved" (Shaw & Wong, 1989, p. 83). Individual human beings may communicate their emotional states, and in general their counterpart is able to interpret the meaning of some such messages correctly, irrespective of group membership and culture (Ekman, 1979, p. 202).

To summarize, it may be suggested that mutual understanding, the focal issue of human communication, depends to a large extent on emotional processes that in turn have a genetic basis. Viewed against the background of evolutionary theory, and taking the notion of scarcity of time and energy into consideration, emotional states may be regarded as investments of comparatively large amounts of time and energy into particular situations. This may be evident both in the case of sexual arousal and defense against an aggressor. Not surprisingly, both cases clearly show their relation to reproduction. The central importance of reproduction for a general understanding of human emotionality is emphasized in Masters's classification of human emotions into "bonding, attack and flight" (1986, p. 234). Evidently, each of these classes of emotional behavior is related to the reproductive problem in some way, either to finding a mate, the rearing of offspring, fighting competitors, or fleeing from imminent attack by a predator. While this classification may be quite superficial from a specialist's point of view, it is entirely sufficient for the present purpose with its emphasis on three major behavioral categories and their corresponding emotional states.

With regard to the individual time-energy budget, it may be suggested that emotional states, as well as their corresponding perceptory thresholds, have been designed by natural selection in such a way that the individual will tend to repeat pleasurable experiences, but avoid painful experiences, thereby enlarging his or her time-energy budget to some degree. To cut a long story short, viewed against the evolutionary background of the general scarcity of resources, the emotional system may be understood as a system of cues that "causes organisms to seek events and repeat acts that tend to increase reproductive success" (Alexander, 1987, p. 110; Badcock, 1991, p. 96). Compared to the quasi-mechanical operations of animal instincts, the human emotional system may not be as reliable in terms of reaction speed and accuracy, but these losses are more than made up by the feasibility of linking up emotions with systems of cognitions. It seems that due to this combination, the peculiarly human mental system compensates for losses in terms of speed, typical of animal instincts, by establishing learned standards of behavioral reaction, thereby enabling human beings to foresee to some extent behaviors of animals, as well as of conspecifics. Due to these learned standards, human beings outperform most animal instincts in reaction speed, as the spread of human populations all over the globe illustrates. Recently, neurobiology has vastly improved our understanding of the neurological and other aspects of the peculiarities of the human behavioral apparatus. Unfortunately, this chapter will have to refrain from pointing to the more general implications of these findings. It will, however, emphasize the role of some neurotransmitters, for instance, serotonin, for operations of the human emotional system.

Returning to the problem of the evolutionary foundations of human cooperation, I suggest that the term "bonding" can be used as a somewhat vague label for a multifunctional emotion underlying the most diverse patterns of cooperative behavior. While the existence of parent-offspring conflicts, as stressed by sociobiology, cannot be denied, numerous studies in child psychology demonstrate that newborn infants actively seek to establish bonds with other individuals. In general, this will be the biological mother, but in cases when the mother is not available, the infant will tend to establish bonds with nonrelated individuals in a similar fashion. To be sure, the infant's search for individuals to bond with is in keeping with sociobiological predictions because successful bonding is synonymous with securing further parental investments in the postnatal phase.

Likewise, from the sociobiological point of view, the parents should be expected to further invest in their child because this is the only way to protect previous investments of time and energy (Badcock, 1991, p. 189). It may be summarized that whereas conflicts of interest are likely to decrease in the course of ontogeny, there is a corresponding increase in the intensity of emotional bonds. Parent-offspring relations will never be entirely harmonious, but they are the major basis for kinship and other types of human social association. This basis has been designed by natural selection in such a way that it is adaptable to a wide variety of environmental conditions, replacing hard-wired genetic determination of behavior with soft-wired emotional processes. As indicated previously, human organisms seek "to repeat acts that tend to increase reproduction" (Badcock, 1991, p. 97); that is, due to the linkage between genetic, emotional, and cognitive processes, human beings, despite the lack of hard-wired genetic determination, adhere to the principles of evolutionary processes. Viewed across the lifetime of several generations, human populations tend to develop selectively tested sets of cognitions and pass them on to the oncoming generation. These sociocultural systems have proven to be superior to any other behavioral system because, due to man's vast soft-wired learning capabilities, they provide chances for adaptation to environmental conditions superior to those of hard-wired animal species.

With regard to yet another set of causes underlying human cooperation, it should be stressed once again that human behavior, despite its unprecedented mental and cultural potentials, is established upon an evolutionarily tested bulwark of genetic dispositions and emotions that supports the preference for cooperation, once it has been established. As indicated previously, there is nearly unanimous agreement among social theorists such as Gouldner and sociobiologists such as Trivers and Alexander that "if you want to be helped by others you must help them" (Gouldner, 1977, p. 37). According to these authors, reciprocity is to be considered one of the mainstays of human social behavior. This view is further endorsed in the following section, which draws the reader's attention to a set of proximate causes of cooperation that so far seem to have escaped most social theorists' notice, namely, the neurobiological causes underlying human emotionality.

I suggest that due to some characteristics of these processes, cooperating individuals gain in terms of neurotransmitters, such as serotonin and benzodiazepine, and given that no major conflict of interest exists between the individuals concerned, this is another proximate cause for their preference for cooperation. Unlike the discussions about the ultimate causes underlying the preference for cooperation, these proximate causes cannot explain why cooperation became a goal in the first place, but they can shed some light on how this preference is brought about in everyday behavior.

Given the background of the Darwinian scarcity assumption, it should be stressed that due to this causal structure, individuals can by their own behavior "create resources through the coordination of activities" (Smillie, 1993, p. 136). As indicated previously, emotions can be regarded as investments of scarce physiological energy, as well as of time, into a certain behavior. With regard to the origin of the evolutionary preference of cooperation, it is important to note that by a certain class of behaviors, namely, cooperation, individuals can in fact increase their supplies of physiological and emotional energies. Although this chapter will put further emphasis on this productive aspect of cooperation, there is no need to dispute the fruitfulness of Darwin's notion of scarcity of resources in its entirety. On the contrary, for the present purpose it is sufficient to stress the survival value of "secure social units" (Smillie, 1993, p. 123), units that by their very existence can provide partaking individuals with larger amounts of neurotransmitters, larger yields of game captured or of crops from the rice paddies, and more cases of enemies successfully repelled. All these benefits are not likely to be gained unless a certain degree of conformity and coordinated action has been secured. In view of these selective advantages of cooperation, it should be expected that individual group members seek to defend the integrity of their social system against free riders and other types of cheaters. Some examples of such defense of group members against certain types of deviant behavior will be given later.

Some patterns of social behavior of schoolchildren, for instance, the tendency to ostracize certain individuals, may subconsciously be motivated by the preference for cooperative over egoistic and purely aggressive behaviors, because, as will be pointed out later, it is usually the overly aggressive individuals who are ostracized by their classmates. A second example is patterns of blood revenge, feuding, and the practice of ostracism among tribal people, as interpreted in some recent studies. In order to focus on the proximate causes underlying these behaviors, a recapitulation of the neurological aspects underlying emotions seems advisable.

Some recent neurobiological studies have pointed out that certain substances cause similar effects in the most different species. According to Gray (1985, p. 103), benzodiazepines play a rather similar role in the neurochemical processes underlying behaviors in such diverse species as rats, *Homo sapiens*, and even some fish species. Moreover, levels of such substances as the catecholamines or serotonin will cause behavioral states similar to feelings of joy or of depression in humans. While the implications of these findings for an understanding of the phylogeny of human emotions cannot be dealt with in the present context, it is interesting to look at the impact of such substances on human social behavior. According to Collins (1986), higher levels of serotonin correspond to high social status of human individuals, whereas lower levels of this substance correlate with lower social status. It should be noted that these insights point to a circular type of causality between social and physiological facts because physiological states can be influenced as much by the social environment as the physiological state itself prepares the individual for a certain type of behavior.

These findings suggest that human individuals may enlarge their respective levels of serotonin, as well as of other neurotransmitters, by holding high social status, as well as by conforming with other people's expectations. Conversely, individuals are likely to diminish their individual supplies of these substances by holding low status and/or by failing to comply with others' expectations. Undoubtedly, human individuals have various behavioral options at their disposal, that is, to meet others' expectations or not, but from the neurobiological point of view one would expect that they would prefer compliance with these expectations in the majority of situations. If this is in fact the case, as is being suggested here, this may be regarded as one of the foundations of the preference for "bonding" over other behavioral options. It may be added from the viewpoint of ultimate causality that "bonding" obviously has been designed by natural selection in such a way that by actually bonding, the individual may expect benefits to chances for survival, because in the type of primeval environmental setting that persisted for the most extensive period of human history, human individuals were under the continuous threat of predators as well as of enemy groups, and any individual would gain in terms of physical safety and ultimately in terms of reproductive value by coordinating his defense with fellow clansmen (Slurink, 1994, p. 464). According to Corning, "The percentage of time each animal spent scanning for predators decreased markedly as the group size increased" (Corning, 1996b, p. 10).

Evolutionary theory should lead us to expect that physiological processes that are likely to enhance the survival of their individual carrier organisms should be favored. Given that bonding may in fact enhance the individual's supplies of serotonin and of other relevant neurotransmitters, it should further be expected that human individuals will actively search for situations in which the physiological concomitants of bonding are likely to be elicited. In many situations of social life, this effect is to be expected when individuals mutually coordinate their behaviors, or when they comply with social rules. Regarding the relation of ultimate and proximate causality, it seems very likely that various levels of proximate causality have evolved, namely, human emotionality, and on a distinct level social rules, the functional integration of which is subservient to the ultimate goals, defined by natural selection. Due to these various levels of causality, individuals tend to prefer bonding and cooperation over purely egoistic behaviors whenever their actual interests allow for such preference.

The infant-mother relation certainly is one of the closest types of bonding feasible between human individuals. Due to this proximity, and leaving aside conflicts of interest, the biorhythms of both individuals are synchronized to the utmost degree. With regard to the infant's levels of neurotransmitters, it may be hypothesized that this rhythmic interaction provides him or her with a feeling of safety, as well as with the kind of stimulation of his or her emotional and cognitive systems that seems to be indispensable for a healthy development of the infant. Viewed against this background, the infant's striving for establishing social ties makes sense because he or she is likely to gain in terms of time and energy from bonding.

According to a study by Barner-Barry on schoolchildren, in an ensuing phase of ontogeny, these children ostracized classmates, particularly those individuals who by their acts "disrupt the cooperative system of a group" (1986, p. 143). Thomas suggests that "natural selection tends, in the long run, to pick as real winners the individuals, and then the species, whose genes provide the most inventive and effective ways of getting along" (Thomas, 1984, quoted by Barner-Barry, 1986, p. 143). This view is further endorsed by two studies on the practice of blood revenge and ostracism among the Pathan tribes of Pakistan and Afghanistan, as well as by the people of Montenegro. According to Mahdi, the Pathan tribes, who are well known for their bellicosity, would ostracize an individual "who is likely to provoke reprisals" (Mahdi, 1986, p. 153). Boehm presents similar findings about the practice of ostracism among Montenegrins. These warlike people were likely to reject individuals who would not comply with rules: "The net effect of clan execution was to enhance the reproductive and therefore the political success of the clan, as a group of closely related males who constituted the primary political subunit within the tribe" (Boehm, 1986, p. 170). Boehm stresses that while chimpanzees also avoid individuals, they lack the moral codes underlying the practice of ostracism and blood revenge in human societies.

With regard to the origins of human cooperation, it may be inferred from these studies that people take an active interest in upholding the cooperative system. Unlike individual members of some animal species, human individuals are prepared to ostracize certain group members even prior to the decisive violation of social rules, as a report on the practice of ostracism and warfare among the Pueblo Indians of the southwestern part of the United States demonstrates. According to Ellis, "Even recently an unusually beautiful woman or successful hunter might be killed—quietly and accidentally—or someone's fine horse be found dead or his big house despoiled" (1951, p. 178). The reason for this practice of ostracism was that "the imposition of village-sanctioned punishment of competition and individualism no doubt has aided tremendously in dissipation of hostilities officially permitted little other expression" (Ellis, 1951, p. 178).

In the sociobiological analysis of human cooperation, the practice of ostracism may be classified in Alexander's category of "indirect reciprocity," which occurs "when interested people observe direct reciprocity between others and use the observations to determine who will be their own future associates and how they will interact subsequently with the observed parties" (Alexander, 1986, p. 107). Since people take an active interest in storing social obligations from which they hope to benefit in the future, it seems plausible that they would avoid having to take over debts from others if this can be avoided at all. It is quite evident that this sort of reasoning underlies the Pueblo Indian practice of ostracism: Since the unusual beauty of a woman may cause trouble, why not get rid of her? While, from a moralistic point of view, this may seem a very high price for upholding cooperation, these people obviously were prepared to pay it.

A second point worth considering from a sociobiological point of view is that the practice of ostracism does not necessarily involve the notion of group selection. Viewed from a group-selectionist point of view, ostracism is practiced solely for the "good of the group." Contrary to this view, it is suggested here that the practice of ostracism, as demonstrated in the examples given here, seems to be accepted by every single individual because he or she expects to gain from it. It is as if the schoolchildren, the exotic warriors, and the Pueblo Indians realized that everybody was better off if he or she would not have to take over the debts caused by other members of the group. According to this view, the individuals accept the practice of ostracism because they expect an increase in terms of time and energy, especially fights they are not going to be involved in, and that is basically why they adhere to ostracism. It must not be assumed, however, that the individual assessment of expected utilities is ever totally rational. On the contrary, in most cases the practice of ostracism, as well as of other forms of cooperation, is accepted in a subrational and more or less emotional manner.

With regard to individual supplies of various neurotransmitters, bonding may be an important foundation for the preference for cooperation to fighting and conflict. It is evident from a Darwinian point of view, with its emphasis on the "struggle for existence," that agonistic phases cannot be totally avoided, and therefore they are a necessary concomitant of social life. Darwin himself used the concept of "struggle for existence" in "a large and metaphorical sense, including dependence of one being on another, and including not only the life of the individual, but success in leaving progeny" (Darwin, 1968, p. 459), thereby avoiding inappropriate emphasis on fighting. It may be suggested that the present interpretation of the origins of human cooperation is in keeping with Darwin's ideas on the foundations of social life.

To summarize, I suggest that sociobiology can improve our understanding of the origin of human cooperation. In fact, this approach avoids some of the pitfalls of classical social theory, as, for instance, Hobbes's notion of continuous competition and conflict. While there is partial agreement between the Darwinian and the Hobbesian approach on the unavoidability of competition and conflict, they disagree on the origins of cooperation. Darwinian theory, particularly sociobiology, does not require the advent of a hypothetical "leviathan," but points to the impact of "kin selection," as well as to the various forms of altruism, as the natural buttresses of human cooperation.

With regard to Smith's concept of "sympathy," there seems to be a considerable harmony with reciprocity, as well as with cooperation. Hobbes's, Smith's, and Darwin's views coincide in taking the human individual as the point of departure for their analyses; that is, they explain social phenomena such as cooperation in terms of individual inclinations rather than in terms of functional contributions to the survival of groups or of systems. Due to this partial agreement on the central role of the human individual for social theory, these views can be discussed against the background of evolutionary theory, which regards individuals and their differential reproduction as the subject matter of evolution.

Our understanding of the origins of human cooperation may be vastly improved by giving due consideration to sociobiological accounts in terms of ultimate causality and linking them up with other biological approaches that emphasize proximate levels of causality. It remains to be seen whether future developments in social theory will seize the advantages of these new disciplines for the explanation of a long-standing problem.

REFERENCES

Alexander, R. (1986). Ostracism and indirect reciprocity: The reproductive significance of humor. In M. Gruter & R. D. Masters (Eds.), *Ostracism: A social and biological phenomenon* (pp. 105–123). New York New York: Elsevier.

———. (1987). *The biology of moral systems*. New York: Aldine de Gruyber.

Badcock, C. (1991). *Evolution and individual behaviour: An introduction to human sociobiology.* Oxford: Basil Blackwell.

Barner-Barry, C. (1986). Rob: Children's tacit use of peer ostracism to control aggressive behavior. In M. Gruter & R. D. Masters (Eds.), *Ostracism: A social and biological phenomenon* (pp. 133–147). New York: Elsevier.

Boehm, C. (1986). Capital punishment in tribal Montenegro: Implications for law, biology, and theory of social control. In M. Gruter & R. D. Masters (Eds.), *Ostracism: A social and biological phenomenon* (pp. 157–173). New York: Elsevier.

———. (1989). Ambivalence and compromise in human nature. *American Anthropologist, 91*(1), 921–939.

Boyd, R., & Richerson, P. J. (1992). Punishment allows the evolution of cooperation (or anything else) in sizable groups. *Ethology and Sociobiology, 13,* 171–195.

Brown D. E. (1991). *Human universals.* New York: McGraw-Hill.

Chagnon, N. A. (1979). Mate competition, favoring close kin, and village fissioning among the Yanomamö Indians. In N. A. Chagnon & W. G. Irons (Eds.), *Evolutionary biology and human social behavior* (pp. 86–132). North Scituate, MA: Duxbury Press.

Cliquet, R. L. (1996, July 22–25). *Below-replacement fertility and gender politics.* Paper presented to the nineteenth Annual Meeting of the European Sociobiological Society, Alfred University, Alfred, NY.

Collins, R. (1986). *How to incorporate sociophysiology into sociological theory.* Paper presented at Stanford University, California.

Corning, P. A. (1983). *The synergism hypothesis. A theory of progressive evolution.* New York: McGraw-Hill, July 25–27.

Corning, P. A. (1996). *Bioeconomics and biopolitics.* Paper presented at the Joint Meeting of the Association for Politics and the Life Sciences and the International Political Science Association, Alfred, NY.

———. (1988, July 22–25). Holistic Darwinism: Synergy and the new evolutionary paradigm. In V. S. Z. Falger, P. Meyer, & J. M. van der Dennen (Eds.), *Research in Biopolitics.* Vol. 6. Stamford, CT: JAI.

Crippen, T. (1994). Toward a new Darwinian sociology: Its homological principles and some illustrative applications. *Sociological Perspectives, 37* (3), 309–335.

Daly, M., & Wilson, M. (1988). *Homicide.* New York: Aldine de Gruyter.

Darwin, C. (1968). *On the origin of species by means of natural selection; or The preservation of favoured races, in the struggle for life.* London: Penguin.

Dawkins, R. (1978). *Das egoistische Gen.* Berlin/Hamburg: P. Parey.

Ekman, P. (1979). About brows: Emotional and conversational signals. In M. v. Cranach, K. Foppa, D. Lepenies, & D. Ploog. (Eds.), *Human ethology: Claims and limits of a new discipline* (pp. 169–203). Cambridge: Cambridge University Press.

Ellis, F. H. (1951). Patterns of aggression and the war cult in southwestern Pueblos. *Southwestern Journal of Anthropology, 7,* 177–201.

Gouldner, A. (1977). The norm of reciprocity: A preliminary statement. In S. W. Schmidt, L. Guasti, C. H. Landé, & J. C. Scott, (Eds.), *Friends, followers, and factions* (pp. 28–43). Berkeley: University of California Press.

Gray, J. A. (1985). A whole and its parts: Behavior, the brain, cognition, and emotion. *Bulletin of the British Psychological Society, 38,* p 99–112.

Gruter, M., & Masters, R. D. (Eds.). (1986) Ostracism, A social and biological phenomenon. *Ethology and Sociobiology* (314). New York: Elsevier.

Haig, D. (1995). Genetic conflicts in human pregnancy. *Quarterly Review of Biology, 68*(4), 495–532.

Hamilton, W. D. (1978). The evolution of altruistic behavior. In T. H. Clutton-Brock & P. H. Harvey (Eds.), *Readings in sociobiology* (pp. 31–33). Reading, MA and San Francisco: W. H. Freeman.

Hobbes, (1968). *Leviathan*. Harmondsworth, Middlesex: Penguin.

Luhmann, N. (1985). *Soziale Systeme*. 2nd ed. Frankfurt/am Main: Suhrkamp.

Mahdi, N. Q. (1986). Pukhtunwali: Ostracism and honor among the Pathan Hill tribes. In M. Gruter & R. D. Masters (Eds.), *Ostracism: A social and biological phenomenon* (pp. 147–157).

Masters, R. D. (1986). Ostracism, voice, and exit: The biology of social participation. In M. Gruter & R. D. Masters (Eds.), *Ostracism: A social and biological phenomenon* (pp. 231–247).

Maynard Smith, J., & Szathmáry, E. (1995). *The major transitions in evolution*. Oxford: W. H. Freeman/Spektrum.

Meyer, P. (1995). The evolution of warfare: Origins of cultural variation. In V. Cauchy (Ed.), *Violence and human coexistence: Proceedings of the IInd world congress of ASEVICO* (pp. 259–265). Vol. 5. Montreal: Editions Montmorency.

Miller, T. C. (1993). The duality of human nature. *Politics and the Life Sciences, 12*(2), 221–241.

Münch, R. (1995). Elemente einer Theorie der Integration moderner Gesellschaften: Eine Bestandsaufnahme. *Berliner Journal für Soziologie, 5*(1), 5–24.

Phillips, C. S. (1987). Politics: An aspect of cultural evolution. *Politics and the Life Sciences, 5*(2), 234–237.

Prigogine, I., & Stengers, I. (1981). *Dialog mit der Natur: Neue Wege naturwissenschaftlichen Denkens*. Munich and Zurich: Piper.

Raphael, D. D. (1991). *Adam Smith*. Frankfurt and New York: Campus Verlag.

Shaw R. P. & Wong, Y. (1988). *Genetic seeds of warfare: Evolution, nationalism, and patriotism*. Boston: Hyman.

Slurink, P. (1994). Causes of our complete dependence on culture. In R. A. Gardner et al. (Eds.), *The ethnological roots of culture* (pp. 461–474): Amsterdam: Kluwer.

Smillie, D. (1995). Darwin's tangled bank: The role of social environments. In P. P. G. Bateson, P. H. Klopfer, & N. H. Thompson (Eds.), *Perspectives in ethology: Vol. 10. Behavior and evolution* (pp. 119–141). New York: Plenum Press.

———. (1995). Darwin's two paradigms: An opportunistic approach to natural selection theory. *Journal of Social and Evolutionary Systems, 18*, (3), 231–255.

Smith, A. (1986). *The theory of moral sentiments*. Düsseldorf Franktfurt am Main: Verlag Wirtschaft und Finanzeu.

Trivers, R. L. (1971). The evolution of reciprocal altruism. *Quarterly Review of Biology, 46*, 35–57.

van den Berghe, P. L., & Whitmeyer, J. (1990). Social class and reproductive success. *International Journal of Contemporary Sociology, 27*, 29–48.

Voland, E. (1996). Konkurrenz in Evolution und Geschichte. *Ethik und Sozialwissenschaften, 7*(1), 93–107.

Wiessner, P. (1982). Risk, reciprocity, and social influences on Kung San economics. In E. Peacock & R. Lee (Eds.), *Politics and history in band societies* (pp. 61–84). Cambridge: Cambridge University Press.

5

Evolution and Culture: The Missing Link

Robin Allott

The relation of evolution and culture has been much debated. There have been many different approaches, of which the most notable have been those of Cavalli-Sforza and Feldman, Lumsden and Wilson, and Boyd and Richerson. Also noteworthy are the views of Durham and Hinde and most recently of the evolutionary psychologists. If these other accounts, or any one of them, seem to cover the subject adequately and to be intellectually satisfying, no new approach would be needed. The first step, then, is to summarize and assess the theories that have been presented.

THEORIES

Cavalli-Sforza and Feldman (1981) adopted a quantitative approach. After pointing out that up to that date cultural transmission had received little attention, they stressed the need for a theory of cultural change; they chose to develop a mathematical theory since the modern theory of biological evolution owed much of its strength to the mathematical background, primarily in population genetics. They sought to deal with the dynamics of the changes within a population of the relative frequencies of the forms of a cultural trait under defined cultural interactions, while recognizing that for humans it is difficult to partition the process of transmission into purely genetic and purely cultural components. Cultural traits vary in significance. There are relatively trivial ones (innovations such as the spread of Coca-Cola or volleyball) where participation in the trait cannot appreciably alter the probability of surviving or having children; in these instances some kind of non-Darwinian selection is involved that they termed "cultural." Other traits are important elements of culture (notably language); these are subject to processes analogous to those in biological evolution to which the concepts of drift and migration can be applied. Cultural selection can act counter to natural selection, although harmony between the two is expected on average, based on the assumption that the neural structures or mechanisms that permit choice evolve under the control of natural selection, which thus indirectly controls the scope of cultural choices made.

Lumsden and Wilson (1981, 1983) postulated that human cultural transmission is ultimately gene-culture transmission. Their aim was the technical development of a theory of gene-culture coevolution, a first attempt to trace development all the way from genes through the mind to culture. The approach, similar to that of Cavalli-Sforza and Feldman though more mathematically ambitious, centered around the concept of the culturgen (producing culture), the basic unit of inheritance. They derived the concept from the operational units of culture in archaeology (artifacts) but extended it to cover all kinds of transmissible behaviors, mentifacts, and artifacts. The transmission of culturgens was governed by epigenetic rules, the genetically determined peripheral sensory filters, interneuron coding processes, and cognitive procedures of perception, learning, and decision making. These together affected the probability of one culturgen being transmitted rather than another. Genetic and cultural evolution drive each other forward; culture is created and shaped by biological processes, while the biological processes are simultaneously altered in response to cultural change.

Boyd and Richerson (1985) presented a Darwinian theory of the evolution of cultural organisms. By culture they meant the transmission from one generation to the next, via teaching and imitation, of knowledge, values, and other factors that influence behavior. The basic point of departure for their dual-inheritance model was the analogy between genes and culture; the relationship between them was, they said, the most interesting scientific problem presented by human evolution. The evolution of the structure of cultural transmission in humans was analogous to the evolution of the genetic system. Genes and culture are mechanically distinct systems of inheritance, but behavior is the product of predispositions resulting from genetic inheritance alone and predispositions resulting from cultural inheritance alone.

Others have approached the relation between cultural and biological evolution from other standpoints, ranging from Rindos (1985), who argued that cultural evolution is explainable by exactly the same processes as those that underlie genetic evolution, with selection as the ultimate determinant, to Hinde (1982), who suggested that natural and cultural evolution can to some extent proceed independently, but many everyday actions may nevertheless depend on propensities selected in another context to promote individual fitness. Most recently the evolutionary psychologists have promoted a novel but still Darwinian view of mental organization and so of cultural behavior. Perhaps one should recall Dobzhansky's (1951, p. 304) observation that human genetics has not been superseded by human culture; the former remains the foundation that enables man to manifest the kinds of behavior that are called social and cultural. The interrelationships between biology and culture are reciprocal.

ASSESSMENT

Ingold comments severely but perhaps justly on the various attempts to give a theoretical or systematic account of the relation of culture and biological evolution and particularly the elaboration of formal models of gene-culture coevolution by Boyd and Richerson, Cavalli-Sforza and Feldman, and Lumsden and Wilson: "Not much can be said for their models in their present state of development; the assumptions on which they rest are either so remote from reality or so ultimately trivial that they do not so much advance our understanding of evolutionary processes as provide an excuse for the exercise of mathematical ingenuity" (1986, p. 364). The authors themselves recognized the necessarily preliminary and speculative character of their pioneering work. Boyd and Richerson said that they were acutely aware that the dual-inheritance model of human evolution rested on less than completely compelling arguments; for the present the purpose of the theory could only be to summarize a state of quite imperfect knowledge about the causes of human behavior in a way that would make further refinement as simple as possible. This comment can equally well be applied to the work of Cavalli-Sforza and Feldman and Lumsden and Wilson.

Unresolved difficulties or lacunae in the different approaches relate to the concept of culture, the lack of empirical evidence to test the various mathematical models, uncertainty about the cultural significance of symbols, the role of language in the evolution of culture, the absence of specific examination of major segments of culture, the hypothesis of unitary cultural traits analogous to genes, emphasis on the transmission of culture and neglect of the more important problem of the creative aspect of human cultural development, and the origins of particular cultural systems. One of the main impressions left by the varying coevolutionary accounts is a certain fogginess about the idea of culture. More than a hundred definitions of culture have been proposed by anthropologists and others; the authors of the coevolutionary theories add to the number. The definition used by Cavalli-Sforza and Feldman is closest to that in Webster's dictionary: "the total pattern of human behavior and its products embodied in thoughts, speech, action, and artifacts and dependent upon man's capacity for learning and transmitting knowledge to succeeding generations; the transfer of abstract instructions and explanations in ways that do not require face-to-face observation and direct imitation." For them, the term "cultural" included all sorts of activities, from technology to entertainment, and all sorts of beliefs, values, and behaviors. Lumsden and Wilson defined culture to include the sum total of mental constructs and behavior, including the construction and employment of artifacts; they acknowledged the preeminence of symbols in human culture but also considered as cultural a substantial class of imitative behaviors. Boyd and Richerson discussed at some length the most appropriate definition of culture, pointing out that in anthropology, the term "culture" is used in many different and only partially overlapping senses; "Culture is information capable of affecting individuals' phenotypes which they acquire from other conspecifics by teaching or imitation" (1985, p. 33). They did not restrict culture to behavior encoded by symbolic constructs such as language, myth, or ritual, nor, unlike other approaches, did they assume the existence of units—"discrete particles"— of cultural inheritance such as memes or culturgens. For the evolutionary psychologists Tooby and Cosmides (1992), culture is the manufactured product of psychological mechanisms situated in individuals living in groups that evolved by natural selection in the evolutionarily stable environment of the Pleistocene.

This chapter is concerned with human culture in the broadest sense (originally adopted by Tylor, 1865, pp. 1–4): essentially the whole complex of behaviors, structures, and processes that make us human. Culture should be approached primarily not in terms of the accumulation of artifacts or of what Lumsden and Wilson call mentifacts but of the behavioral potential which underlies the production of the cultural artifacts and mentifacts in the individual and in human society. The key question is the evolutionary source of the potential. What is needed is a theory that deals adequately and specifically with the origin, transmission, and change of the central nontrivial aspects of human culture: language, morality, social systems, science, the arts, and religion. While what distinguishes cultural evolution from genetic evolution is transmission by nonbiological means, cultural origins and the source of variations are as important as the mechanisms of transmission.

In *The Descent of Man* (1871) Darwin considered at some length such major aspects of human culture as morality and language but did not attempt to deal with high culture. In his autobiography he described how he became blind or deaf to music, poetry, mathematics, philosophy, and art: "Now for many years I cannot endure to read a line of poetry: I have tried lately to read Shakespeare and found it so intolerably dull that it nauseated me. I have also almost lost any taste for pictures or music. . . . I am so utterly destitute of an ear, that I cannot perceive a discord, or keep time, or hum a tune correctly"; he spoke about "not being able to see any meaning in the early steps in algebra." At the university he was what Matthew Arnold would have termed a Philistine: "My time was sadly wasted there and worse than wasted [drink, dissipation, riding]. . . . How I did enjoy shooting." Nevertheless, "I cannot help looking back on that time with much pleasure" (1876, pp. 58, 60–61, 138).

CULTURE AS A SYMBOLIC SYSTEM?

For the most part, later writings on coevolutionary theory also fail to deal specifically with the major aspects of human culture. Partly this may have been the result of overemphasis on cultural units, traits, culturgens, and memes rather than cultural systems. The other contributing factor seems to have been a limiting, and perhaps mistaken, view of the general character of culture as an aggregation of symbols, a consequence of taking language as the model, paradigm, or prime example of cultural evolution generally. Though there are differences of emphasis between the authors whose coevolutionary theories have been summarized earlier, all of them comment on the symbolic character of culture while rejecting some of the more extreme views of anthropologists. Thus Boyd and Richerson said that some of the most strenuous objections to human sociobiology came from the symbolic anthropologists who believed that symbols are the essential feature of human culture, often defining culture as a set of symbols whose meanings are shared by members of a human society, but such a sweeping view was unsatisfactory unless it could be made consistent with the argument from natural origins: "The key defining feature of symbols is that they are arbitrary; in language it usually does not matter what sound pattern or what series of letters are used to signify a particular thing or concept; it only matters that the members of a speech community agree on some convention. Other cultural symbol systems are similar to language. Much of human behavior consists of the use and production of symbols" (1985, p. 272). Lumsden and Wilson identified the central importance in cultural evolution of a process they called reification for which the enabling device was symbolization. Human language is largely the manipulation of symbols to convey the reified concepts of the mind; language is the means whereby culturgens are labeled and juxtaposed to assemble and communicate more complex knowledge structures; gene-culture coevolution was only possible because of the invention by the human species of reification, symbolization, and language.

The treatment of culture as a symbolic process is not new; Sol Tax (1960, p. 280) said that cultural behavior has a quality of arbitrariness because it does not flow through the genes and is therefore not anchored in the individual, as is seen most clearly in the arbitrariness of the symbols of language. Cavalli-Sforza and Feldman also developed their account of cultural evolution using the development and differentiation of languages as prime examples of cultural phenomena, with considerable parallelism between genetic and linguistic evolution: the formation of two or more different languages from a single predecessor as an analogue of speciation. Language and its components (words, rules, and sounds) could be regarded as cultural "objects." Other aspects of language touched on by Cavalli-Sforza and Feldman included the correlation of linguistic and genetic variation between populations (as extensively developed in Cavalli-Sforza's later publications), the distribution of surnames, vowels as examples of continuously variable traits, and the acquisition of language by children of foreign parents. In the learning of language, they commented, genetics might seem a priori to be unimportant, but given the difference between man and animals, it was unavoidable to conclude that language has a genetic basis despite the present impossibility of proof.

The concept of the symbol, in language or otherwise, can lead one far astray and can be even more misleading if one assumes that a symbol must necessarily be arbitrary. The character and origin of the symbol have been the subject of much controversy; the relation between symbol and the external world is not simple. A symbol may be arbitrary, or it can be the physical product of the process or object that it symbolizes—or a symbol can be structurally related to the process or object that it symbolizes, or a symbol can be generated by unconscious processes of association or imitation ultimately dependent on brain organization and function. To assume that language is symbolic, that language is the prime example of a human cultural phenomenon, and that therefore cultural traits are also essentially symbolic and arbitrary makes a bad starting point for any theory of the relation of cultural and biological evolution.

In summary, one is forced to conclude, with Ingold, that the approaches or theories of cultural evolution presented by these authors are not convincing, satisfactory, or adequate. They deal essentially with cultural transmission, and the arguments tend to be circular; culture for them is what is transmitted culturally. They say little or nothing about the human potential for the creation of culture, about the major aspects of culture in human evolution, and about the evolutionary source of the potential and of the content and form of the major cultural systems. Even in terms of their account of the transmission of culture, the idea (which most of them espouse) that culture is atomistic, composed of culturgens, cultural traits, memes, and the like, seems mistaken or at least misleading. Their accounts of cultural transmission, heavily mathematicized in ways derived directly from population genetics, remain at an abstract level, with little or no attempt to apply them to real cultural content. But if these approaches are unsatisfactory, what new or different approach to the relation of biological evolution and culture is possible?

LANGUAGE AS THE LINK

The new approach proposed is a revived stress on language as the foundation of culture, as the source of the human cultural potential, as the mode of transmission and of change of cultural systems, and as the form through which cultural systems are originated and stabilized. Language has a much greater importance in relation to cultural evolution than the probably mistaken use of it as the archetypal symbolic, arbitrary cultural system. Much of what is classified as culture is preserved in language, transmitted over time by language, and conveyed from parent to child by language. At the level of society, virtually every major aspect of human culture is language dependent; social systems, kinship classifications, morality, custom, science, group organization, and religion. For human culture, imitation without any linguistic context is a much less important mode of cultural transmission; even the transmission of technologies is heavily dependent on language. At the level of the individual, patterns of behavior and thought are also language dependent. Darwin stressed the wider importance of language in the development of human culture. After drawing attention to the curious parallels between the formation of different languages and of distinct species, he stated, "If we possessed a perfect pedigree of mankind, a genealogical arrangement of the races of man would afford the best classification of the various languages now spoken throughout the world; and if all extinct languages, and all intermediate and slowly changing dialects, were to be included, such an arrangement would be the only possible one" (1871, p. 402). More significantly, he commented on the fundamental importance of language in human evolution: "It is not the mere power of articulation that distinguishes man from other animals . . . but it is his large power of connecting definite sounds with definite ideas . . . the relation between the continued use of language and the development of the brain has no doubt been far more important [than language simply as a means of communication]. . . . it may well be that [self-consciousness, abstraction, and the like] are incidental results of other highly-advanced intellectual faculties; and these again are mainly the result of the continued use of a highly-developed language" (p. 105). "A great stride in the development of the intellect will have followed, as soon as, through a previous considerable advance, . . . language came into use; for the continued use of language will have reacted on the brain and produced an inherited effect; and this again will have reacted on the improvement of language. The large size of the brain in man . . . may be attributed in chief part . . . to the early use of some simple form of language—that wonderful engine which affixes signs to all sorts of objects and qualities" (p. 390). Besides the evolutionary contribution of language to the development of mind, as suggested by Darwin, language has made human culture in its totality possible. The significance of language is not only what it contributed to human evolution hundreds of thousands of years ago, but its contribution now both in the social maintenance of culture and in advancing understanding of brain function. Lashley (1951) said that language is the only window we have on the mind, and though we have modern scanning technologies such as positron emission tomography (PET), magnetic resonance imaging (MRI), and magnetic encephalography (MEG) that allow us to inspect the working of the brain, language remains the most specific and discriminating mode of insight into the mind. The current importance of language is obvious.

THE SOURCE OF THE POWER OF LANGUAGE

Rather than considering the appearance of language in human evolution as the product of some mysterious general symbolizing capacity, we need a genetic (physiological and neurological) theory of language origin and language function; such a theory would constitute the missing link in the relation between evolution and culture. The new approach leads necessarily to the question of the source of the culture-forming and culture-transmitting power of language, both as a current phenomenon and as the key development in the advance of the human species. To understand the current role of language in human individual mentality and in human social organization, one has to attempt to go back to the origins, the vital but neglected issue of the evolutionary origin of language. This is a topic that, after neglect during the nineteenth century and continuing neglect by mainline linguists, has come to assume a much greater importance; there is a rapidly growing literature on the subject, with accounts that differ both as regards the evolutionary mechanisms proposed and the theories of language adopted.

In "mainline linguistics," if this is an appropriate term for the Chomskyan approach, Chomsky accepts that language acquisition is the result of innate factors but rejects any possibility of giving a classical Darwinian account of language origin by natural selection. His linguistic theories have changed radically over the years; by slow steps he has moved away from his long obsession with syntax as the essence of language and has come to appreciate more fully language as the interaction of syntax and lexicon. His account of universal grammar has been based very heavily on the peculiarities of English, in which word order plays the central role; this approach is less appropriate for other languages that rely on highly differentiated lexical structures, with word order as a subsidiary process. There is nothing remarkable about word order; words can only be uttered serially, just as one can only perform actions serially, and syntax is very largely implemented through words. Chomsky suggests that the unordered superrules (principles) are universal and innate, and that when children learn a particular language, they do not have to learn a long list of rules, because they were born knowing the superrules. All they have to learn is whether their particular language has the parameter head-first, as in English, or head-last, as in Japanese, as if the child were merely flipping a switch to one of two possible positions. This general conception of grammar is called the "principles and parameters" theory and is the latest in a line of theories from syntactic structures in 1957, moving from transformational-generative grammar centered on the idea of "deep structure," to government and binding in 1980. Many would not accept Chomsky's current account described above as a useful basis for considering the evolution of language.

PINKER AND OTHERS

After these introductory remarks about Chomskyan linguistics, one can turn to Steven Pinker's book *The Language Instinct* (1994), which is the most recent, comprehensive, and ambitious attempt to account for the origin of language. It approaches the topic from within the Chomskyan framework, which is unusual since Chomskyan linguists have generally remained silent about language evolution. Pinker offers an evolutionary account of language based on the idea that it evolved, step by minimal step, adaptively by natural selection, in exactly the same sort of way as that proposed by Darwin (and refined by Dawkins) for the evolution of the eye, or, also as proposed by Darwin, for the evolution of animal instincts. Language, Pinker (1994, p. 18) says, is not a cultural artifact but "a distinct piece of the biological make up of the brain"; the task is to explain the evolution as a brain module or organ of what Chomsky terms of "universal grammar." Chomsky's theory of universal grammar is well known: the brain must contain a recipe or program that can build an unlimited number of sentences from a finite list of words; the program may be called a mental grammar; children—"grammatical geniuses"—must innately be equipped with a plan common to the grammars of all languages that tells them how to extract the syntactic patterns from the speech of their parents. Pinker does not share Chomsky's skepticism about whether Darwinian natural selection can explain the origins of the language organ. At a number of important points, Pinker's account seems unsatisfactory, for example, the idea that language could have developed, like the eye, by minute steps under the pressure of natural selection, the idea that eventually neuroscientists will be able to locate a "language organ" or behavioral geneticists to discover a grammar gene, the postulation of a uniform distinct language of thought, mentalese, to be translated into any particular spoken language, his discussion of the arbitrariness of the sign, and his account of the acquisition of language by children.

These criticisms both of the Chomskyan account of language and of Pinker's attempt to present a gradualistic account of the evolution of language as a distinct function, instinct, or organ are dealt with at greater length in a paper for the meeting of the Language Origins Society at Pecs, Hungary (Allott, 1995b). Piattelli-Palmarini (1994, p. 339), an enthusiast for Chomsky, says that Pinker's account (developed with Paul Bloom) is the best adaptationist reconstruction, but is still unconvincing. In Pinker's book there is much little-examined speculation about the time when the first rudiments of language might have emerged, about the manner in which children can acquire lexicon, about the anthropological basis for improved language capacity, about the role of genetic mutation in bringing about changes in language structures, and about the possibility of survival benefit for the individual flowing from mutations affecting language competence and performance. The principal error is the failure to treat adequately the social character of language as a possession of the group, of the speech community, and not simply of the individual. Language development and change as increasing inclusive fitness, that is, serving the long-term reproductive success of the individual, is simply not a plausible proposition. The other major error is concentration on the evolution of syntax and grammar and treating superficially the vital role of the development of lexicon, both as a representation of the perceived world and as an instrument for syntactic manipulation of utterances through function words, inflections, and the like. The final and perhaps most important error is a mistaken view of natural selection as limited to gradualistic change in a complex structure serving a specified function; natural selection also operates through serendipitous transfer of complexity developed for one function to a new function, for example, the move from swim bladder to lung, from webbed foot to wing, from gill to structures of the ear, and so on.

Nor is it necessary to spend much time on the treatment of language in Maynard Smith and Szathmáry (1995). For them, language is the fifth major transition in evolution. They disclaim any special knowledge of linguistic theory; in their account of the evolution of language, they say that they rely heavily on two sources, Bickerton (1990) and Pinker's (and Bloom)'s scenario of language evolving by minimal steps. While they adopt the evolution of language and of the eye as parallel processes, they recognize the difficulties of accounting in this way for the step-by-step evolution of Chomskyan grammar. The criticisms already mentioned in relation to Pinker's account apply equally to that produced by Maynard Smith and Szathmáry.

A DIFFERENT APPROACH

If both the Chomskyan approach to language and the gradualistic account of language evolution are rejected, what alternatives are there? Any theoretical approach to language has to go beyond phrase structure and cope with the elaborated systems of grammar and lexicon found in many world languages. Other approaches follow a quite different tack: Reject the idea that language could have emerged gradually by minute steps following the pattern of individual natural selection with inclusive fitness related to minor improvements in the language capacity. Suggest instead another type of verified evolutionary origin: the construction of new biological functions on the basis of systems that had evolved for other purposes. Darwin and others have given accounts of this kind of evolution, for example, the swim bladder serving a new purpose as an air-breathing lung, the evolution of gills into the structures of the ear, the conversion of the forefoot for locomotion into the hand for grasping, and the conversion of webfeet for movement in water into surfaces for gliding or flying. Evolutionary accounts of language following this different approach predicate a transfer from previously developed perceptual or motor systems, requiring changes in brain organization, to form a capacity for the development of languages and the complexities seen in languages.

The idea that language may have been modeled on or directly derived from preexisting brain systems has been explored by a number of writers. The possibilities include modeling on tool use (Greenfield, Parker, Gibson, and others), modeling on the visual system (Givón), modeling on throwing action (Calvin), and modeling on motor control (Studdert-Kennedy, Lieberman, and Allott). The earliest suggestion on these lines was by Karl Lashley (1951), who discussed the generality of the problem of syntax and drew attention to the parallels between the syntax of language and the syntax of action. There has since been considerable discussion of the grammar of action and of the grammar of vision by Richard Gregory (1976). It is not possible in this chapter to present these alternatives at any length, but the following paragraphs briefly describe them.

Greenfield's 1991 article "Language, Tools, and Brain: The Ontogeny and Phylogeny of Hierarchically Organized Sequential Behavior" postulated an evolutionary homologue of the neural substrate for language production and manual action that provided a foundation for the evolution of language before the divergence of the hominids and the great apes. The role of toolmaking as a precursor for or as coevolving with language has been extensively discussed. Perhaps it should be treated as the first approach to investigating the relation between language and the cerebral motor-control system.

Studdert-Kennedy (1983) suggested that linguistic structure may emerge from, and may even be viewed as, a special case of motoric structure, the structure of action. Later he added that for language, the goal is to derive its properties from other, presumably prior, properties of the human organism and its natural environment: "We should try to specify the perceptual and motor capacities out of which language has evolved" (1983, p. 329). Evidence from brain stimulation (notably the work of Kimura, Ojemann, and Mateer) almost forced the hypothesis that the primary specialization of the left hemisphere is motoric rather than perceptual; language would be drawn to the left hemisphere because the left hemisphere already possessed the neural circuitry for control of the fingers, wrists, and arms, precisely the type of circuitry needed for control of the larynx, tongue, velum, and lips and of the bilaterally innervated vocal apparatus. Ojemann and Mateer (1979) and Ojemann (1991) identified common cortical sites for sequencing motor activity and speech. Language arises at least in part in brain areas that originally had a predominantly motor function; the development of language seems to have incorporated brain mechanisms originally developed for motor learning.

Givón in a paper for the 1994 Berkeley meeting of the Language Origins Society took the system of visual perception as the basis on which language emerged in a process of coevolution. In this theory the evolution of language was linked directly to the development of the visual system. He discussed the correspondences between visual and linguistic information and suggested that language processing piggybacked on visual processing. In evolution there had been an early coexistence of auditory-vocal and visual-gestural codes; the rise of visual-gestural coding provided a neurocognitive preadaptation for a shift to audio-oral coding because of the adaptive advantages it offered, freeing the hand and body for other activities transcending the immediate visual field. He developed these ideas in the light of recent evidence from PET scans and other research of brain localization of particular aspects of language processing in relation to visual and auditory brain organization.

Lieberman (1984, 1991) has presented a motor theory of the origin of syntax. According to this theory, the evolution of speech and language follows from Darwinian processes; organs that were originally designed to facilitate breathing air and swallowing food and water were adapted to produce human speech. The development of language was an instance of the mechanisms of preadaptation that, besides examples such as swim bladders and lungs, produced the sometimes surprising preadaptive bases of various specialized organs, for example, milk glands from sweat glands and the bones of the mammalian middle ear from the joint of the lower jaw. The initial stage in the evolution of the neural bases of human language appears to have involved lateralized mechanisms for manual motor control, facilitating precise one-handed manual tasks. Brain mechanisms that allow the production of the extremely precise complex muscular maneuvers of speech, the most difficult motor-control task that humans perform, may have provided the preadaptive basis for rule-governed syntax that may reflect a generalization of the automatic schema first evolved in animals for motor control in tasks like respiration and walking. A change in brain organization that allowed voluntary control of vocalization is the minimum condition for vocal communication.

Calvin (1992) argues the case for an even more specific preadaptation for the neural machinery underlying language in the neural circuitry required for planning sequential hand movements such as hammering and throwing. Since hand-arm sequencing circuitry in the brain has a strong spatial overlap with the location of language circuitry in the left brain, perhaps the same massively serial architecture can do double duty for language and planning ahead. The well-formed sentence and the reliable plan of action have some strong analogies to more familiar Darwinian successes, a matter of what Darwin called "conversion of one function to another" or metamorphosis of function. To describe the original function from which the conversion of function was made, the better word is "exaptation" because of the "preconceived" connotations of preadaptation. A given piece of anatomy can have more than one function. The conversion of function is an excellent candidate for how beyond-the-apes language abilities originated. Hominid-to-human language is a "free" secondary use of neural sequencing machinery that was primarily shaped by the food-acquisition uses of ballistic movement skills.

In a number of works (Allott 1989, 1992, 1994, 1995a, 1995b, 1995c) I have presented a comprehensive motor theory of language evolution and function. The theory proposes as a universal principle that the structures of language (phonological, lexical, and syntactic) were derived from and modeled on the preexisting complex neural systems that had evolved for motor control, the control of bodily activity. Motor control at the neural level requires preset elementary units of action that can be integrated into more extended patterns of action—neural motor programs. These in turn have to be linked to and integrated with one another by "syntactic" neural processes and structures. In this theory, given that speech is also essentially a motor activity, language made use of the elementary preset units of motor action to produce the equivalent phonological units (phonemic categories); the neural programs for individual words were constructed from the elementary units in the same way as motor programs for bodily action are formed from them (in both cases a neural program is formed in direct relation to the perceived structure of the external world); the syntactic processes and structures of language proper were modeled on the "syntactic" rules of motor control.

Chomsky, Pinker and Bloom, and Piattelli-Palmarini argue against preadaptation on the basis of the visual or motor systems on grounds that are directly related to their perhaps idiosyncratic formal analysis of language, with its emphasis on syntax. Thus Piattelli-Palmarini (in a 1994 article that argued strongly against Piaget's view that language was derived from or related to motor schemata) said that "The form of linguistic principles is very specific (e.g., c-command, X-bar, PRO, projection of a lexical head, trace of a noun-phrase)" (1994, p. 339) and went on to say that "there is no hope, not even the dimmest one, of translating these entities, these principles, and these constructs into generic notions that apply to language as a "particular case." Nothing in motor control even remotely resembles these kinds of notions; concrete linguistic examples (drawn from Chomskyan theory) make it vastly implausible that syntactic rules could be accounted for in terms of sensorimotor schemata. Chomsky in *Language and Problems of Knowledge* (1988) said that the visual system is unlike the language faculty in many crucial ways; though there are some similarities in the way that the problems can be addressed, in relation to vision and language, the visual faculty does not include the principles of binding theory, case theory, structure dependence, and so on. The two systems operate in quite different ways. Pinker, Chomsky, and Piattelli-Palmarini, in rejecting a preadaptive or exaptational basis for the evolution of language in the visual or motor systems of the brain because it is impossible to see how such a basis could accommodate the formalisms of transformational-generative grammar, government and binding, or principles and parameters, ignore the unwelcome possibility that there is something fundamentally wrong with the linguistic theories, not with the Darwinian process by which there can be conversion of function from an already-existing complex neural system for perception or action to serve as the basis for speech and language function. Chomsky is left in the awkward position of being unable to conceive of a Darwinian origin for language even though he asserts that it must have a biological basis; this leads Pinker to propose a gradualistic account of language evolution as the product of a series of minimal genetic and language changes, which is implausible in accounting for the step-by-step accretion of the elements required for Chomskyan phrase-structure theory, and even less plausible in accounting for the development of other complex grammatical and lexical features of world languages.

The way out of the impasse is to see the evolution of language as a system founded on, reflecting, and expressing the preexisting complexities of the perceptual and motor systems of the brain. If the evolution of language is viewed in this way, the relation between biological and cultural evolution loses a great deal of its mystery. Language is the biological link between culture and noncultural aspects of human evolution both in its role in the development of the brain and cognition and in its continuing role, as part of brain organization and function, as the instrument for the preservation and transmission of culture from generation to generation.

REFERENCES

Allott, R. (1989). *The motor theory of language origin*. Lewes: Book Guild.
———. (1992). The motor theory of language: Origin and function. In Jan Wind, Brunetto Chiarelli, Bernard Bichakjian of Alberto Nocentini (Eds.), *Language origin:*

A multidisciplinary approach. NATO ASI Series Advanced Science Institute Series. Series D: Behavioural and Social Sciences, Vol. 61. Dordrecht: Kluwer Academic Publishers.

————. (1992). *Gestural equivalence (equivalents) of language.* Paper for meeting of the Language Origins Society, Berkeley, CA.

————. (1995a). Motor theory of language origin: The diversity of languages. In Jan Wind, Abraham Jonker, Robin Allott, & Leonard Rolfe (Eds.), *Studies in language origins,* Vol. 3 (pp. 125–160). Amsterdam: John Benjamins.

Allott, R. (1995b). *Pinker's language instinct: Gradualistic natural selection is not a good enough explanation.* Paper for meeting of the Language Origins Society, Pecs, Hungary.

Allott, R. (1995c). Syntax and the motor theory of language. In M. E. Landsberg (Ed.), *Syntactic iconicity and linguistic freezes: The human dimension* (pp. 307–329). Berlin: Mouton de Gruyter.

Bickerton, Derek. (1990). *Language and species.* Chicago: University of Chicago Press.

Boyd, Robert, & Richerson, Peter J. (1985). *Culture and the evolutionary process.* Chicago: University of Chicago Press.

Calvin, William. (1992). Evolving mixed-media messages and grammatical language: Secondary uses of the neural sequencing machinery needed for ballistic movements. In Jan Wind, Brunetto Chiarelli, Bernard Bichakjian & Alberto Nocentini *Language origin: A multidisciplinary approach* (pp. 163–179). NATO ASI Series. Series D: Behavioural and Social Sciences, Vol. 61. Dordrecht: Kluwer Academic Publishers.

Cavalli-Sforza, L. L., & Feldman, M. W. (1981). *Cultural transmission and evolution: A quantitative approach.* Princeton, NJ: Princeton University Press.

Chomsky, N. (1957). *Syntactic structures.* The Hague: Mouton.

————. (1988). *Language and problems of knowledge.* Cambridge, MA: MIT Press.

Darwin, Charles. (1871 [1981]). *The descent of man.* Reprint with Introduction by J. T. Bonner and R. M. May. Princeton, NJ: Princeton University Press.

————.(1876 [1958]). *The autobiography of Charles Darwin 1809–1882.* Nora Barlow, London: Collins.

————. (1882 [1971]). *The origin of species.* Introduction by L. Harrison Matthews. Reprint of 6th edition. London: Dent.

Dobzhansky, T. (1951). *Genetics and the origin of species* (3rd ed.). New York: Columbia University Press.

Durham, William H. (1991). *Coevolution: Genes, culture, and human diversity.* Stanford: Stanford University Press.

Givón, T. (1994). *On the co-evolution of language, cognition, and neurology.* Paper for the Tenth Meeting of the Language Origins Society, Berkeley.

Greenfield, P. M. (1991). Language, tools, and brain: The ontogeny and phylogeny of hierarchically organized sequential behavior. *Behavioral and Brain Sciences, 14,* 531–595.

Gregory, R. L. (1976). *Concepts and mechanisms of perception.* London: Duckworth.

Hinde, Robert A. (1982). *Ethology: Its nature and relations with other sciences.* Oxford: Oxford University Press.

Ingold, Tim. (1986). *Evolution and social life.* Cambridge: Cambridge University Press.

Lashley, K. S. (1951). The problem of serial order in behavior. In L. A. Jeffress (Ed.), *Cerebral mechanisms in behavior* (pp. 112–135). New York: John Wiley.

Lieberman, P. (1984). *The biology and evolution of language*. Cambridge, MA: Harvard University Press.

———. (1991). *Uniquely human: The evolution of speech, thought, and selfless behavior*. Cambridge, MA: Harvard University Press.

Lumsden, Charles J., & Wilson, Edward O. (1981). *Genes, mind, and culture: The coevolutionary process*. Cambridge, MA: Harvard University Press.

———. (1983). *Promethean fire: Reflections on the origin of mind*. Cambridge, MA: Harvard University Press.

Maynard Smith, J., & Szathmáry, Eors. (1995). *The major transitions in evolution*. Oxford: W. H. Freeman.

Ojemann, G. A. (1991). Cortical organisation of language and verbal memory based on intraoperative investigation. In D. Ottoson (Ed.), *Progress in sensory physiology* (Vol. 12, pp. 193–230). Berlin: Springer-Verlag.

Ojemann, G., & Mateer, C. (1979). Human language cortex: Identification of common sites for sequencing motor activity and speech. In O. Creutzfeldt, H. Scheich, & C. Schreiner (Eds.), *Hearing mechanisms and speech* (pp. 192–220). Berlin: Springer-Verlag.

Parker, S. T., & K. R. Gibson. (1975). A developmental model for the evolution of language and intelligence in early hominds. *Brain and Behavioral Services, 3*, 367–408.

Piattelli-Palmarini, Massimo. (1994). Ever since language and learning: Afterthoughts on the Piaget-Chomsky debate. *Cognition, 50*, 315–346.

Pinker, S. (1994). *The language instinct*. London: Penguin.

Pinker, S., & Bloom, P. (1990). Natural language and natural selection. *Behavioral and Brain Sciences, 13*, 707–784.

Rindos, David. (1985). Darwinian selection, symbolic variation, and the evolution of culture. *Current Anthropology, 26*, 65–88.

Studdert-Kennedy, M. (Ed.). (1983). *Psychobiology of language*. Cambridge, MA: MIT Press.

———. (1986). Development of the speech perceptuo-motor system. In B. Lindblom & R. Zetterstrom (Eds,) *Precursors of early speech* (pp. 206–217). New York: Stockton Press.

Tax, Sol. (1960). The celebration: A personal view. In Sol Tax & Charles Callender (Eds.), *Issues in evolution*. Vol. 3 of *Evolution after Darwin* (pp. 271–282). Chicago: University of Chicago Press.

Tooby, J., & Cosmides, L. (1992). The psychological foundations of culture. In J. H. Barkon, L. Cosmides, & J. Tooby (Eds.), *The adapted mind: Evolutionary psychology and the generation of culture* (pp. 19–136). New York: Oxford University Press.

Tylor, E. B. (1865). *Researches into the early history of mankind*. London: John Murray.

6

Evolution: Implications for Epistemology and Cultural Variation

Dennis Werner

REALITY AND KNOWLEDGE

How can I swear to tell the truth? I can only say what I know. How can I know if this is the truth?
            —Cree Indian testifying about a dam being built on his territory,
                                          cited in Richardson (1975)

Perhaps there are individuals in all cultures who stop to think about the relationships between truth and knowledge, reality and the mind, facts and reason, object and subject. In order to illustrate how contemporary evolutionary theory affects these reflections, I would like to contrast an *evolutionary* view of reality with three other views, which may be called *naïve realism, idealism*, and *phenomenalism*. Other works on evolutionary epistemology (Lorenz, 1973; Wuketiks, 1990; Campbell, 1970) have addressed themselves mainly to the natural sciences or to other social sciences in which the views of reality contrasted here are less common. Since my notions of these views may differ somewhat from those of others, I would like to clarify briefly what I mean by each, discussing implications for science, morality, and esthetics.

NAÏVE REALISM

A rose is a rose is a rose

                                                            —Gertrude Stein

For those who have never reflected much about philosophical issues, it seems a great waste of time to question the existence of things. It just seems so obvious that "a rose is a rose is a rose." But of course when we start to think a little, we see that things are not so easy. We may be pretty clear about what we mean by Mrs. Smith's house on Elm Street, but when we read that a census found 403 houses in our neighborhood, we might wonder how researchers counted the children's playhouse in Mr. Jones's backyard, or whether they counted as one or two houses the place where Mrs. Thompson resides with her married daughter and grandchildren. When we move on to phenomena like "war," "love," or "causality," we find it even more difficult to establish "facts," and we realize that we need to think about what we mean by each of these terms. We need to talk about "concepts"—that is, something in our own minds rather than "out there" in "reality."

In terms of esthetics, naïve realists usually assume that some things really are more beautiful than others. Likewise, with regard to morality, they have no problems declaring that some acts are really wrong and others right. Lists of sins or rules of behavior drawn up by authority figures such as churches, governments, or professional organizations promote this idea and may help some people to remain naïve realists throughout their entire lives.

It would be nice to think that cultural anthropologists have given up any naïve notions of reality they may once have held, but there are still many moments when naïve realism creeps into anthropological thought. This is especially evident in exhortations to "free oneself of one's biases" and accept the "facts as they are" when doing fieldwork.

IDEALISM

> Particular natural facts are the symbols of particular spiritual facts. Nature is the symbol of spirit. . . . Every appearance in nature corresponds to some state of mind.
> —Ralph Waldo Emerson (1983)

Although we are confronted in our personal lives with a great diversity of houses, dogs, roses, and the like, we are still able to form images of what we mean by "house," "dog," "rose," and so on. These images correspond to something that has never been seen, but that combines all of the essential qualities for forming a concept. For Plato and other idealists, the "things" "out there" are only imperfect shadows of the ideal forms we hold in our minds. To know the world, then, we need first to know ourselves. Learning consists in "remembering" or "raising consciousness."

Aristotle was less confident about this "inside" route to finding ideal forms and insisted on a more empirical approach to reality. He used experimentation to clarify how the imperfect things in this world manage to achieve their ideal forms (their "essences"), for example, his famous experiment (described in Harré, 1981) in which he opened twenty chicken eggs on successive days to describe the development of the chick embryo. As Harré (1989) points out,

Figure 6.1
Aristotle's Four Causes and Ideal Potentials

	Explains the actual	Explains the ideal potential
Explains the current state of things	Material cause	Formal cause
Explains change	Efficient cause	Final cause

Aristotle's famous "four causes" were also conceived in light of how things realize their potential ideal, as shown in figure 6.1.

These different causes (I prefer to call them explanations) are all relevant to anthropological research. For example, a study of war in a given society might look first at the formal explanation, which would clarify what either the participants or the ethnologists mean by war (its "essence") in terms of rules, structures, and the like. The particular war under study may not conform to this ideal, and the material explanation—the inventory of soldiers, arms, and other resources that make up the fighting force—might explain why the ideal is not achieved. Why this particular war broke out at this moment would be explained by the efficient cause—a fight over women, or an insult suffered by someone, for example. The final cause may consist in examining why warfare exists at all in this or other societies (some effect of general social structure, or adaptation, for example).

In the idealist view, moral and esthetic questions are very closely related to the question of truth. Everything is achieved when perfection is reached. In the words of Emerson (1983, p. 36), "The true philosopher and the true poet are one, and a beauty which is truth, and a truth which is beauty, is the aim of both." "Sensible objects conform to the premonitions of reason and reflect the conscience. All things are moral."

For centuries idealism dominated Western philosophy and science. In biology, it was found in Aquinas's "scale of nature" that organized animals in a hierarchy going from the most imperfect to the most perfect—from plants to simple animals, primates, humans, angels, and finally God—and could also be found in Lamarck's evolutionary ideas involving progress. It has also appeared in more recent ideas such as Odum's (1963) view of the "climax" stage in an ecological sere—the stage in which an ecological system achieves its full potential—and in comments about upsetting the natural "balance of nature" (see Reichholf, 1992, for a critique).

Idealist notions have also been quite common in the social sciences. When popular psychologists exhort people to realize their potential, for example, they are demonstrating an idealist worldview. In anthropology the idealist view is reflected in the attempts by Maine, Edward Tyler, Lewis Henry Morgan, and Friedrich Engels to describe the inevitable progress of humankind (even if, as with Engels, the progress occurs only after some regression). I would also classify Margaret Mead as an idealist. This is reflected in her interest in cultural norms, which do not refer to what people actually do, but rather to what their culture sets out as ideal forms of behavior. Thus not all Mundugumor may follow the aggressive norms of their culture, and the practical functioning of society may even depend on the existence of deviants. But a culture's norms are not totally arbitrary; they need to be understood in terms of what they offer a society. Mead's daughter, Mary Catherine Bateson (1985), explains the relationships between Mead's ideas of norms and higher values. Margaret Mead often deposited her daughter in different households while she went off to study different cultures or deliver lectures. Mary Catherine was instructed to observe the particular norms of each household where she was placed, and she was given ideas about why the households maintained different rules. In preparation for her wedding, Mary Catherine and her mother reviewed the rules of etiquette one by one in order to understand what each rule was for, so that they could decide which rules they wished to follow.

PHENOMENALISM

Que toda la vida es sueño, y los sueños, sueños son (all of life is a dream, and dreams are but dreams).

—Pedro Calderón de la Barca

Although they agree that the things of this world, at least as we perceive them, may be the product of our minds, many thinkers do not share the idealist's confidence in ideal forms. What we have in our minds may also be just a shadow, pure illusion, only a dream. In ancient Greece the Sophists already rejected the idea that we could arrive at a universal truth. In his declaration that "man is the measure of all things," Protagoras emphasized that humans live only according to the laws of their own culture, which can vary from one place to the next. There are no greater laws. As Malefijt (1983, p. 16) points out, this is the doctrine of cultural relativism accepted by many, if not most, contemporary anthropologists.

For the phenomenalists, the material world also lacks any greater meaning. The aphorism of Heraclitus that "a man never steps twice into the same river" is often cited to illustrate the fleetingness of phenomena and, presumably, the fragility of our concepts with regard to these phenomena. How can we talk of a "river" when the water is always changing, and also the sunlight's reflection and the observer as well? Nothing stays the same. We can only sense the fleeting moments. The "essence" of reality, if there is such an "essence," consists simply in that passing sensation that never repeats itself. Any attempt to join disparate phenomena into a single concept (like "river") is arbitrary and has no greater justification.

Phenomenalists react differently to these reflections. For some Sophists, they led to the nihilism that so disturbed Socrates and Plato. On the other hand, the

Epicureans decided that since humans exist only in the here and now, the best one can do is to enjoy what life has to offer and avoid suffering. Existentialists like Sartre, Camus, and the characters in Ingmar Bergman's films reacted with anguish to the absurdity of life. Genet managed to achieve a mystic ecstasy while contemplating his own insignificance, especially when he was provoked by humiliation. In these works, the ideas of right and wrong and beauty and ugliness are often inverted in order to point out that they depend on the eye of the beholder.

Since phenomenalists emphasize how reality is a product of thought, they tend to concentrate principally on consciousness itself. As Cupani (1985, p. 32) observed, for Husserl, "The world is made up of feelings or meanings that depend on consciousness." Everything is "subject," even the "object." Rather than speaking of "facts," the phenomenalist prefers to speak of "concepts."

Ruth Benedict (e.g., 1934) was one of the first anthropologists to take seriously the phenomenalist point of view. That people in different cultures gave their lives different objectives or meanings did not require additional explanation in terms of psychology or other higher ideals, as Mead thought. I would also classify Benedict's one-time classmate, Gilberto Freyre (e.g., 1933), in the phenomenalist school. He turned Brazilians' view about the country's racial and ethnic mixture on its head, finding virtues where others found only problems. But instead of the existentialist's angst over life's meaning, Freyre demonstrated always the Epicurean's delight in everyday pleasures. More recently, social historians like Foucault and Ariès have emphasized how different cultures construct their own realities, while postmodernists anguish over the ethics of cultural relativism.

EVOLUTIONISM

> Geschrieben steht, am Anfang war das Wort.
> Hier stock ich schon. Wer hilft mir weiter fort? . . .
> Der Geist mir hilft. Auf einmal seh'ich Rat
> Und schreibe getrost: Am Anfang war die Tat.
> (In the beginning was the Word, the Scriptures read.
> But here I balk. Who'll help me now proceed?
> The Spirit helps. For once I'll take its heed
> and write consoled: In the beginning was the deed.)
> —Goethe's Faust translating the Bible

> The miracle is not that the world has laws, but that we are able to understand them.
> —Albert Einstein

Although initially evolution was conceived in an idealist framework, today, following Darwin, it has come to refer to evolution via natural selection. Natural selection affects the way we view the relationship between mind and reality. Naïve realists assume that there is a reality out there, independent of our minds, and give little thought as to how we come to know this reality. With idealism, natural "facts" are symbols of spiritual "facts," so there is a perfect correspondence between reality and mind, at least when both achieve perfection. For phenomenalists, reality is a construction of our minds, with only the most fragile connections to anything "outside." Evolutionists admit the existence of an external reality, but like the phenomenalists, they assume that reality, as we perceive it, is indeed a construction of our minds. For both evolutionists and phenomenalists, any attempt to find a greater meaning to life is pure illusion. But there is a basic difference. While the phenomenalists see the mind as relatively independent of "reality," or indeed as the constructor of reality, evolutionists place limits on how far the mind can stray. If our minds were too independent of reality, we would never have survived natural selection. But this does not mean that the mind is perfectly adapted to reality, as the idealists thought. Natural selection is pragmatic, not perfect. For example, although all animals must care for their own offspring if they are to pass on their genes, the ways they recognize these offspring may vary tremendously—by location, smell, or a particular peep. Even in their own environments, none of these recognition devices is foolproof, and parents are often deceived (Allport, 1997).

Natural selection has allowed us to survive and reproduce, but there is no guarantee that our way of thinking can arrive at the "Truth." This is because our notions of "true" or "false" are also the products of natural selection. They may have great pragmatic value, but they are not absolute. As Hofstadter (quoted in Méro, 1990) stated this idea: "Most philosophers and logicians are convinced that truths of logic are 'analytic' and a priori; they do not like to think that such basic ideas are grounded in mundane, arbitrary things like survival. They might admit that natural selection tends to *favor* good logic—but they would certainly hate the suggestion that natural selection *defines* good logic." Various researchers have begun to trace the origins of our notions of truth and falsehood in the social interactions of primates and other animals, especially in studies of deceit and play (Sommer, 1992; Byrne & Whiten, 1988). In logic it is propositions (not reality) that are "true" or "false." Likewise, in social relations it is the signals from other animals that must be judged as "true" or "false." Psychological studies of the "mistakes" Westerners make in their use of formal logic also reveal that these "mistakes" are much more easily explained if we assume that they are based on a "Darwinian cheating detector," rather than other factors, such as lack of familiarity with the materials or situations (Cosmides & Tooby, 1992; Gigerenzer & Hug, 1992).

Just as evolutionary theory can help us clarify the origins and nature of our logic, it can also help us clarify the origins and nature of our morality or esthetics. Alexander (1987) suggests that our notions of morality may have evolved in conjunction with systems of social reciprocity, especially indirect reciprocity. However, as Alexander and others (Wright, 1994; Rachels, 1990) point out, we cannot conclude that traits that are adaptive (including our notions of morality) are also moral. This would be a case of concluding from what is to what ought to be, which is possible only within an idealist vision of reality. Social thinkers are often guilty of making this connection in one or more of its forms. With the *naturalist fallacy* people conclude that what is "natural" is good; with the *relativist fallacy* they conclude that whatever any given group thinks is good *should* be considered good; with the *moralist fallacy* they conclude that what "ought to be" is what "is" (for example, "the value of an object *ought* to represent the work invested in it; therefore, the value of an object *results* from the work invested in it"), or they conclude that what "ought not to be" "cannot be" (for example, Lysenko's conclusion that natural selection cannot be true, because it is immoral). The study of how our ideas about morality evolved cannot tell us what is right, but can be useful in pointing out biases and cheating strategies that we should look out for. In particular, we should be wary of the selfish motives behind self-righteousness. The ability of evolutionary theory to alert us to these biases is one of its most important heuristic aspects.

EPISTEMOLOGICAL IMPLICATIONS OF EVOLUTIONISM

The epistemological implications of evolutionism are based on how it perceives the relationship between mind and matter. Naïve realists do not think much about how mind relates to matter, except, perhaps, to comment that our biases or faulty perceptions may at times blind us from seeing reality as it is. For idealists, there is a true and perfect correspondence between mind and matter. Plato thought that we possess a priori very specific ideal forms in the mind, for example, the notion of "dog" or "house." Later philosophers (such as Locke with his "blank-slate" view of mind) argued that these concepts are constructed from experience. Although Locke recognized that we also need something else in order to build concepts, he basically ignored what this something else might be (Losee, 1979). It was left to Kant to think through how a limited number of a priori "categories" (like space, time, number, and cause) might suffice for us to construct all of our other concepts. But Kant did not explain where these a priori categories might come from and depended instead on the idealistic notion of a basic correspondence between mind and matter. Later scholars, most notably Piaget (e.g., 1974), expanded on Kant's ideas and attempted to clarify how basic concepts are constructed.

For anthropologists, doubts about the a priori character of Kant's categories stem primarily from Durkheim (1915), whose critiques have "definitively been incorporated into anthropological thought" (Oliveira, 1988, p. 4). Durkheim asked where Kant's categories might have come from and concluded that they could only come from society. Our notions of contradiction, causality, time, and space come from our first social experiences and are transmitted via language, which is a social product. Since societies and languages vary, so must these general categories.

Evolutionists can welcome Durkheim's questions. After all, the categories, if they exist, must have come from somewhere. But Durkheim's answer cannot explain how an infant could learn the very first social categories if it did not already come to the world with a few tools to incorporate this social experience. Where did these tools come from? As was pointed out by Lorenz (1941), the answer must lie in the whole process of biological evolution. Researchers from areas as diverse as neurology (e.g., the special issue of *Scientific American, 267* (3), 1992), neuroethology (Camhi, 1984), cognitive ethology (Ristau, 1991; Byrne & Whiten, 1988; Parker & Gibson, 1990; Cheney & Seyfarth, 1990), cognitive psychology (Anderson, 1990: Gigerenzer & Hug, 1992), and artificial intelligence (Drescher, 1992; Méro, 1990) have all been busy trying to figure our how sensory systems, perceptions, concepts, reasoning, and other mental phenomena evolved.

This research has epistemological implications because it helps clarify the limits to human thought. These limits may be absolute or relative. As Alexander (1987) points out, there is not a whole lot we can do about absolute limits to our imagination. Physicists sometimes talk about things we cannot imagine (like four dimensions, curved space, or the beginning of time itself) and may give the impression that they can imagine these things. But what really happens is that they extend a mathematical logic that begins with something we can imagine (like a cubic meter) to things we cannot imagine (like a quartic meter).

We can do a lot more about our relative limitations. These limitations include things like the number of pieces of information we can hold in short-term memory or biases with regard to how we perceive things—biases that we can see once they have been pointed out to us, and that we can to some extent control. We can construct research designs that include controls for a few standard biases, like the tendency to perceive only evidence that confirms our ideas. We can use double-blind tests, for example. We can sometimes carry out computer simulations to see if our reasoning about an issue is as clear as we think it is. We can perhaps also stimulate the generation of a diversity of ideas by guaranteeing that people of diverse backgrounds and interests all participate in the context of discovery.

Like phenomenalists, evolutionists also recognize that we cannot deal with "facts," defined as phenomena that correspond exactly to reality (a reality either independent of our minds, according to the naïve realists, or with exact correspondence to our minds, according to the idealists). For both phenomenalists and evolutionists, then, "objectivity" simply refers to "intersubjectivity." That is, we accept something as more "objective" if there is more agreement among those who study the phenomenon. But there is a basic difference between the phenomenalists and the evolutionists on what they mean by "intersubjectivity." Phenomenalists are more likely to emphasize that intersubjectivity refers to "consensus." Zilman (1979) suggests that "science" be defined as "the study of theses for which we have arrived at a universal consensus," although in practice Zilman thinks that "universal" could be toned down to "almost universal." Some postmodernists, like Lyotard (in Oliveira, 1988, pp. 98–99), argue that "consensus" would inevitably be false, given biases and power differences, and emphasize instead a "dialogue" between speakers (Tedlock, 1991). In this case, intersubjectivity is achieved through communication rather than consensus.

For evolutionists, intersubjectivity can take on a more concrete character because universal or common elements in the human mind guarantee that there will be some convergences in the ways different people think of or perceive the world. We can agree about some things—about logic, for example, or about certain perceptions. This permits *replicability*. On the occasion of a cheese-tasting contest, Sokolov (1993) provides a good example of how agreement about common perceptions can be achieved. It is consensus at these more concrete levels that provides the *data* for testing more abstract theories. Unlike "facts," "data" are also mental constructions, but unlike the consensus at more abstract levels, data are replicable. This changes the game between different speakers. Winning an argument consists not in gaining consensus among the listeners about the argument itself, but rather in coming up with an argument that best accounts for the data. Einstein understood this problem well. When he was warned by a friend about a publication entitled "A Hundred against Einstein," he replied simply, "One would do" (Calder, 1979, p. 142).

Some critics argue that replicability may be valid for the natural sciences, but not for the social sciences, because human phenomena are unique. Here it is relevant that we are always replicating *aspects* of phenomena and not the phenomena themselves. For example, we compare apples and oranges on aspects like price, sweetness, or vitamin C content.

This discussion also has implications with regard to Popper's (1982) view that theories can never be shown to be true, but can be shown to be false if they disagree with the data. As others have pointed out, data are also mental constructions and so cannot be shown to be true. If we cannot affirm data, then we cannot disconfirm theories. Thus it is not possible to show that theories are false. In concrete terms, we see this problem when statistical data are used to evaluate a hypothesis. A failure to find a correlation may mean simply that the data are not good, not that the theory is bad. To carry Popper's analogies between science and natural selection one step further, it is noteworthy that natural selection does not eliminate "wrong" animals. It only selects among alternatives. The same is likely true with ideas. Natural selection equipped us with the capacity to decide between alternative courses of action, but not to determine if a given course of action is "correct" or not. In terms of research strategies, such a bias means that we cannot evaluate one hypothesis by itself, but can only compare alternatives.

Besides these implications with regard to how we evaluate explanations, evolutionary theory may also affect other intellectual pursuits, such as how we come up with new ideas and structure them, how we persuade others of our ideas, how we comprehend the routines and thought structures of others, and how we reinterpret the ideas of others in order to come up with new solutions to problems. These topics are all benefited by evolutionary theory's power to give us specific clues as to the structure and content of the human mind.

THE STRUCTURE AND CONTENT OF MIND

Few, if any, anthropologists or evolutionists today would argue that major differences between ethnic groups can be explained by biological differences in their brains. However, anthropologists are not at all agreed on what "psychic unity" or "human nature" consists of. For Malinowski, cross-cultural differences are only superficial. The Trobrianders, for example, may begin with different premises and cite different types of evidence, but the logic of their argumentation is the same as that of Westerners (see the discussion of the male's role in procreation in Malinowski, 1929). Like Westerners, the Trobrianders and other peoples are also motivated to acquire prestige and wealth, although the forms of this prestige and wealth may vary from society to society (Malinowski, 1929). For Benedict (1934), the differences go much deeper. Depending on their culture, people are motivated by entirely different objectives. For example, in many North American Indian groups people seek intense emotional experiences, but among the Zuni, people are motivated to avoid these experiences. The Kwakiutl may long for prestige and wealth, but not the Zuni. For Ariès and Foucault, the differences go much deeper still. While Benedict admitted that people in different cultures think about and treat their children and their homosexuals differently, Ariès (1981) and Foucault (1978) argue that "children" and "homosexuals" themselves are recent constructions of Western cultures. In other societies there may be "miniature adults" or there may be a multitude of individuals who have sexual relations with others of their sex, but there is no reason to think, for example, that a Sioux *winkte*, if brought up in today's United States, would become an American gay. Just how similar and how different are humans?

Thinking about how the mind evolved can help us clarify these questions. Major differences in points of view go back to a distinction that defines hostile camps in at least the three fields most familiar to me: cultural anthropology, psychology, and evolutionary biology. This is the distinction between "structuralist" versus "atomistic" views. It is a distinction that also seems to define national rivalries—French versus Anglo-Saxon. Briefly stated, the structuralist view sees different phenomena as inextricably linked, so that their separation is meaningless, much as a musical note is meaningless when it is extracted from the symphony it belongs to. The "atomistic" view considers such "decontextualization" as possible.

Rieppel (1992) traces this distinction back to the ancient Greek debates between "idealists" like Plato and Aristotle, with their emphasis on formal explanations, and the Sophist "materialists" who saw things as constructed out of smaller divisible units. Although this distinction appears similar to my distinction between idealists and phenomenalists, it is not the same. The "idealist/phenomenalist" dimension deals with the belief or disbelief in ultimate ideal potentials, while the "structuralist/atomist" dimension deals with the importance of form versus specific content. Intellectuals may combine these beliefs in different ways. For example, I would classify Margaret Mead as both "idealist" and "atomistic," Ruth Benedict as both "phenomenalist" and "atomistic," Claude Lévi-Strauss as both "idealist" and "structuralist," and Michel Foucault as both "phenomenalist" and "structuralist."

For evolutionary biology, Rieppel (1992) distinguishes between "atomists" like Darwin, Richard Dawkins, E. O. Wilson, and the nineteenth-century followers of "natural theology," on the one hand, and "structuralists" like Cuvier, Lamarck, Stephen J. Gould, Niles Eldridge, and the nineteenth-century *Naturphilosophen*, on the other. "Atomists" emphasize the functional or adaptive significance of specific characteristics. They are more likely to search for parallel developments across species and explain them in terms of adaptation to similar environmental constraints. "Structuralists" emphasize "design constraints" rather than environmental constraints in evolution. This implies greater attention to how different characteristics join together to form a whole, or how an organism develops throughout its life span. Structuralists emphasize how a specific change (e.g., hormone release) at a given moment affects a multitude of features that develop subsequently (see Gould, 1984b, and Money & Ehrhardt, 1972, for examples). Cross-species comparisons emphasize the structural similarities in closely related species. While "atomists" see evolution primarily as an adaptive process, "structuralists" see evolution as a cumulative process requiring the study of how complex forms can develop out of simpler forms.

In psychology, structuralist viewpoints are represented by many forms of psychoanalysis (e.g., Lacan) and by Piaget's "constructivist" view of psychological development. "Atomism" is represented by behaviorist learning theories or by more recent sociobiological attempts to explain specific human or animal cognitive processes. In cultural anthropology, the contrast is between French structuralism versus British or American "culturalism," including functionalism, relativism, and cultural ecology (Descola, Lenclud, Severi, & Taylor, 1988; Schulte-Tenckhoff, 1985).

At first glance it might seem that adopting a more structuralist position in biology would also imply a more structuralist position in psychology or cultural anthropology. However, the connection is not so clear. In fact, acceptance of the very "design constraints" that are so important to structural biology may imply a less structuralist view of mind. This is because these constraints mean that the human mind is unlikely to have evolved as a whole piece. Natural selection can only act on already-existing structures. It is like a *bricoleur* (a tinkerer), constructing new forms out of preexisting pieces. A house built by a *bricoleur* has a history. A kitchen is added on. A closet becomes a bathroom. Such a house has many "unnecessary" supports to hold it up. It is not like the more efficient house built by an engineer in which all the pieces are planned ahead of time, and no extra columns are included to support the structure. But structural psychologists often view the human mind as if it had been built by an engineer. Indeed, Piaget was heavily influenced by engineers. Like Kant (on whom his theory is based), Piaget begins with the construction of a few elementary categories that are then used to construct, via interaction with the environment, all the more abstract and differentiated concepts humans possess. Today Piaget's theories may be especially applicable to engineering attempts to build artificial intelligence (e.g., Drescher, 1992). But on theoretical grounds we might well question their applicability to human intelligence. A mind constructed by a *bricoleur* would have a hodgepodge of many more independent and sometimes relatively "hard-wired" concepts and routines—perhaps not as many as Plato imagined, but certainly more than Kant or Piaget admitted.

Gould (1984a) cited the case of Mozart to illustrate how the mind is constructed in relatively independent modules. When he was only eight years old, Mozart showed so many musical talents that an English intellectual, a certain Mr. Barrington, felt obligated to verify his birthdate in the Salzburg records in order to assure himself that this was not a trick. But what most impressed Barrington was that, outside of music, Mozart was an absolutely typical eight-year-old boy. Barrington could only conclude that the human mind is organized in relatively independent modules. How else could we explain the enormous difference between Mozart's musical comprehension and other aspects of his mind?

Most researchers would agree that some aspects of our cognition are closely related to other aspects, but that there are also relatively independent modules. The question is to figure out just what is most related to what. Here empirical studies must come into play.

Many cultural anthropologists would agree that cognitive abilities and personality can vary somewhat independently of each other, but they would insist that "concepts" can only be understood in terms of their relationship or contrasts with other concepts, and this is why structural analyses are so important. In part this insistence stems from a confusion between concepts and words. No one doubts that sememes, morphemes, phonemes, and other aspects of speech are defined by their structural relationships. However, there are many grounds for distinguishing between concepts and words. First, as shown by Damasio and Damasio (1992), there are different regions of the brain associated with concepts and words. Second, numerous studies of animal cognition (see Ristau, 1991; Byrne & Whiten, 1988; Parker & Gibson, 1990; Cheney & Seyfarth, 1990) make it clear that animals possess concepts, but not words to express these concepts. Finally, studies of human development also show that babies and children possess concepts long before they are able to express them (e.g., Bowerman, 1981; McKenzie, 1990).

But even if concepts cannot be equated with words, concepts themselves may be based on contrasts with meanings derived from relationships with other concepts. It is useful to distinguish between the evolution of the *structural* complexities of speech—syntax, morphology, phonology, and the like—and the *conceptual/semantic* aspects of speech. Richman (1993) notes that we may find the answer to the evolution of structure in song. On the other hand, abstract concepts may have been built up differently, via analogy. They do not depend on contrasts to give them meaning, because they are rooted in universal, inborn concepts. Lorenz (1973) suggests that the first concrete concepts to evolve dealt with spatial movements such as forward, backward, up, down, left, and right. This is because these were the first things animals needed to represent mentally. Building on these concepts, we produce abstractions like "righteousness," "downtrodden," and "backwardness." Givón (1989) follows Charles S. Peirce in arguing that etymologies generally "track" how analogies are used to construct abstract concepts. Thus, the contemporary debates between de Saussure's *semiologie* and Peirce's semiotics may simply reflect the different amounts of attention given to linguistic structures and to concepts, respectively.

What does all of this have to do with cultural anthropology? The main implication is that there may be aspects of thought that are relatively independent of each other, and that we can "decontextualize" some things. We do not need to see everything as embedded in a culture's symbolic system. We do not need to speak of "total social facts," as Mauss and later structuralists thought. This means, for example, that it is reasonable to expect that people with different symbolic systems may come up with similar solutions to similar problems.

We may also need to rethink how we define "culture." If culture refers only to symbolic systems, or only to what can be transmitted via language from one person to the next, then we may miss out on much that is important. We will have excluded from cultural anthropology the study of shared ideas, values, attitudes, and the like that have originated independently in different people in response to some similar problem or social circumstance. It is foolish for cultural anthropology to simply define these phenomena out of the field. An overly symbolic definition of culture amounts to a type of "love it or leave it." The field of cultural anthropology can only lose prestige, personnel, and students by such an attitude.

EVOLUTIONARY THEORY AND EXPLANATIONS FOR CROSS-CULTURAL VARIATION

Evolutionary theory is most relevant for cultural anthropology because of its implications for epistemology and for our theories about mind. However, specific theories about evolution may also help explain particular aspects of cultural variation. It is these more specific theories that are most controversial in cultural anthropology. For example, van den Berghe and Mesher (1980) argued that incest makes genetic sense for men and their genetically related wives where the men enjoy polygyny, and where the male offspring of the incestuous couple can reliably count on enjoying polygyny as well. This may represent an extreme case of the age-old tendency to invest in those offspring that are genetically most like oneself. The authors suggest that this explains why socially instituted incest relations occurred only among the royal families of societies like ancient Egypt, Korea, Hawaii, the Incas, and several African states. While this type of argument may be interesting, there are also other ways to link evolutionary ideas to cross-cultural variations.

Although it may sound paradoxical, it is the universals implied by evolutionary theory (and that include more than ideas about kin selection) that can best be used to explain cross-cultural variation. Indeed, one of the best ways to check if something really is universal is to see if it predicts cross-cultural variation. For example, in his study of adult human speech errors in French and Hungarian, Fónagy (1988, p. 184) suggested that "phonetic as well as syntactic and semantic rule transgressions, far from being 'noise' are generated by a paralinguistic iconic code." Fónagy suggested that this iconic code can be traced to "a communication system phylogenetically prior to language." For example, "the increased maxillary angle in anger may suggest the threat of oral aggression, whilst the tender lip rounding that transforms /i/ into [y], and /e/ into [oe] may allude to a kiss," and "the violent tension of the articulatory organs may be interpreted as a residue of the general muscular contraction preparatory to fighting. Heavy stresses might represent blows" (Fónagy, 1988, p. 184). Fónagy could attempt to verify these ideas by checking whether the same speech errors are associated with tenderness and aggression in one language after another. Of course Fónagy could never check all human languages, since many of these languages no longer exist, and he thus could never really show that this association is universal. But it may be possible to transform this idea into a hypothesis about cross-cultural variation. We might check, for example, whether heavy accents or tense vocalizations are more common in societies where aggressiveness is more highly valued. In fact, Lomax and his associates (1968) have shown that this is true with regard to singing styles.

Cultural anthropologists have been slow to check ideas in this way, but practically all ideas about human universals imply similar associations. Do certain facial expressions mean the same thing everywhere? Then maybe we can predict the kinds of masks produced by different peoples around the world based on the emotions that are most valued (see Aronoff, 1975, for a study of masks). Are the same colors most salient all over the world? Then maybe we can predict cross-cultural variation in color categorizations (see Boster, 1986; Berlin & Kay, 1969). In fact, although this point is not always well appreciated, any explanation for cross-cultural variation implies the assumption of some universal in human thought, whether this be a specific process of psychological development and learning, a rational way to solve practical problems, or some inborn tendency to associate one thing with another.

In sum, although there are also many other theories that might help to explain cross-cultural variation, there is good reason to look at specific evolutionary theories as well. Of course, following my own epistemological advice, these arguments should be contrasted with alternative arguments. Also, of course, researchers need to remember that phenomena are never explainable by a single "cause," and this brings us to another source of great confusion. Psychological studies suggest that our concept of "cause" (also a product of evolution) really applies only to *differences*, or more exactly to "co-variation within a context" (Werner, in press; Cheng & Novick, 1991; Hewstone, 1989). Thus, for example, we can provide one cause (different genotypes) to explain why the Pima Indians have more diabetes than Anglos with similar lifestyles, and another cause (diet and other environmental factors) to explain why contemporary Pima Indians have more diabetes than their nineteenth-century ancestors. But we cannot talk of *the* cause or even of the causes of "diabetes" without specifying some type of comparison.

CONCLUSIONS

This chapter has emphasized how an understanding of evolutionary theory is important for cultural anthropology. This is not so much because evolution offers suggestions about how or why cultures vary, but more due to its epistemological implications. Evolutionary theory helps clarify how we think, what we are capable of "knowing," and how we may go about learning. The implications of the theory of natural selection are also important in understanding how the human mind evolved and how it is structured. This affects how we decontextualize aspects of culture and how we study cross-cultural variation. Since I have not seen these issues discussed in anthropological publications, it seems that most cultural anthropologists have not thought much about them. If I am wrong on this point, perhaps this chapter will at least serve to draw these thoughts out of the woodwork and into public debate.

ACKNOWLEDGMENTS

The original version of this chapter was written while I was on sabbatical leave at the Institut für Humanbiologie, Universität Hamburg, with the generous support of a grant from the Brazilian Research Foundation (CNPq). I thank both of these institutions and my colleagues at the Federal University of Santa Catarina and at the Institut for their encouragements.

REFERENCES

Alexander, Richard D. (1987). *The biology of moral systems*. New York: Aldine de Gruyter.
Allport, Susan. (1991). *A natural history of parenting*. New York: Harmony Books.
Anderson, John. (1990). *Cognitive psychology and its implications*. 3rd ed. New York: W. H. Freeman & Company.
Ariès, Philippe. (1981) *História social da família e da criança*. Rio de Janeiro: Zahar.

Aronoff, Joel. (1975). *Threat characteristics in masks*. Paper presented at the meetings of the Society for Cross-Cultural Research, New York.

Bateson, Mary Catherine. (1985). *With a daughter's eye*. Stephens City, VA: PB Whitney.

Benedict, Ruth. (1950). *Patterns of culture*. New York: Mentor Books.

Berlin, Brent, & Kay, Paul. (1969). *Basic color terms: Their universality and evolution*. Berkeley: University of California Press.

Boster, James. (1986). Can individuals recapitulate the evolutionary development of color lexicons? *Ethnology, 25*(1), 61–74.

Bowerman, Melissa. (1981). Language development. In C. Triandis & A. Heron (Eds.), *Handbook of cross-cultural psychology* (Vol. 4, pp. 93–185). Boston: Allyn & Bacon.

Byrne, Richard W., & Whiten, Andrew. (Eds.). (1988). *Machiavellian intelligence: Social expertise and the evolution of intellect in monkeys, apes, and humans*. Oxford: Clarendon Press.

Calder, Nigel. (1979). *Einstein's universe*. New York: Viking Press.

Camhi, Jeffrey M. (1984). *Neuroethology: Nerve cells and the natural behavior of animals*. Sunderland, MA: Sinauer Associates.

Campbell, Donald T. (1973). Natural selection as an epistemological model. In R. Naroll & R. Cohen (Eds.), *Handbook of method in cultural anthropology* (pp. 51–88). New York: Columbia University Press.

Cheney, Dorothy L., & Seyfarth, Robert M. (1990). *How monkeys see the world: Inside the mind of another species*. Chicago: University of Chicago Press.

Cheng, Patricia W., & Novick, Laura R. (1991). Causes versus enabling conditions. *Cognition, 40*, 83–120.

Cosmides, Leda, & Tooby, John. (1992). Cognitive adaptations for social exchange. In J. H. Barkow, L. Cosmides, & J. Tooby (Eds.), *The adapted mind: Evolutionary psychology and the generation of culture* (pp. 163–228). New York: Oxford University Press.

Cupani, Alberto. (1985). *A crítica do positivismo e o futuro da filosofia*. Florianópolis: Imprensa da Univ. Fed. de Santa Catarina.

Damasio, Antonio, & Damasio, Hanna. (1992). Brain and language. *Scientific American, 267*(3), 63–71.

Descola, Philippe, Lenclud, Gerard, Severi, Carlo, & Taylor, Anne-Christine. (Eds.). (1988). *Les idées de l'anthropologie*. Paris: Armand Colin.

Drescher, Gary L. (1992). *Made-up minds: A constructivist approach to artificial intelligence*. Cambridge, MA: MIT Press.

Durkheim, Émile. (1965). *The elementary forms of the religious life*. New York: Free Press.

Emerson, Ralph Waldo. (1983). *Nature: essays and lectures*. New York: Library of America.

Fónagy, Ivan. (1988). Live speech and preverbal communication. In M. E. Landsberg (Eds.), *The genesis of language: A different judgement of evidence* (pp. 183–204). New York: Mouton de Gruyter.

Foucault, Michel. (1980–1988). 3 vols. *The history of sexuality*. New York: Vintage Books.

Freyre, Gilberto. (1943). *Casa Grande e Senzala*. Rio de Janeiro: José Olympio.

Gigerenzer, Gerd, & Hug, Klaus. (1992). Domain-specific reasoning: Social contracts, cheating, and perspective change. *Cognition, 43,* 127–171.

Givón, T. (1989). *Mind, code, and context: Essays in pragmatics.* Hillsdale, NJ: Lawrence Erlbaum Associates.

Gould, Stephen J. (1984b). Mozart and modularity. *Natural History, 93*(6), 6–14.

———. (1984b). A short way to corn. *Natural History, 93*(3), 12–20.

Harré, Rom. (1981). *Great scientific experiments: Twenty experiments that changed our view of the world.* Oxford: Oxford University Press.

———. (1989). *The philosophies of science.* Oxford: Oxford University Press.

Hewstone, Miles. (1989). *Causal attribution: From cognitive processes to collective beliefs.* Cambridge, MA: Basil Blackwell.

Lomax, Alan. (1968). *Folk song style and culture.* Washington, DC: American Association for the Advancement of Science.

Lorenz, Konrad. (1941). Kants Lehre von apriorischen im Lichte gegenwärtiger Biologie. *Blätter für Deutsche Philosophie, 15,* 94–125.

———. (1973). *Die Rückseite des Spiegels: Versuch einer Naturgeschichte menschlichen Erkennens.* Munich: R. Piper & Co.

Losee, John. (1979). *Introdução histórica à filosofia da ciência.* Bel. Horizonte: Ed. Itatiatia.

Malefut, Annemarie de Waal. (1983). *Imágenes del hombre.* Buenos Aires: Amorrorta editores, S.A.

Malinowski, Bronislaw. (1929). *The sexual life of savages in North Western Melanesia.* New York: Harcourt, Brace World.

McKenzie, Beryl E. (1990). Early cognitive development: Notions of objects, space, and causality in infancy. In C. A. Havert (Ed.), *Development psychology: Cognitive, perceptuo-motor, and neuropsychological perspectives* (pp. 43–61). Amsterdam: North-Holland.

Méro, Láslo. (1990). *Ways of thinking: The limits of rational thought and artificial intelligence.* Teaneck, NJ: World Scientific.

Money, John, & Ehrhardt, Anke A. (1972). *Man and woman, boy and girl: The differentiation and dimorphism of gender identity from conception to maturity.* Baltimore: Johns Hopkins University Press.

Odum, Eugene. (1963). *Ecology.* New York: Holt, Rinehart & Winston.

Oliveira, Roberto Cardoso de. (1988). *Sobre o pensamento antropológico.* Rio de Janeiro: Tempo Brasileiro.

Parker, Sue Taylor, & Gibson, Kathleen Rita. (Eds.). (1990). *"Language" and intelligence in monkeys and apes: Comparative developmental perspectives.* New York: Cambridge University Press.

Piaget, Jean. (1974). *A epistemologia genética.* São Paulo: Editora Victor Civita.

Popper, Karl. (1982). *Conjecturas e refutações.* Brasilia: Universidade de Brasilia.

Rachels, James. (1990). *Created from animals: The moral implications of Darwinism.* Oxford: Oxford University Press.

Reichholf, Josef H. (1992). *Der schöpferische Impuls: Eine neue Sicht der Evolution.* Stuttgart: Deutsche Verlags-Anstalt.

Richardson, Boyce. (1975). *Strangers devour the land.* Toronto: Macmillan.

Richman, Bruce. (1993). On the evolution of speech: Singing as the middle term. *Current Anthropology, 34*(5), 721–722.

Rieppel, Olivier. (1992). *Unterwegs zum Anfang*. Munich: Deutscher Taschenbuch Verlag.

Ristau, Carolyn A. (Ed.). (1991). *Cognitive ethology: The minds of other animals*. Hillsdale, NJ: Lawrence Erlbaum Associates.

Schulte-Tenckhoff, Isabelle. (1985). *La vue portée au loin: Une histoire de la pensée anthropologique*. Lausanne: Editions d'en bas.

Scientific American. (1992). [Special edition on the brain.] *Scientific American, 267*(3).

Sokolov, Raymond. (1993). A cheddar aesthetic. *Natural History, 102*(11), 88–91.

Sommer, Volker. (1992). *Lob der Lüge: Täuschung und Selbstbetrug bei Tier und Mensch*. Munich: Beck.

Tedlock, Dennis. (1986). A tradição analógica e o surgimento de uma antropologia dialógica. *Anuário Antropológico*, 183–202.

Van den Berghe, Pierre L., & Mesher, Gene. (1980). Royal incest and inclusive fitness. *American Ethnologist, 7*(2), 300–317.

Werner, Dennis. (1997). *O pensamento de animals e intelectuais: Evoluçao e Epistemologia*. Florianópolis: Editora da Universidade Federal de Santa Catarina.

Wright, Robert. (1994). *The moral animal*. New York: Pantheon.

Wuketits, Franz M. (1990). *Evolutionary epistemology*. Albany: State University of New York Press.

Zilman, John. (1979). *Conhecimento público*. Belo Horizonte: Liv. Itatiaia.

7

Culture and the Darwinian Heritage: Implications for Literary Research in the University

John Constable

THE HUMANITIES AND THE SCIENCES

It is not immediately evident that Darwin has any significance for those studying literature beyond typifying, for the conservative appreciative critic, scientific philistinism, and for others, a naïve trust in the value of the scientific method. On the first point we have Darwin's own self-sentence: "Up to the age of thirty, or beyond it, poetry of many kinds, such as the works of Milton, Gray, Byron, Wordsworth, Coleridge and Shelley, gave me great pleasure, and even as a schoolboy I took intense delight in Shakespeare, especially in the historical plays. . . . But now for many years I cannot endure to read a line of poetry: I have tried lately to read Shakespeare, and found it so intolerably dull that it nauseated me" (Darwin, 1958, p. 138). To make things still worse, he claimed to prefer novels, "if they do not end unhappily," and provided that they contained "some person whom one can thoroughly love" (Darwin, 1958 pp. 138–139). On the second point, Matthew Arnold acts as undertaker: "I have heard it said that the sagacious and admirable naturalist whom we lost not very long ago, Mr. Darwin, once owned to a friend that for his part he did not experience the necessity for two things which most men find so necessary to them—religion and poetry; science and the domestic affections, he thought, were enough" (Arnold, 1883, pp. 336–337).

Hardly anyone would agree to live in accord with this asceticism, but the course by which Darwin arrived at it is not unlike that undergone by many students of literature, and even by some who go on to become professional literary scholars. After a period of infatuation or addiction comes a deepening sense of the futility of literary and critical study, and after that, for the academics at least, a desperate attempt to convert their criticism into something that will allow them to hold up their heads in common room and lecture hall. This often takes the form of an exaggerated insistence on the value of a humanistic education, or an equally insistent denial of this position combined with an attempt to erect a criticotheoretical structure describing literature, the study of which will constitute the education that reading Shakespeare, for example, so obviously is not. These two parties have much in common, and overall, one might say that criticism squares up to its subject material rather as Paleyite natural theologians related to the Creation, but there is an important distinction between the revolutionaries and the conservatives. Whereas those who still insist on the importance of reading literature for its not very clearly specified human values are obvious analogues of clerics who find evidence of the exquisite handiwork of God in every biological structure, every perfect orbit, radical critics might be better termed natural diabologists. Their mission is to expose literary texts as the ramshackle contrivances of a predatory demiurge buried deep either in the political structure of the society that produces the literature or, mystically, in language itself. In the last twenty years the Manicheans have had much the best of the battle with their pious colleagues, but these local triumphs have been unable to conceal the process of protracted failure afflicting the entire project of which they are a part. The study of "literature as literature" wilted in the university when it became clear that though it was an agreeable hobby, its intellectual rewards were slight. "Literary theory," as it stands currently, is now going the same way, and it must, for literature does not explain the world very well, and literary theory does not even explain literature.

The reason for this general decline is simple: both radical and traditional literary criticism are incompatible with the rest of human knowledge and are therefore superfluous. Within the university, and arguably elsewhere, we do not need a religion of poetry or its inversion, because science is enough. For literature and nonprofessional discussion of literature, this matters less, since the public world will probably maintain both, but for academic literary theorists and critics, incompatibility will ultimately be professional death. The root cause of this isolation is readily identified: workers in the humanities have clung to the natural theological and diabological attitudes long after the majority of the intellectual world has embraced the major alternative, naturalism. It is not too late to tag along, and I hope in this chapter to sketch some of the reorientations that are needed to bring about worthwhile change, and to outline, if only for researchers in other fields who may wish to encourage its first falterings, some of the ventures that might be undertaken with the new pilot.

NATURALIZING LITERARY STUDY

We might begin by saying a literary study that was correctly naturalized would provide physicalistic, causal explanations of cultural objects, by which we would mean not only ideational objects, brain states, but also consequent behaviors and objects external to our bodies, such as books. It would thus be able to relate the results of its work to other areas of science. For such an attempt to succeed, the overall government for this process of naturalization must come from a field in which these principles are already widely accepted, and biology, and Darwinian thought in general, is on two counts a clearly appropriate source. First, Darwinized cognitivism (Tooby & Cosmides, 1992) provides a fully materialist psychology on which may be based a satisfying generational theory of culture. Second, Darwinism introduces population thought to the study of cultural objects, as exemplified in the epidemiological theories of Dan Sperber (1985, 1990, 1994, 1996a, 1996b) and the work of Pascal Boyer on the distribution of religious ideas (1994).

There are already, as it happens, numerous signs that biology is having an effect in literary study, and that this process of naturalistic theorization is taking place. Joseph Carroll's work, including his monumental *Evolution and Literary Theory*, is an inescapable, and in many ways an estimable, sign of this change (Carroll, 1995a, 1995b; for discussion of Carroll's position, see Constable, 1996). However, just as Darwinism led some thinkers in the nineteenth century to abandon religion for an imperfect naturalism in which the process of natural selection became an authority to be venerated and obeyed, there seems every reason to suppose that the impact of the biological sciences on the humanities in the coming years may amount to little more than a revision of current practices. The recent history of literature departments and the way that fields as diverse and diversely creditable as psychoanalysis and linguistics have been used show us how Darwinism will be employed. This new inrush of thought will be taken up in order to retool the engine of critical commentary and so enable critics to work over their chosen texts once more, this time producing evolutionary readings and Darwinized interpretations. In short, it will be used to reinforce a biologized criticism. To continue the metaphor, since departments of literature regard the conceptual product of other disciplines as a mere component part of the real business, just cogs for the motor of literary discussion, contemporary critics see their own work as an end product. They do not deny that other disciplines also have products, but they reserve the right to pass critical judgment on them. Thus products in the sciences become subject to the ethical criticism of the university literary world, which becomes the final arbiter of value. So, not only do critics regard themselves as the end users of all other products within the intellectual world, but they regard their own production as the end product to end all other products and the medium through which the intellectual activities of all other thinkers are to be represented to students and to the public. An example may make this danger seem more immediate. This is the opening statement from a recent study of romantic poetry by Karl Kroeber:

My purpose in this book is to encourage the development of an ecologically oriented literary criticism. This criticism, escaping from the esoteric abstractness that afflicts current theorizing about literature, seizes opportunities offered by recent biological research to make humanistic studies more socially responsible. Biologists have arrived at the frontier of revolutionary new conceptions of humanity's place within the natural world. Humanists willing to think beyond self-imposed political and metaphysical limits of contemporary critical discourse can use these scientific advances to make literary studies contribute to the practical resolution of social and political conflicts that rend our society. Humanists could help to ensure, for example, that the effects on our world of new biological research are beneficent rather than malign. (1994, p. 1)

In this case of retooling, the writer "seizes" ideas from the sciences, wields them in conflicts with other schools, and then attempts to turn them on the sciences from which they were taken. When literary critics speak of taking an interest in science or of becoming integrated with it, this is what they envision. They wish to benefit from association with the immense and deserved prestige of the sciences without forgoing any authority. Literary researchers are, in the devastating phrase of the eminent historian, Frank Sulloway, scholars who think that science is a "topic rather than a method" (quoted in Alexander, 1995, p. 1). The problem is not a simple one, even for those who recognize that it must be solved, since criticism is deeply entrenched in the working habits of most researchers in the humanities (see Carroll, 1995a; Cooke, 1995; Dissanayake, 1992, 1995; Fox, 1995; Nesse, 1995; Storey, 1996), and no evaluative approach, however subtle the moral connoisseurship it embodies, can be a science of culture. Criticism articulates conflicts between individuals and groups within a cooperative society and is purposely built to register such differences. By contrast, scientific method is designed to facilitate agreement and to minimize interference from the divergent social, economic, political, and sexual interests of the researchers. The point can be illustrated by observing that scientists in two warring nations can readily agree upon the validity of a piece of work by either one of them, and governments expend large sums of money in espionage to extract scientific secrets from hostile states, whereas discussion of cultural materials, literature, or even music becomes still more contentious in times of open conflict than it was before. If these points are translated onto the personal and social planes, they apply equally well. Conflicts in criticism are conflicts of personal or personal and sectional interest, and critical discussion is an area in which such conflicts take place, thus assisting individuals in negotiating settlements. This proposition is neither novel nor extremely controversial and is widely recognized amongst Marxist critics, such as Terry Eagleton, who has persuasively described the history of criticism in England as coextensive with conflict, though his description is largely disabled by being limited to class conflicts (Eagleton, 1984).

The extremely broad variety of mutually antagonistic approaches to literature is a registration of social friction and the fact that these approaches cannot be brought into any kind of harmony, and that teachers have been compelled, in order to make educational sense of this question, to "teach the conflicts," is inevitable, for there are no concluded agreements, only temporary alliances. On this view, then, criticism cannot be reformed, and its introduction into the university as a discipline during this century is to be regretted. The rejection of academic criticism will not only enable a scientific study of literature, freeing those who, in history, linguistics, stylistics, and poetics, are at present often forced to camouflage their work as subservient to critical projects, but would also benefit public critical debate, because university criticism is clearly neither sufficiently responsive nor capacious to be able to meet public needs. That is, in large societies with numerous interest groups and rapidly changing interests, a self-elected elite of literary specialists cannot, even when equipped with an expanded canon and polysemantics with which to parse its texts, adequately represent and debate these conflicts. Such matters are already better handled through the more adequate electoral democracy of periodical publications and Internet newsgroups.

INTEGRATED SCIENCE AND THE STUDY OF LITERARY DATA

Evolutionary thinkers need not and perhaps should not concern themselves with this general political problem. However, they must ensure that literary researchers are prevented from bringing criticism into an evolutionarily oriented study of culture. This is not to advocate a return to some earlier position in the "scholars versus critics" debates of the past, for, from the perspective outlined here, "scholars," as they have been defined in the university, are deeply permeated with "criticism," and their scholarship is driven by critical imperatives that may, as radical critics have correctly claimed, be only more or less obscured or denied. Nor is it a way of rejecting "theory" and returning the literary humanities to some more orthodox commonsense position underwritten by the authority of science. Insofar as the drive to theory has been motivated by a desire to meet the problems inherent in a university incarnation of evaluative criticism, there is much with which to sympathize, though the solutions have not been fortunate. Indeed, since they fail to provide powerful causal explanations of cultural phenomena, to organize known data, or to define research programs, these solutions hardly deserve to be called theories at all. Where there are insights, for example with regard to the use of cultural objects to manipulate others, these lack any clear definitions of what the interests are and how the manipulation takes place, points on which Darwinian science has much to say (see Cronk, 1995; Eibl-Eibesfeldt, 1988, p. 61; Scalise Sugiyama, 1996). The aim, then, must be to produce an agreed-upon and robust theory of culture, and the best chance of achieving this is to base the attempt on theories that are already agreed upon, that is, the sciences, and to aim for a degree of conceptual integration that will allow university researchers working on cultural objects to avoid the pitfall of criticism and to become part of a cooperative scheme. This means that literary workers have to be shown where they fit into an integrated science and how they contribute to its conceptual product.

The most important step in this positioning is to accept that an integrated science must be physicalist and should subscribe to ontological reductionism. Working out the implications of this initial statement entails locating the humanities in the hierarchy of reduction, or, to use terms from Tooby and Cosmides again, to place the phenomena under study in the "integrated causal model" (Tooby & Cosmides, 1992, p. 23). Biology reduces ontologically to chemistry, and chemistry to physics, and, within biology, behavior reduces to physiology. It is common to speak of these areas as levels of analysis, each with its own modeling principles (theory reduction is not assumed), and this seems unobjectionable, at least on pragmatic grounds. However, it is important to recognize that within the field of behavior itself the only levels of which it makes sense to speak are defined by the number of individuals involved in any particular act. A group action reduces to individual actions and thus to physiology. There is no level of "food cultivation," or of "shelter building," "sexual activity," "tool manufacture," or, for that matter, "linguistic activity," though these of course have an appropriate level at which they may be studied. "Linguistic activity," and thus the writing of novels and plays and poems, like "shelter building," is a data field available to many levels, on which we may draw when we are studying a particular act. Thus we arrive at the point of most interest for literary scholars, the point at which books, printed pages, have to be examined within this scheme. It is initially implausible to think that they reduce to physiology, and certainly if we place them in a level of their own, it is awkward to so reduce them, but if we regard them as components in and products of an act, the difficulty vanishes, and we are left with material objects, bound pieces of paper covered with ink, as elements in the extended causal consequences of the organized matter that constitutes the generating person or persons. For example, if I want to examine a courtship act, I can specify its level according to the numbers of individuals involved in it and then turn to the relevant component data fields for that act, among which could be linguistic data.

This distinction between levels of analysis and data fields is primarily of use because it will prevent the isolation of those who, inevitably, specialize in particular kinds of data. If literary specialists are allowed to think of their work as constituting a level, then they will tend to insist on some degree of autonomy, and before long they will be demanding independence. If, on the other hand, they are encouraged to regard their work as the study of a data field, then the risk of dissociation will be greatly reduced, since anyone working on literary data will have to recognize that constant reference must be made to other data concerning a particular act. Furthermore, by insisting on the unity of the behavioral level and the multiplicity of data fields within it, the question of integrational compatibility is greatly simplified. Rather than worrying as to whether the "literary level" should integrate with history or with linguistics, or with some other field or combination of fields, we can say that human behavioral studies itself integrates along only one frontier; that is, it must integrate with the study of the psychological mechanisms that generate behavior. Thus working in the literary data field means studying literary objects as evidence of human acts and producing intelligible, which means abstract and economical, descriptions of those acts for other workers at the behavioral level and for psychologists. In order for this to appear feasible, some clarification of the ontological character of the data field under study is clearly needed, and then a clarification of the way that study of these objects relates to other elements in a unified scheme.

POPULATION THOUGHT AND CULTURAL STUDY: THE EPIDEMIOLOGY OF REPRESENTATIONS

Ernst Mayr has observed that the most significant lesson of Darwinism for our general philosophy is the replacement of essentialism by population thinking (1991, pp. 40ff.). Certainly we can transform our approach to literature when we determine ourselves to regard printed materials as physical elements in a population of such and other related elements, and an individual's reading of a text as one of a population of readings. The clearest statement of this line of thought and its implications is found in the writings of the anthropologist Dan Sperber, whose proposed epidemiological program aims to explain the distribution of "representations," where this term is defined as an "object [that] is a representation *of* something, *for* some information processing device" (1996a, p. 61). Thus when we are liberated to "talk of representations as concrete, physical objects located in time and space," still more illuminating distinctions can be introduced: "At this concrete level, we must distinguish two kinds of representations: there are representations internal to the information processing device—*mental representations*; and there are representations external to the device and which the device can process as inputs—that is, *public representations*" (Sperber, 1996a, p. 61).

The work that Sperber outlines is the study of the causal chains connecting members of these classes; that is, by making a mental representation we may be motivated to change our external environment, perhaps by constructing a public representation that is then processed by another individual to form a mental representation. This individual may then be motivated to form a new mental representation, perhaps including elements of the older one, and thus to make a new public representation. To put it colloquially, someone may have an idea and write a piece of text about the idea, which another person may read and then, in turn, write about. This model has considerable advantages over others based on the "replication" of cultural particles (Dawkins, 1989) and the coevolution of these with genes, since it does not require us to assume that when a reader processes a text, something is, in a very obscure way, being copied or transmitted (see Sperber, 1996a, pp. 100ff.; and Boyer, 1994, pp. 283–284, for more detailed discussions of this point, and also Tooby & Cosmides, 1992,

p. 118). If we try, for example, to say that when I read a sonnet, the ''meaning,'' or some sort of ''information,'' is copied, then it becomes necessary to show how this is physically instantiated or coded and why this deserves the term ''copying.'' This does not appear to be possible, and so to talk of meaning or information transfer in a causal account is little better than invoking a philosophy of spirit. Sperber's approach, on the other hand, holds that communication is not principally a process of transfer, but the provision of encoded cues from which the recipient can make inferences, the bulk of communication resulting from the inferential stage. Thus successful communication involves predicting what sort of cues will provoke a desired inference (Sperber & Wilson, 1995).

The value of this approach is that its materialist foundations prevent us from mistaking abstract descriptions of these causal chains for causally efficacious objects, as writers about literature routinely do. Errors of this type are common even among those who, on paper at least, spurn idealism, since the identification of ''materialism'' with the links between economic structure and ideology has precluded a more satisfactory causal theory based on materialism as it is understood in the natural sciences (for examples, see Jenks, 1993; the papers in During, 1993; and Williams, 1981). This approach also short-circuits fruitless discussion of whether there is a one true immanent meaning for any particular text, for clearly there cannot be, and yet equally clearly there is much to be said for trying to ensure that we are aware of the inferential patterns planned and expected by the cue generator. Similarly, melodramatic skepticism concerning the absence of a shared meaning or a shared text is defused by recasting its points in a physically precise fashion that does justice to the anti-idealist case concerning meaning, but does not oblige us to feign anaesthesia with regard to the communication that does occur. Our attention is concentrated, first, on actual physical instances of a representation and their actual psychological consequences when they are processed as inputs by a particular reader, and, second, on the study of the representation's abstract, formal properties as ''potential psychological properties'' (Sperber's terms). More crudely, we are forced into seeing how texts in fact work on readers and how they might work on readers (and of course on the representations of this potential entertained by writers). Sperber's own example of this is oral, but we may take it as the brief for a rigorous study of literature: ''Potential psychological properties are relevant to an epidemiology of representations. One can ask, for instance, what formal properties make Little Red Riding Hood more easily comprehended and remembered than, say, a short account of what happened today on the Stock Exchange'' (Sperber, 1996, p. 63).

Equally, we can ask ourselves why certain formal properties seem to recur in certain populations of representations. For example, it might be asked why all linguistic communities have a set of utterances and formal structures that are restricted in extent and certain other ways and are mostly considered special or poetic, as opposed to utterances that are unrestricted in extent and are mostly regarded as undistinguished prose. A psychologically motivated discussion might be able to provide some answers to this question (Constable, 1997 and 1998, attempts this). For if, as Sperber puts it, "culture is the precipitate of cognition and communication in a human population" (1990, p. 42), then knowing something of the cognitive mechanism and the terms of the communication may enable us to explain why it is that we get the precipitate that we do get and not some other. Most importantly, great emphasis is thrown onto the significance of history. Epidemiological studies must always be taken as speaking about a particular human population at or during a specified time, for the potential psychological properties of a public representation depend crucially on the state of the device that processes it as an input, and this state depends in large part on the mental representations that already inhabit the device. For instance, extremely restricted forms of verse appear to have been much more widespread and prestigious in the past than at present, and a causal explanation for this probably does not involve genetic change and consequent changes in brain modularity, but almost certainly involves competition with other representations, for example, with unrestricted forms that have, on average, higher pragmatic value.

Literary study, then, is not simply a business of processing texts or exposing students to appropriate external representations, cultural background, as we say, in order that they too may process texts and generate approved internal representations, negative or approbatory or bemused. Or rather it should not be so, though that at present is what it is. Both conservative and radical critics might reply that to evade these matters is to ignore all the features that make cultural representations, and texts in particular, of interest. My own view is that Pascal Boyer's rejection of this type of remark with regard to his study of religion is correct, that "lack of humanistic 'significance' or interest is often the price to pay for causal relevance" (Boyer, 1994, p. 295), and that this is a price worth paying. Indeed, I would go further, by reemphasizing the points made against criticism earlier in this chapter, and suggest that causal explanations are the only ones worth disciplinary pursuit in the university (Sperber, 1996a, p. 98, it should be noted, does not accept this view). Taking our materialism and our commitment to causal explanations seriously means directing attention away from the muddled discussion of essentialist meaning, a soft target in any case, and focusing instead on an examination of actual instances in which public representations are processed and form mental representations, and so on the reasoned and nonevaluative examination of the potential psychological properties of public representations. In other words, the study of literary objects and other cultural objects is to be justified by the light it sheds on the cognitive systems that precipitate them, and on the process of this precipitation.

USING LITERARY MATERIALS

At present, researchers have little experience in using literary materials as data in this way, and the making of psychological inferences from them is liable to be rejected as intentionalism, or confused with Freudian attempts to pathologize the writer, or ridiculed as an outdated belief in the formative importance of the individual author. Some combination of the first and third of these points is liable to be extremely popular and requires notice because it in fact contains an element of interest to be salvaged, namely, the suggestion that not all the order in a text is generated by the author. This point can easily be accepted within the terms of the argument presented here, but without therefore rejecting the concept of human agency altogether. The challenge for those who do not adopt a physicalist ontology is to explain how the complex order of texts can be causally explained other than by reference to a human agency, an author, or authors, or editors. At present, the only source of such order that we know of is the order in human minds (computer-generated texts are not yet autonomous of external instruction, yet and even when they become so, they will long bear their human heritage). This order has two sources: (1) the order resulting from the action of the evolved modular structure of the mind, a modular structure that also makes the learning of order possible, and (2) learned order.

The order in a particular text, however, may not be generated by the named author of that text, as is obvious from quotations. If, for example, I quote from Shakespeare, "To be or not to be," the order in that text fragment is to be explained by reference to the order generation in the brain of a person, thought to be William Shakespeare, but its occurrence in my text is to be explained by reference to the order generation in my brain. Similarly, the order of the natural-language English text that you are now reading is not to be explained wholly as an order generated by my brain. Much of it is to be explained as an order learned by my brain, with its specially adapted module for the purpose, a module that also gives a considerable amount of ordering to that learned material. With appropriate qualifications, then, there is not only no objection to discussing the role of authors in text generation, but it seems there is no alternative. Moreover, many inferences are as unavoidable as seeing something when we open our eyes.

With these very basic levels of inference legitimated, we can safely go further into more dubious territory, and in practice we will rarely be confronted with materials that are not accompanied by a context that prevents the formation of misleading conclusions. The range of uses for literary data is obviously very large, and there is no need to attempt even a vague outline of the possibilities, but there is some point in reminding ourselves that the degrees of complexity will be extremely variable. On the one hand, relatively simple questions concerning structural properties—verse form might be a representative case—can be approached with confidence. For example, any inference that supposed that the order found in a sonnet was to be explained as the invention of a twentieth-century poet would be unsound, since the sonnet form can be found in the literature of previous centuries. The fact that a certain writer wrote in a sonnet form does allow one to make inference, however, about exposure to that form. We can safely say that any poet who has written a sonnet in the last twenty years was exposed to an external representation of a sonnet and formed an internal representation that led him or her to create another external representation with similar formal properties. Such banalities form a secure basis for more adventurous activities. We might, for example, wonder why this form has remained in circulation for so long, and perhaps make inferences about the mind's susceptibility to sonnets, as compared to other verse forms. Alternatively, we might construct hypotheses relating to the functions of verse forms as compared to nonverse (Constable, 1997). Such questions can be approached more or less independently of other properties and are amenable to simple counting surveys. Analysis of narrative techniques would only be marginally more difficult, and a good deal of the work in that area has already been done by theorists of fiction.

On the other hand, the construction of psychological hypotheses relating to complex formal properties in the local texture of a work presents a theoretical headache. No agreed-upon techniques for content analysis exist, and there is always the risk of becoming bogged down in futile critical debate over interpretive differences. One possible way of cutting through this Gordian knot would be to use published criticism as the source of surveyed interpretation, but this, though helpful, only transfers the interpretive problem from one text to a multitude. More problematic still is the difficulty of sorting out which of the many properties of a novel, say, causes it to become epidemically published, and why its author gave it these formal qualities and not others, though some headway has already been made here by those using the content of fiction as data for psychological research on sexuality (Ellis & Symons, 1990; Whissell, 1996). Optimistically, we might say that this rich data field holds many challenges, but it might be as well to admit at the outset that the complex potential psychological properties of texts constitute a very noisy data field, and that the resulting information, even when filtered, may not be of a very high quality.

Setting aside these doubts, we may happily take up the tools developed by stylistics and the linguistically grounded parts of poetics and turn to the work of inference construction. The skill involved in this sort of project rests in determining whether a line of reasoning is of sufficient general interest. Here again we come up against one of the habitual working assumptions of the critic, that the most suitable phenomena for research are those that are unique or rare, hence the traditional emphasis on writers rather than readers, a bias that has to some degree been corrected in recent years. Moreover, it is assumed that the fascinating thing about writers is that they are unusual in some profound way, whereas the approach recommended here tends to assume that what really makes writers unusual is that they write so much. This might well itself become a matter for investigation, but it should not be allowed to overwhelm study of other features shared with nonwriters, that is, with readers. Hence we can conclude that, for example, inferences about the exposure of individual poets to individual ordering principles, in other words, traditional influence studies, are not worth pursuing in themselves, but would properly form an element in a larger project that attempted to widen the database available for psychological speculation. One way of ensuring that an inferential research project of this type is not wasted effort is to return to the principle described earlier, that literary data are typically an element in a commonly occurring behavior.

An example may make this clearer. Let us say that I decide to study the psychology of human judgment, because literary texts abounding in all sorts of judgments are plentiful. After preparing a theoretical framework to enable me to categorize, quantify, and discuss various sorts of judgments, no mean feat, as I have indicated earlier, I turn to the material at hand. Do I then confine my research to one judgment, found in a poem, say, or to one author's judgments? Obviously not, since, first, it is a working principle that literary material is only an element in a larger phenomenon, and, second, discussion of an individual psychology alone is unlikely to produce reliable and general hypotheses. Therefore, although I might indeed study individual judgments and the careers of authors, the inferences made at this level will constitute minor, ancillary work leading up to larger-level inferences.

At this point the data attitude and the current of population thought join to force yet another novelty on the literary researcher, the principle of quantitative evidence. Literary argument is normally essentialist in its assumptions; that is, it regards choice, salient evidence as sufficient and conclusive to support a thesis, and the canons of rhetorical elegance in our journals are built around this principle. Contrary evidence, on the other hand, is excluded in the hope that it will escape the attention of opponents. Moreover, the range of cases, typically individual authors, is usually small. To achieve any degree of reliability, however, an epidemiological inferential study would require large numbers of examples. At present, we have no way of making such a presentation sit prettily upon the page. Given that this extensive epidemiological study is in hand, what is it, psychologically, that is under consideration? At this point we can note that in terms of Sperber's theory the mind may be seen as "susceptible" to culture, and that the student of cultural representations is therefore more or less indirectly studying the susceptibilities or insusceptibilities of mental modules. Clearly, this applies most strongly to readers, and we can reasonably ask what evolved feature of the mind, operating in specified environmental conditions, was susceptible to this cultural representation, this formal property. Let us presume, for example, that, as I think is in fact the case, judgment styles vary across a historical period. We can then ask, "Why this style at time A, but that style at time B?" The same approach can also be used with regard to a single author, regarding the author as susceptible to his or her own production. Briefly, then, cultural objects can be regarded as evidence of preferences, both authorial and readerly, and the task of the researcher will be to relate these preferences and the environment in which they were manifested to what is known of the evolved modular structure of the mind, in the hope that this knowledge may be extended. No assumption need be made about the adaptiveness of either producing or hosting a particular representation, though this question is unlikely to be irrelevant in many cases, and, of course, for a great number of representations we may need to explain their prevalence by the pragmatic power they give to those who hold them, or to those who force or persuade others to hold them.

THE DARWINIAN HERITAGE: ENTAILMENTS FOR
LITERARY RESEARCH

The importance of Darwin for students of culture is that the theory of evolution supports physicalism and opens the way for a powerful antidualist psychology. The theorization of culture that takes place under this aegis can reject criticism, the dominant mode of discourse in departments of literature, and regards cultural materials, whether they are books or brain states, as data to be incorporated into a causal model of human activity through a physicalist psychology and a physicalist ontology. The most important general consequence of this theoretical orientation is that unlike criticism, which is an autonomous activity using other fields but not requiring reciprocal relationships with them, this approach demands integration. An integrated causal model of cultural study regards that study as deriving its importance from its contributory relationships to other fields, principally psychology. Interpretation and explication, traditional and hitherto apparently self-sufficing activities, are seen as subordinate to these relations.

The studies taking place within this framework would be in many ways familiar, both in their use of detailed history and in their use of fine-grained stylistic analysis. Where they would not be so, where the theory breaks decisively with the critical stance, is in the rejection of the suggestion that literature or, to use the current substitute term, textuality is a transcendent category. On this view a poem, a novel, a play, a sentence, a phrase, or any cultural object is a physical object with consequences for the human brain. Since these representations cannot compete with those generated by science in allowing us to understand what they represent, the only reason for studying such cultural objects within a university is that they shed light on the psychologies that generate, use, and host them, and on the history of these activities. Most important, no attempt is made to intervene directly in the public debate over the ideologico-moral implications of cultural objects. Just as there is no place in zoology for the qualitative grading of organisms, there should be none in a naturalized study of cultural objects, where any representation is of potential interest as a topic of study. Equally, the criteria by which the importance of a topic is determined are revised and rendered dependent on the concepts arising from that study, rather than on some ill-defined sense of social prestige. Currently, to work on a minor author is considered somewhat shameful, and scholars are still, despite recent attempts to broaden the scope, concentrated around a few big names, with researchers on the periphery trying to make their own chosen authors into bigger names. If biological research were run on similar lines, we should have a plethora of studies of lions, birds of paradise, and alligators, but very few of naked mole rats or bacteria, and this would entail, as its analogue does in literature, an immense loss of understanding. Naked mole rats and botulism are not worth comprehension intrinsically, though detailed knowledge may have considerable pragmatic utility, but of interest because such examples enable us to develop a detailed causal account of the overall pattern of organic evolution. Similarly, marginal or despised cultural representations may, when placed into an appropriate theoretical context, illuminate general psychological principles as well as or better than apparently central texts of high status. Nevertheless, it is far from certain that a study of cultural materials revised in his name would reconcile Darwin to Shakespeare, but at least, if the plays themselves fail Hume's rigorous tests, thought about Shakespeare, if it were put into an ancillary relationship to history and evolutionary psychology, would be sufficiently theoretically integrated to generate "abstract reasoning concerning quantity and number" and "experimental reasoning concerning matter of fact and existence" (Hume, 1975: 165). It might then be as teachable and rewarding to study as today it is otherwise.

REFERENCES

Alexander, Richard. (1995). The view from the president's window. *Newsletter: Human Behavior and Evolution Society, 4*, 1.

Arnold, Matthew. (1883). Literature and science. In *The works of Matthew Arnold* (Vol. 4, pp. 317–348), London: Macmillan.

Boyer, Pascal. (1994). *The naturalness of religious ideas: A cognitive theory of religion.* Berkeley: University of California Press.

Carroll, Joseph. (1995a). *Evolution and literary theory.* Columbia: University of Missouri Press.

———. (1995b). Evolution and literary theory. *Human Nature, 6*, 119–134.

Constable, John. (1996). Reviews. *Studies in English Literature*, English Number, pp. 91–99.

———. (1997). Verse form: A pilot study in the epidemiology of representations. *Human Nature, 8*, 171–203.

———. (1998). The character and future of rich poetic effects. In S. Sakurai (ed), *The view from Kyoto: Essays on twentieth-century poetry* (pp. 89–108). Kyoto: Rinsen Books.

Cooke, Brett. (1995). Microplots: The case of Swan Lake. *Human Nature, 6*, 183–196.

Cronk, Lee. (1995). Is there a role for culture in human behavioral ecology? *Ethology and Sociobiology, 16*, 181–205.

Darwin, Charles. (1958). *Autobiography* (Nora Barlow, Ed.). New York: Norton.

Dawkins, Richard. (1989). *The selfish gene* (new ed.). Oxford: Oxford University Press.

Dissanayake, Ellen. (1988). *What is art for?* Seattle: University of Washington Press.

———. (1992). *Homo aestheticus: Where art comes from and why.* New York: Free Press.

———. (1995). Chimera, spandrel, or adaptation. *Human Nature, 6*, 99–117.

During, Simon (Ed.). (1993). *The cultural studies reader.* London: Routledge.

Eagleton, Terry. (1984). *The function of criticism.* London: Verso.

Eibl-Eibesfeldt, Irenäus. (1988). The biological foundations of aesthetics. In Ingo Rentschler, Barbara Herzberger, & David Epstein (Eds.), *Beauty and the brain: Biological aspects of aesthetics* (pp. 29–68). Basel: Birkhauser.

Ellis, Bruce J., & Symons, Donald. (1990). Sex differences in sexual fantasy: An evolutionary psychological approach. *Journal of Sex Research, 27*, 527–555.

Fox, Robin. (1995). Sexual conflict in the epics. *Human Nature, 6*, 135–144.

Jenks, Robin. (1993) *Culture.* London: Routledge.

Kroeber, Karl. (1994). *Ecological literary criticism: Romantic imagining and the biology of mind.* New York: Columbia University Press.

Mayr, Ernst. (1991). *One long argument: Charles Darwin and the genesis of modern evolutionary thought.* Cambridge, MA: Harvard University Press.

Hume, David. (1975). *Enquiries concerning human understanding and concerning the principles of morals.* 3rd ed. L. A. Selby-Bigge and P. H. Nidditch (Eds.). Oxford. Clarendon Press.

Nesse, Margaret H. (1995). Guinevere's choice. *Human Nature, 61*, 145–163.

Rentschler, Ingo, Herzberger, Barbara, & Epstein, David. (Eds.). (1988). *Beauty and the brain: Biological aspects of aesthetics.* Basel: Birkhauser.

Scalise Sugiyama, Michelle. (1996). On the origins of narrative: Storyteller bias as a fitness-enhancing strategy. *Human Nature, 7*, 403–425.

Sperber, Dan. (1985). Anthropology and psychology: Towards an epidemiology of representations. *Man, 20*, 73–89.

———. (1990). The epidemiology of beliefs. In Colin Fraser & George Gaskell (Eds.), *The social psychological study of widespread beliefs* (pp. 25–44). Oxford: Clarendon Press.

———. (1994). The modularity of thought and the epidemiology of representations. In Lawrence A. Hirschfeld & Susan A. Gelman) (Eds.), *Mapping the mind: Domain specificity in cognition and culture* (pp. 39–67). Cambridge: Cambridge University Press.

———. (1996a). *Explaining culture: A naturalistic approach.* Oxford: Blackwell.

———. (1996b, December 27). Learning to pay attention. *Times Literary Supplement*, pp. 14–15.

Sperber, Dan, & Wilson, Deirdre. (1995; 1st ed., 1986). *Relevance: Communication and cognition* (2nd ed.). Oxford: Blackwell.

Storey, Robert. (1996). *Mimesis and the human animal: On the biogenetic foundations of literary representation.* Evanston: Northwestern University Press.

Tooby, John, & Cosmides, Leda. (1992). The psychological foundations of culture. In Jerome Barkow Leda Cosmides, & J. Tooby (Eds.), *The adapted mind: Evolutionary psychology and the generation of culture* (pp. 9–136). New York: Oxford University Press.

Whissell, Cynthia. (1996). Mate selection in popular women's fiction. *Human Nature, 7*, 427–447.

Williams, Raymond. (1981). *Culture.* London: Fontana.

PART III

SOCIOBIOLOGY AND POLITICAL SCIENCE

The chapters in this part focus on the political activities of humans, which are certainly important for the understanding of human interactions within and, perhaps even more dramatically, between cultures. Starting with E. O. Wilson's books on human sociobiology and the criticisms leveled against him by Marxists, Lucio Ferreira Alves paints a picture of Darwin's evolutionary theory applied to humans in contrast to that of his contemporary Karl Marx. Alves finds that there is no real justification for the Marxist view, while Darwin's work lives on as a vital link to the present.

Pouwel Slurink looks at the possible social features of the ancestor of the human lineage prior to the appearance of culture to gain a sense of our biological foundations inherited from the past. Patterns of ecological instability in Africa may have led to the emergence of new hunting patterns, say, in australopithecines, requiring a division of labor and the first appearance of cultural learning systems that ultimately meant conflicts between groups. This may account for the beginning of war as a human trait. Johan M. G. van der Dennen also considers sources in the literature that attempt to account for the human propensity for war, starting from the work of Darwin and his view of intergroup competition. Other possibilities found in this rich literature involve the appearance of male hunting, which necessitates male bonding. Finally, he looks at the accounts of ethnocentrism as leading to warlike conflicts, a theme explored in subsequent chapters in this volume.

Sociobiology offers the possibilities of a breakthrough in this area of human social conflict, a problem area that continues to receive attention. Tatu Vanhanen reflects this continuing line of investigation, offering a fascinating empirical study of 183 countries where ethnic cleavage helps us to gain a clearer idea of the roots of ethnic conflict of various kinds. In a broad sense Meyer's "kin selection" is represented in ethnic differences within and between societies. Vanhanen shows a surprisingly high correlation between ethnic cleavage and ethnic conflict, certainly one possible clue to the widespread existence of war among humans. A. J. Nabulsi deals with a similar theme in focusing on tribal groups within Jordan. Here again, looking at this pattern of intracultural differences in a specific setting, he finds that tribal groups are held together by inbreeding, relationships of intermarriage that aid social solidarity. These studies support one another and reveal the promising possibilities of a new scientific integration of political life understood from an evolutionary perspective.

Finally, the work of J. Philippe Rushton gives an additional view of biological patterns in different cultural and ethnic settings, a view that includes measured brain size and different levels of response to intelligence tests between ethnic groups. This, of course, is a sensitive area since we are coming in contact with contemporary efforts to modify intracultural conflict between groups and to emphasize equality rather than difference. We will not easily resolve these issues of the way we live and the way we understand the evolutionary roots behind our conflicts, but it is important that we do not hide from the facts even as we attempt to heal differences. Rushton brings the factual side to the fore.

8

Marx, Darwin, and Human Nature

Lucio Ferreira Alves

In 1975, Harvard biologist Edward Wilson published *Sociobiology: The New Synthesis*, a book of encyclopedic proportions that was defended and attacked with equal enthusiasm by biologists and social scientists. Wilson defined sociobiology as the systematic study of the biological basis of all social behavior in all kinds of animals, including humans. Concerning humans, it refers only to the application of the evolutionary theory to the study of human social behavior. To put it another way, the role of sociobiology is to examine how the diversity of human societies reflects the adaptation of individuals to their social and ecological environments. Thus the social sciences are placed within a framework constructed from a synthesis of evolutionary biology, genetics, population biology, ecology, animal behavior, psychology, and anthropology (Wilson, 1975, 1976).

Because Wilson was merely extending the neo-Darwinian theory into the study of social behavior and animal societies, the criticism the book received from other biologists was overwhelmingly favorable. However, social scientists (here to be understood in the academic sense) and Marxists (here to be understood in their ideological sense) attacked the book vigorously. Wilson had not expected much reaction from the former, and he did not even think about the latter. As he admitted, he underestimated the Durkheim–Boas tradition of the autonomy of the social sciences and the ideological character of Marxist criticism. These criticisms surprised him, but he was unprepared to deal with ideological arguments (Wilson, 1978b).

Three years later, Wilson (1978a) wrote *On Human Nature*, a book totally dedicated to human social behavior. At the end of the book, he said that the dream of social theorists to devise laws of history that can foretell something of the future of humankind died because of their basic misconception of human nature. Karl Marx was certainly the most influential of such theorists.

In the 1970s, altruism was the central problem of sociobiology, but in the 1990s, attention began to shift in human sociobiology to gene-cultural evolution (Wilson, 1998; Lumsden & Wilson, 1983). In the light of modern evolutionary biology, the nature–nurture controversy is obsolete. Humans and animals are prepared to learn certain behaviors and predisposed to avoid others. They form a subclass of *epigenetic rules* (Wilson, 1998), defined as "a sum of all the interactions between the genes and the environment that created the distinctive trait of an organism" (Lumsden & Wilson, 1983, p. 71). Epigenetic rules are adaptive; they confer Darwinian fitness on organisms by improving their survival and reproduction (Wilson, 1998).

Although sociobiology and Marxism have opposite points of view in almost all aspects of human social behavior (such as religion, nationalism, the origin of the family, ethics and moral behavior, and even human pleasures and desires), comparative analyses between them have been extensively explored by Feuer (1977), Zhang (1987, 1994), Flew (1992), Frolov (1986), Lobão (1984), and Paastela (1991). Feuer stresses that Marx's language and method from 1859, when Marx published *Contribution to the Critique of Political Economy*, to 1867 and 1873, when he published *Capital*, became "biologized." While Flew rejects the idea that Marx was one of the greatest men of science, Zhang views Marxism and human sociobiology as two models of understanding human beings. In his analysis, he compares both theories to see if and to what extent they complement rather than compete. Frolov points out some contrasting and converging aspects between them. Like Zhang and Frolov, Lobão focuses her paper on substantive areas where sociobiology and Marxism contrast and converge, but she also contends that "Marxism and sociobiology are theories of progress with individuals envisioned in historical advancement toward ever more states of being" (Lobão, 1984, p. 4). Paastela discusses three basic questions that are relevant for Marxism and sociobiology: namely, the question of materialism, the question of what is inherited and what is learned, and the question of human nature.

Except for the question of materialism, sociobiology and Marxism are two theories that are polar extremes. Contrary to Lobão's assumption, evolution does not necessarily lead to progress "with humans involved in historical advancement toward more satisfactory states of being" (p. 23). It means only change and adaptation. The nature of human "progress" in Marx's and Darwin's thought is, thus, fundamentally different.

Marx and Darwin have only three points in common. First, they were contemporaries. Marx and Engels wrote the *Manifesto of the Communist Party*, an apocalyptic view of the capitalist society and a visionary one of its communist counterpart, in 1848, and Darwin published *On the Origin of Species* in 1859. Second, both shared the view of philosophical materialism. Third, as a result of the first two, both shook the establishment with their works. But the coincidences end here. Therefore, if Marxism and Darwinism are two incompatible theories, why should one continue to compare them?

The present chapter explores Darwinian explanations to account for the fall of communism in Eastern Europe in 1989. To do this, I argue that Marx and Engels (it is quite difficult to take their works separately) not only had a misconception and overoptimistic view of human nature, but also misinterpreted Darwin's theories.

UTOPIAS OLD AND NEW

The idea of an altruistic, classless, and egalitarian society based on the principles of voluntary membership, absence of private property and "to each according to his needs, from each according to his abilities" is much older than the *Manifesto of the Communist Party*. It has been proposed by many thinkers from Plato to Tommaso Campanella, Thomas More, Jean-Jacques Rousseau, and Karl Marx. What distinguishes Marx from the other authors is his belief that his method was inexorable, scientific, and thus historically inevitable.

Freud (1929, p. 304), who knew human nature much better than Marx did, remarked that communism as the final aim of a society was "an untenable illusion." Indeed, given our Darwinian heritage, it is not surprising that all attempts to build such a society, with the exception of the Hutterites of North America and the kibbutzniks of Israel, were unsuccessful. Pierre van den Berghe and Karl Peter (1988) suggested that the success achieved by the Hutterites and the kibbutzniks in establishing and perpetuating economically thriving and biologically self-reproducing settlements based on egalitarian communalism lies in the numerous points of independent convergence they have in common: small-scale, self-effective social control through diffuse sanctions rather than centralized authority, agricultural production with a division of labor based on age and sex, recognition and accommodation of affinal and consanguineal ties, institutionalization of the family as the main rearing child unit, dealienation of labor through the provision of self-paced, nonrepetitive, autonomous jobs, provision of cradle-to-grave security, and a pragmatic adoption of modern productive technology combined with a quasi-peasant social structure.

Anthropological and sociological data show that people with power will always try to maximize both their reproductive and productive advantages. Such data include the Yanomamö from Venezuela (Chagnon, 1979), the Nambikuaras from Brazil (Lévi-Strauss, 1944), the K'ekchi' farmers from Belize (Berté, 1988), the Turkmen from Persia (Irons, 1979), the Kipsigis from Kenya (Borgerhoff Mulder, 1987), the Fijians, the Noma Hottentotes of southern Africa, the Inca of Peru, the Shavante of Brazil, and the Ifaluk from the Western Caroline Islands (Betzig, 1988; Turke & Betzig, 1985). Such data show that even in stateless societies chiefs and leaders are present and, of course, identifiable because, as in any other human groups, there are men who enjoy prestige for its own sake. To put it briefly, when different reproductive and material success are involved, egalitarianism does not prevail, but quite the opposite occurs: not everyone is in fact equal. Therefore, the absence of egalitarianism is not cultural but rather is part of those psychological raw materials out of which any given culture is made (Chagnon, 1979; Lévi-Strauss, 1944).

These patterns are all consistent with the Darwinian perspective that predicts that people with prestige, rank, and power will exploit and use the available resources as a material and reproductive gain and contradicts the Marxist thesis that exploitation of others' productive labor is absent in stateless societies (Engels, 1884; Terray, 1972). Thus, if an unequal distribution of resources exists in primitive societies, how could Marx's concept of a communist society be put in practice in a large, stratified, state-level society? Is it possible to build a society in the terms imagined by Marx and Engels? Given our Darwinian heritage, it is quite improbable that such a society could come into being. But before we seek an explanation for the collapse of communism, it is worth comparing the Marxist and Darwinian concepts of human nature.

MARXISM AND HUMAN NATURE

Marx, Engels, and Lenin were aware of the bourgeois character of the working class. They remarked that the lower middle class, the small manufacturer, the shopkeeper, the artisan, the peasant, and the democratic petty bourgeoisie were not revolutionaries, but reactionaries, because they were fighting against the bourgeoisie not to change the whole society, but to save from extinction their own existence as fractions of the middle class (Marx & Engels, 1848, 1850).

In a letter to Marx, Engels (1858) wrote that the English proletariat was becoming more and more bourgeois, so that the ultimate aim of England was the possession of a bourgeois aristocracy and a bourgeois proletariat. Twenty-four years later, he remarked how the English workers were profiting from the English monopoly in the world market and colonies (Engels, 1882). Lenin (1902) also admitted that through its own efforts the working class was able to develop only a trade-union consciousness (or ''economicism''); that is, they were only interested in economic ends, fighting for higher salaries, better work conditions, and social improvement. The task of a revolutionary party was, therefore, to awake the masses and divert them from ''economicism.''

It is true that Marx sent Darwin an inscribed copy of the German edition of the first volume of *Das Kapital* and professed to be a ''sincere admirer'' of Darwin's, but there is no evidence that Marx tried to dedicate the second volume of his influential book to Darwin and even less that Darwin declined the offer. Darwin admitted that *Das Kapital* was a ''great book,'' although its content was ''so different'' from his own work. Politely, he wrote to Marx that he wished that he was ''more worthy to receive it, by understanding more of the deep and important subject of political economy'' (Desmond & Moore, 1992, p. 602). Gould (1978) suggested that Darwin was not a devotee of the German language, but in his autobiography, Darwin (1882) himself realized that during his whole life he was singularly incapable of mastering any foreign language. The fact, as I have had the opportunity to see in Darwin's library at Down House, is that except for the first few dozen pages, Darwin's copy of this book is uncut, so it is clear that he never read it.

Marx and Engels held Darwin in very high regard. They recognized the importance of Darwin's *On the Origin of Species*, which Engels (1859) found "absolutely splendid" (p. 551), and tried to apply its conclusion to their own writings, in which they remarked that Darwin's theory of natural selection provided an important basis for the study of class struggle (Engels, 1859; Marx, 1860, 1861, 1862). In a letter to Engels, Marx (1862) remarked how Darwin rediscovered, among the beasts and plants, the English society with its division of labor, competition, opening up of new markets, and Malthusian struggle for existence. Engels wrote two essays to integrate Marx's and Darwin's works. In the *Dialectics of Nature* (Engels, 1873), he argued that Darwin's *Origin* was a bitter satire on mankind, and especially on his own countrymen, because free competition and the struggle for existence, celebrated by the economists as the highest historical achievement, are the basic rule of the animal kingdom. Later, in the *Part Played by Labor in the Transition from Ape to Man*, which quite incidentally is published jointly with *Dialectics*, he stressed that labor created man, but that most naturalist scientists of the Darwin school were unable to form any clear idea of the origin of man because of their idealistic world outlook (Engels, 1896). Likewise, many European socialists regarded Marx's historical materialism as a corollary of Darwin's evolutionary theory (Feuer, 1977).

Nonetheless, Darwin considered totally unsound and ridiculous any link between socialism and evolution. He was quite aware of the misconception his theory could invoke, and, in *On the Origin of Species*, he stressed that he used the term "struggle for existence" in a large and metaphorical sense. It included not only the dependence of one being on another but also success in leaving progeny. Moreover, it was only near the end of *Origin*, and even then in a short paragraph, that he dared to mention human evolution.

Thus, it can be argued that the conclusions Marx drew from Darwin were erroneous and dogmatic, but to assume that Marx's thought is void of any claim about human nature is fundamentally false. Marx (1844) saw man as a natural being endowed with natural powers, which existed in him as tendencies and abilities, that is, as instinct. Human beings were led astray by private property, from which arose egoism, wars, competition, domination, and the like. Communism, on the other hand, was the political transcendence of private property, the complete emancipation of man, the return of man to himself. In *The Poverty of Philosophy*, Marx (1847, 192) remarked that "all history is nothing but a continuous transformation of human nature," but it was in his *Economic and Philosophic Manuscripts of 1844* that he expounded more systematically his view of human nature. In a similar way, Engels (1845a) remarked that in a communist society everyone would freely develop their human nature and live in positive human relations with their neighbors. Accordingly, the question is not whether Marx and Engels accepted the idea of human nature; rather, it lies in their misconception about it, which they associated with history and specific modes of production. This position led them to associate law, morality, and religion as forms of ideological and bourgeois prejudice that only served class interests (Marx & Engels, 1845, 1848; Engels, 1878).

Such an economic and reductionist position did not prevent Marx and Darwin from being placed on the same scientific level. Two examples are illustrative. At Marx's graveside, Engels traced an analogy between Marx's and Darwin's theories by saying, "Just as Darwin discovered the law of development of organic nature, so Marx discovered the law of development of human history" (1883, p. 467). Almost ten years later, Lenin (1894, p. 140) assumed an identical position. He asserted that the tremendous success enjoyed by Marx's *Capital* was due to the fact that the book showed the whole capitalist social formation as a living thing; therefore, he concluded, the comparison between Marx and Darwin "is perfectly accurate." However, while Darwin did discover the forces of mutation and natural selection, Marx did not discover any law of historical development. Possibly there is no such law.

Moreover, many of Marx's and Engels's writings are vague, contradictory, incoherent, and even mystical. Marx (1866a, 1866b, 1866c) was an admirer of Pierre Trémaux, a French naturalist who proclaimed that the differences among races and peoples should be founded on the corresponding differences in the soil on which they build their lives. But Marx was also convinced that Trémaux represented a significant advance over Darwin. On the other hand, Engels (1896) advanced a hypothesis according to which a meat diet had "crucial importance" in the development of the brain from ape to man.

Finally, Darwin did not write a satire but a scientific work. Moreover, he did not rediscover anything when he developed the theory of natural selection. Although Malthus's book *Essay on the Principle of Population* played a key role in Darwin's work, this does not mean that he transferred the competition he saw among animals and plants to English society. "Malthusian struggle for existence" and "struggle for existence" have different meanings, and Darwin (1859) himself put this very clearly in *Origin*, while in *The Descent of Man* (1871), he stressed that the corporeal strength of man is more than counterbalanced by his intellectual power and by his social qualities, which led him to give aid to his fellows and to receive it in return.

Therefore, when one compares what Darwin really said and what Marx and Engels wrote, it is possible to see how they misunderstood Darwin's theory and distorted it as well as how dogmatic they were. Therefore, Marxism should be discredited as science, and any attempt to put Marx on the same scientific level as Darwin is, as Anthony Flew (1992, p. 224) pointed out, a "sheer indecency."

DARWINISM AND HUMAN NATURE

The answer to the question, "What is man?"—that is, how man differs from other animals, his place in nature, and even the existence of a human nature— has absorbed the human mind almost since the dawn of humankind. This has always been a central question to any system of philosophy, theology, and, more recently, biology. The biologist George Gaylord Simpson (1966) argued that this question was possibly being asked by the most brilliant australopithecine two million years ago. Aristotle saw man as a political animal, and the seventeenth-century British philosopher John Locke viewed man as an independent and rational agent, but it was Charles Darwin who first gave a scientific answer to the question, "What is man?" As Simpson (1966) pointed out, theology, art, metaphysics, and other nonbiological sciences can also help us to understand what man is, but unless they accept his biological nature, they will be no more than fancies or falsities. Ernst Mayr penned: "Every modern discussion of man's future, the population explosion, the struggle for existence, the purpose of man and the universe, and man's place in nature rests on Darwin (1992, p. 7).

Darwinism is well established as the main scientific theory to explain the biological evolution of all living species, including humans. However, when Darwin sailed on the *Beagle* on September 27, 1831, he was a quite orthodox man who believed in every word of the Bible as an unanswerable authority, as he frankly admitted in his autobiography (1882). Five years on board changed Darwin's religious idea and provoked a revolution about the evolution of the species and of man. After all, how could a good and benevolent God permit the abominable practice of slavery, and how could He be responsible for earthquakes and volcanic eruptions that killed thousands of innocent people?

In his autobiography, Darwin (1882) made it clear that it was Malthus's *Essay*, which he read only "for amusement," that struck him and gave him the theoretical foundation he was looking for. Nonetheless, he was so apprehensive about avoiding prejudice that he decided not to write even the briefest sketch of his theory of natural selection (Darwin, 1882). The prejudice that Darwin mentioned was not concerned with evolution itself but with the fact that his conclusions were based on philosophical materialism, that is, that all mental processes are a byproduct of matter. That, much more than evolution, was a heresy.

In 1836, when Darwin returned to England, he already had a reputation he did not want to put at risk with a heretical theory he could not prove; he therefore waited more than twenty years before publishing *On the Origin of Species*. His aim was to demonstrate that all individuals had a single ancestor and that all the closely allied species of most genera were descended from one parent that had migrated from some single birthplace. But it was only at the end of the book that he dared to mention, and even so in a few enigmatic words, the origin of man: "In the distant future I see open fields for far more important researches. Psychology will be based on a new foundation, that of the necessary acquirement of each mental power and capacity by gradation. Light will be thrown on the origin of man and his history" (1859, p. 458). Darwin waited another twelve years to publish *The Descent of Man and Selection in Relation to Sex* (1871), a book in which he developed his theory of natural selection in relation to man and the continuity between human beings and other animals. In this seminal work, Darwin disdained those who maintained that man's origin could never be known and asserted that our moral and ethical predisposition also evolved by natural selection. Otherwise, his theory would be gravely threatened. In so doing, he established for the first time the scientific foundation for the study of human nature. He also assumed that the future would decide whether he had exaggerated the importance of natural selection.

Seven years before the publication of Darwin's *Descent of Man*, Karl Marx wrote in a letter to his uncle: "since Darwin demonstrated we are all descended from apes there is scarcely any shock whatever that could shake 'our ancestral pride' " (1864, p. 542). Nevertheless, man never was an ape, a chimpanzee, or a gorilla. What Darwin's theory says is that man might have a common ancestor with primates that differed from any of us. This does not prevent the Russian philosopher Ivan Frolov from asserting that "it was Karl Marx who first offered a scientific answer to the question: 'What is Man?' " (1986, p. 90). There is no doubt that Darwin's theory on the origin of man represented a shock to human pride, but to assume that "Darwin demonstrated we are all descended from apes" is not merely a misunderstanding of what Darwin really said, but also complete nonsense. All that Darwin proposed was that all forms of life had a common ancestor.

Human beings have indeed suffered several blows to their narcissism and self-esteem. Freud (1917) described three of these blows. The first was a cosmological one: in the sixteenth century Copernicus demonstrated that the earth was not the center of the universe. The second blow was Darwin's theory of natural selection, a biological blow. The third was Freud's theory of human unconsciousness, a psychological blow and in his view the most wounding. Today it is possible to add a fourth blow: that human beings are not only a product of their own environment, but a result of the interaction between culture and genes, which implies the existence of a human nature. This blow seems now to be the most wounding.

THE COMMUNIST SOCIETY

"A spectre is haunting Europe—the spectre of communism" (Marx & Engels 1848, p. 496). The very first sentence of the *Manifesto of the Communist Party* fostered the dreams, illusions, and hopes of several generations until October 25, 1917, when the Bolsheviks launched a coup, overthrew the provisional government in Russia, and promised to build a new man, a new era, a new civilization. After World War II the Soviet experience was extended to Eastern

Europe. Then, nearly 140 years after the publication of the *Manifesto of the Communist Party*, the Marxist regimes collapsed.

A basic premise of Marx's doctrine is that, capitalism, defeated by its own inner contradiction, would be replaced by the inexorable laws of history, by a communist society in which nationalism, religion, the bourgeois family, antagonism between classes, and even personal conflicts between individuals would disappear. In a communist society the numerous national and local literatures would give rise to a world literature (Marx, 1844; Engels, 1845a; Marx & Engels, 1848).

Marx and Engels saw their specific mission as the development of socialism from its utopian stage to its scientific stage, liberating it from its sentimental, moralist, and visionary background. Then, for Marx and Engels, it should be possible through "scientific socialism" to transform the existing mode of production and with it to transform not only the society but also human nature, which they believed could be molded into virtually any form. Marx and Engels were reviving Locke's empiricism, a doctrine according to which all our knowledge comes from experience. Decades of communist experience showed not only that human nature is not infinitely malleable but also that "scientific socialism" is just nonsense. After all, what happened to the promises of the Bolshevik revolution? What happened to the provisions Marx and Engels asserted in many of their writings?

With the victory of the Bolshevik revolution, Lenin could put Marx's doctrine to practice. In a speech delivered at the Third All-Russia Congress of the Young Communist League, Lenin defined communist society as "a society in which all things—the land, the factories—are owned in common and the people work in common. That is communism. Is it possible to work in common if each one works separately on his own plot of land? Work in common cannot be brought about all at once. That is impossible. It does not drop from the skies. It comes through toil and suffering" (1920, p. 296). Indeed, communism, and all attempts to build it, produced toil and suffering. Lenin's definition of communism sounds like the utilitarian's morality, which preaches that a sacrifice is considered wasted unless it increases or tends to increase the total sum of happiness (Mill, 1861). Imagined to be the redemption of man, as Marx (1844) wrote, today communism represents more a historical phenomenon than a viable political, economic, or social alternative. If one compares how things were predicted and how they really developed, one sees how the theoreticians misunderstood the complexity of human nature.

Several interpretations have focused on the death of communism. Some Marxists argue that backward Russia with its huge feudal peasantry would be the least appropriate place for a successful transition to a genuine communism. Others stress that the despotic character of the Bolshevik revolution was an aberration that violated Marx's fundamental doctrine and that even to talk about the collapse of socialism is a grotesque misinterpretation of the facts, because socialism was never really tested. According to these views, what crumbled in

Eastern Europe was neither socialism nor communism, and much less Marxism, but Leninism-Stalinism and state capitalism, which were not only a degenerate form of Marxism, but its very negation (Burawoy, 1990; Harris, 1992; Medvedev, 1989; Nagels, 1991).

Nonetheless, such interpretations did not explain anything. Although none of the conditions that Marx considered essential to the development of a communist society (a conscious, enlightened, free, and united proletariat, a democratic constitution, a developed industry, and a considerable means of production) were present in Russia in 1917, Marxism failed as a social, political, and economic system. It failed not because it was inept, illegitimate, inflexible, bureaucratic, and repressive but because human nature cannot be molded in virtually any form as Marx imagined.

The greatest defect of Marx was his inability to ask, *"And then what?"* He never asked what would be the ultimate consequences of organizing a society according to the aphorism "From each according to his ability, to each according to his needs!" (Hardin, 1977). Marx's famous aphorism is a part of his *Critique of the Gotha Programme* (1875), in which he remarked that in a future communist society individuals would receive back from the society exactly what they gave to it. But who will decide what an individual has given to society? And who will decide each individual's needs?

Marxism was unable to satisfy the growing material aspiration of the citizens where it was implemented because it is quite impossible to know what individual fulfillment is and to what extent it may be expanded or to answer the question, "How much is enough?" The fact, missed by Marx, is that, as Aristotle observed twenty-three centuries ago, human avarice is insatiable; as each of our desires is satisfied, a new one appears in its place (Durning, 1991). Data collected from the former Soviet Union and other Eastern European countries revealed that the number of private cars grew fivefold between 1970 and 1985 (Renner, 1989). Thus, the compulsion to consume seems to be embedded in human nature. Psychologist Michael Argyle shows that the satisfaction derived from money does not come from simply having it, but from having more money than others do and from having more of it this year than last (cited in Durning, 1991). Such data contradict Marxist dogma, according to which our pleasures and desires spring from society (Marx, 1849). One may argue that propaganda and social pressure encourage acquisitive impulses, but nothing could explode in us if it had not been there already, as Jung (1937) remarked. We can only repress such desires. Like any other secular religion, Marxism tried to repress our most inner instinct. As a matter of fact, the success of Marxism lies in its religious appeal. After all, did not Engels (1847) outline what the future communist society would be in a paper called "Draft of a Communist Confession of Faith"? The analogy between Marxism and religion is unmistakable: Messianism, absolute sacrifices, rites, canonical texts, liturgical formulas, creed, an infallible hierarchy, and sacred and profane elements are constantly found in both of them (Alves, 1998).

Aristotle was also aware that "that which is common to the greatest number has the least care bestowed upon it" (quoted in Hardin, 1993, p. 152). Marx missed this reality too. But today ecologists agree that it is both mathematically and biologically impossible to provide the nest for the greatest number of persons, as Hardin (1968) put it in his now classical article "The Tragedy of the Commons."

Louis Wasserman (1979) drew an interesting analogy between Marx's doctrine and a young doctor who could not cure his patient's common cold. The doctor advised her to go out in the rain and catch pneumonia, because he did know how to treat pneumonia. Marx's prescription, Wasserman argued, is like that. He was aware of the economic anomalies of society, but he chose to deal with them in catastrophic terms.

In a classic book, Engels (1845b) described the bourgeois society as the most complete expression of the war of all against all in which man has neither time nor initiative to think about the fact that, in the end, each personal interest coincides with the interest of another. He maintained that only a communist society could eliminate the unpleasant consequences of this egoistic behavior.

Because of our Darwinian heritage, it is doubtful that personal antagonism between individuals will disappear, as Marx (1844) and Engels (1845a) predicted. However, despite the fact that human beings strive first for their reproductive success and that of their kin, our society is based on cooperation, and we do not need to wait for the appearance of a future and utopian communist society to overcome our egoistic behavior. But how can cooperation emerge without a central authoritarian government? An elegant way to solve this question is through the Prisoner's Dilemma game analyzed by the American political scientist Robert Axelrod (1990). The central point of this game is that when two individuals play it once or a known number of times, both players will try to exploit each other. But in a computer simulation Axelrod showed that cooperation will emerge when the players interact an indefinite number of times. A Darwinian perspective can help us to understand what we do in terms of what we have done. This means that if we are more democratic and cooperative today than we once were, it is for the simple reason that we need each other today more than we did before (Betzig, 1991).

CONCLUSION

For many people and even for many social scientists, it is unrealistic, if not immoral, to connect biology and the social sciences. At the end of the nineteenth century, Émile Durkheim (1895) claimed that sociology was a distinct and autonomous science rather than an auxiliary one; therefore, one social act could be explained only by another social factor. A century later, most sociologists still react negatively to any attempt to link biology with sociology (Bock, 1980; Menzies, 1985; Sahlins, 1976). They argue that social rules are created by humans. But human decisions are physiological products of the brain, which evolved by natural selection over thousands of years (Lumsden & Wilson, 1983; Ruse & Wilson, 1986). Therefore, it is reasonable to suppose that human decisions and social rules (including ethics, moral behavior, and the law) also evolved by natural selection, as Darwin (1871) suggested when he wrote: "the intellectual and moral faculties of man . . . are variables . . . and we have every reason to believe that the variations tend to be inherited. Therefore, if they were formerly of high importance to primeval man and to his ape-like progenitors, they would have been perfected or advanced through natural selection" (Vol. I, p. 159).

Although social scientists could not stop the progress of neo-Darwinian theory toward the understanding of human social behavior, they continue to claim that a biological approach to human behavior implies biological determinism and loss of freedom. However, science does not support any of these claims, although some people do claim supportive arguments in some scientific findings. Evolutionary biology argues that the so-called nature–nurture controversy has no meaning. Human social behavior is a product of both. Such hostility, therefore, is ideologically motivated, devoid of any rational argument, and based on lack of sound understanding (van den Berghe, 1991).

History has no laws, much less "inexorable laws." Historicism is not science, but understanding the past can illuminate the future. As George Santayana put it: "Those who forget history are condemned to repeat its failures." Lessons of history, like laws of biology, are, therefore, to be learned. Our future depends on whether political leaders and common citizens have enough knowledge about them.

ACKNOWLEDGMENTS

This chapter was supported by CNPq (Conselho Nacional de Desenvolvimento Científico e Tecnológico) grant no. 204.204.88/7. I am indebted to Benjamin Gilbert (Fundação Oswaldo Cruz, Rio de Janeiro), Octavio Antunes (Instituto de Quimica, Universidade Federal do Rio de Janeiro), Alphonse Kelecom (Instituto de Biologia, Universidade Federal Fluminense), and Carlos Lima (Departamento de Letras, Universidade Estadual do Rio de Janeiro), and to Robin Allott and the organizing Committee of the XVIII Meeting of the European Sociobiological Society for reading the manuscript. I am extremely grateful to Monique Borgerhoff Mulder (Department of Anthropology, University of California), Laura Betzig (University of Michigan), Napoleon Chagnon (Department of Anthropology, University of California at Santa Barbara), Anthony Flew (Reading, England), Ivan Frolov (Department of Philosophy, University of Moscow), Williams Irons (Department of Anthropology, Northwestern University), Linda Lobão (Department of Biology, University of Michigan), Jan Paastela (Department of Political Science, University of Tampere), Michael Ruse (Department of Philosophy, University of Guelph), Paul Turke (Department of Anthropology, University of Michigan), Pierre van den Berghe (Department of

Sociology, University of Washington), and Edward Wilson (Department of Zoology, Harvard University). The articles and reviews they sent to me were essential in preparing this chapter. Without their participation, patient discussion, and constructive comments, this chapter would have been quite different and certainly much inferior.

REFERENCES

Alves, L. F. (1998). Sociobiology, Marxism and religion. *Ciência e Cultura, 50*, 416–425.

Axelrod, R. (1990). *The evolution of cooperation*. New York: Penguin Books.

Berté, N. A. (1988). K'ekchi' horticultural labor exchange: Productive and reproductive implications. In L. Betzig, M. Borgehoff Mulder, & P. Turke (Eds.), *Human reproductive behavior: A Darwinian perspective* (pp. 83–96). Cambridge: Cambridge University Press.

Betzig, L. (1988). Redistribution: Equality or exploitation? In Betzig, Borgerhoff Mulder, & Turke (Eds.), *Human reproductive behavior: A Darwinian perspective* (pp. 49–63).

———. (1991). History. In M. Maxwell (Ed.), *The sociobiological imagination* (pp. 131–140). Albany: State University of New York Press.

Bock, K. (1980). *Human nature and history: A response to sociobiology*. New York: Columbia University Press.

Borgerhoff Mulder, M. (1987). Resource and reproductive success in women with an example from the Kipsigis of Kenya. *Journal of Zoology* (London), *213*, 489–505.

Burawoy, M. (1990). Marxism as science: Historical challenges and theoretical growth. *American Sociological Review 55*, 775–793.

Chagnon, N. (1979). Is reproductive success equal in egalitarian societies? In N. A. Chagnon & W. Irons (Eds.), *Evolutionary biology and human social behavior: An anthropological perspective* (pp. 374–401). North Scituate: Duxbury Press.

Darwin, C. (1859) [1985]. *The origin of species by means of natural selection*. London: Penguin Books.

———. (1871) [1981]. *The descent of man and selection in relation to sex*. Princeton, NJ: Princeton University Press.

———. (1882) [1993]. *The autobiography of Charles Darwin, 1809–1882*. New York: W. W. Norton.

Desmond, A., & Moore, J. (1992). *Darwin: The life of a tormented evolutionist*. New York: W. W. Norton.

Durkheim, E. (1895) [1964]. *The rules of sociological methods*. London: Free Press.

Durning A. (1991). Asking how much is enough? In L. Brown (Ed.), *State of the world* (pp. 153–169). New York: W. W. Norton.

Engels, F. (1845a) [1975]. Speeches in Eberfeldt. In *Collected works of Marx and Engels*, Vol. 4 (pp. 243–264). London: Lawrence & Wishart/Moscow: Progress Publishers.

———. (1845b) [1975]. The conditions of the working class in England. In *Collected works of Marx and Engels*, Vol. 4 (pp. 295–583). London: Lawrence & Wishart/Moscow: Progress Publishers.

———. (1847) [1984]. Draft of a communist confession of faith. In *Collected works of Marx and Engels*, Vol. 6 (pp. 96–103). London: Lawrence & Wishart/Moscow: Progress Publishers.

———. (1858) [1983]. Letter to Marx. In *Collected works of Marx and Engels*, Vol. 40 (pp. 343–345). London: Lawrence & Wishart/Moscow: Progress Publishers.

———. (1859) [1983]. Letter to Marx. In *Collected works of Marx and Engels*, Vol. 40 (pp. 550–551). London: Lawrence & Wishart/Moscow: Progress Publishers.

———. (1873) [1987]. Dialectics of nature. In *Collected works of Marx and Engels*, Vol. 25 (pp. 311–587). London: Lawrence & Wishart/Moscow: Progress Publishers.

———. (1878) [1987]. *Anti-Dühring*. In *Collected works of Marx and Engels*, Vol. 25 (pp. 5–309). London: Lawrence & Wishart/Moscow: Progress Publishers.

———. (1882) [1971]. Brief zu Kautsky. In *Marx-Engels-Gesamt-Ausgabe*, Band. 35, (pp. 356–358). Berlin.

———. (1883) [1989]. Karl Marx's funeral. In *Collected works of Marx and Engels*, Vol. 24 (pp. 467–471). London: Lawrence & Wishart/Moscow: Progress Publishers.

———. (1884) [1985]. *The Origin of the family, of private property and the state*. New York: Penguin Books.

———. (1896) [1987]. The Part played by labor in the transition from ape to man. In *Collected works of Marx and Engels*, Vol. 25 (pp. 452–464). London: Lawrence & Wishart/Moscow: Progress Publishers.

Feuer, L. (1977). Marx and Engels as sociobiologists. *Survey, 23*, 109–136.

Flew, A. (1992). *Atheistic humanism*. New York: Prometheus.

Freud, S. (1917) [1964]. A difficulty in the path of psycho-analysis. In *The standard edition of the complete psychological works of Sigmund Freud*, Vol. 17 (pp. 135–144). London: Hogarth Press.

———. (1929) [1991]. Civilization and its discontents. In *The Pelican Freud Library*, Vol. 12 (pp. 243–340). London: Penguin.

Frolov, I. (1986). Genes or culture? A Marxist perspective on humankind. *Biology and Philosophy, 1*, 89–107.

Gould, S. J. (1978). *Ever since Darwin*. New York: Penguin Books.

Hardin, G. (1968). The tragedy of the commons. *Science, 162*, 1243–1248.

———. (1977). What Marx missed. In G. Hardin & J. Baden (Eds.), *Managing the Commons* (pp. 3–7). New York: W. H. Freeman.

———. (1993). *Living within limits. Ecology, economics and population taboos*. New York: Oxford University Press.

Harris, M. (1992). Distinguished lecture: Anthropology and the theoretical and paradigmatic significance of the collapse of Soviet and East European communism. *American Anthropologist, 94*, 293–305.

Irons, W. (1979). Cultural and biological structures. In N. Chagnon & W. Irons (Eds.), *Evolutionary biology and human social behavior* (pp. 257–272). North Scituate, MA: Duxbury Press.

Jung, C. G. (1937) [1977]. Psychology and religion. In *Collected works of Carl Gustav Jung*, Vol. 11 (pp. 3–105). London: Routledge & Kegan Paul.

Lenin, V. (1894) [1977]. What the "friends of the people" are and how they fight the social democrats? In *Collected works of Lenin*, Vol. 1 (pp. 129–332). Moscow: Progress Publishers.

————. (1902) [1977]. What is to be done? In *Collected works of Lenin*, Vol. 5 (pp. 347–529). Moscow: Progress Publishers.

————. (1920) [1982]. The tasks of the youth leagues. In *Collected works of Lenin*, Vol. 31 (pp. 283–299). Moscow: Progress Publishers.

Levi-Strauss, C. (1944). The social and psychological aspects of chieftainship in a primitive tribe: The Nambikuara of Northernwest Mato Grosso. *Transactions of the New York Academy of Sciences, 7,* 16–32.

Lobão, L. R. (1984). Marxism and sociobiology: Contrasts and theoretical implications. Paper presented at the annual meeting of the American Sociological Association, San Antonio, TX.

Lumsden, C., & Wilson, E. O. (1983). *Promethean fire. Reflections on the origin of mind.* Cambridge, MA: Harvard University Press.

Marx, K. (1844) [1975]. Economic and philosophic manuscripts of 1844. In *Collected works of Marx and Engels*, Vol. 3 (pp. 229–346). London: Lawrence & Wishart/ Moscow: Progress Publishers.

————. (1847) [1984]. The poverty of philosophy. In *Collected works of Marx and Engels*, Vol. 6 (pp. 105–212). London: Lawrence & Wishart/Moscow: Progress Publishers.

————. (1849) [1977]. Wage-labour and capital. In *Collected works of Marx and Engels*, Vol. 9 (pp. 197–228). London: Lawrence & Wishart/Moscow: Progress Publishers.

————. (1860) [1985]. Letter to Engels. In *Collected works of Marx and Engels*, Vol. 41 (pp. 231–233). London: Lawrence & Wishart/Moscow: Progress Publishers.

————. (1861) [1985]. Letter to Ferdinand Lassale. In *Collected works of Marx and Engels*, Vol. 41 (pp. 245–247). London: Lawrence & Wishart/Moscow: Progress Publishers.

————. (1862) [1985]. Letter to Engels. In *Collected works of Marx and Engels*, Vol. 41 (pp. 380–381). London: Lawrence & Wishart/Moscow: Progress Publishers.

————. (1864) [1985]. Letter to Lion Philips. In *Collected works of Marx and Engels* Vol. 41 (pp. 542–544). London: Lawrence & Wishart/Moscow: Progress Publishers.

————. (1866a) [1987]. Letter to Engels. In *Collected works of Marx and Engels*, Vol. 42 (pp. 303–305). London: Lawrence & Wishart/Moscow: Progress Publishers.

————. (1866b) [1987]. Letter to Kugelman. In *Collected works of Marx and Engels*, Vol. 42 (pp. 325–327). London: Lawrence & Wishart/Moscow: Progress Publishers.

————. (1866c) [1987]. Letter to Engels. In *Collected works of Marx and Engels*, Vol. 42 (pp. 321–322). London: Lawrence & Wishart/Moscow: Progress Publishers.

————. (1875) [1989]. Critique of the Gotha programme. In *Collected works of Marx and Engels* Vol. 24 (pp. 75–99). London: Lawrence & Wishart/Moscow: Progress Publishers.

Marx, K., & Engels, F. (1845) [1976]. The German ideology. In *Collected works of Marx and Engels*, Vol. 5 (pp. 19–539). London: Lawrence & Wishart/Moscow: Progress Publishers.

————. (1848) [1984]. Manifesto of the communist party. In *Collected works of Marx and Engels*, Vol. 6 (pp. 477–519). London: Lawrence & Wishart/Moscow: Progress Publishers.

————. (1850) [1978]. Address of the central authority to the league. In *Collected works of Marx and Engels*, Vol. 10 (pp. 277–287). London: Lawrence & Wishart/Moscow: Progress Publishers.

Mayr, E. (1992). *One long argument. Charles Darwin and the genesis of modern evolutionary thought*. New York: Penguin Books.

Medvedev, R. (1989). *Let history judge*. Oxford: Oxford University Press.

Menzies, R. (1985). Genetic ideology: Observations on the biologicization of sociology. *Canadian Journal of Anthropology and Sociology, 22*, 202–225.

Mill, J. S. (1861) [1991]. Utilitarianism. In J. Gray (Ed.), *On liberty and other essays* (pp. 131–201). London: Oxford University Press.

Nagels, J. (1991). *Du socialism perverti au capitalisme sauvage*. Bruxelles: Editions de l'Université de Bruxelles.

Paastela, J. (1991). *Sociobiology and Marxism*. Paper prepared for presentation at the Fifteenth World Congress of the International Political Science Association. Buenos Aires.

Renner, M. (1989). Rethinking transportation. In L. Brown (Ed.), *State of the world* (pp. 95–112). New York: W. W. Norton.

Ruse, M., & Wilson, E. O. (1986). Moral philosophy as applied science. *Philosophy, 61*, 173–192.

Sahlins, M. (1976). *The use and abuse of biology: An anthropological critique of sociobiology*. Ann Arbor: University of Michigan Press.

Simpson, G. G. (1966). The biological nature of man. *Science, 152*, 472–478.

Terray, E. (1972). *Marxism and primitive society*. New York: Monthly Review.

Turke, P. W., & Betzig, L. L. (1985). Those who can do: Wealth, status, and reproductive success on Ifaluk. *Ethology and Sociobiology, 6*, 79–87.

van den Berghe, P. (1991). Sociology. In M. Maxwell (Ed.), *The sociobiological imagination* (pp. 269–282). Albany: State University of New York Press.

van den Berghe, P., & Peter, K. (1988). Hutterites and kibbutzniks: A tale of nepotistic communism. *Man, 23*, 522–539.

Wasserman, L. (1979). Alienation incident. *Humanist, 39*, 4–10.

Wilson, E. O. (1975). *Sociobiology the new synthesis*. Cambridge, MA: Belknap Press.

———. (1976). Sociobiology: A new approach to understanding the basis of human nature. *New Scientist, 70*, 342–345.

———. (1978a). *On human nature*. Cambridge, MA: Harvard University Press.

———. (1978b). What is sociobiology? In M. S. Gregory, A. Silvers, & D. Sutch (Eds.), *Sociobiology and human nature*, (pp. 1–12). San Francisco: Jossey-Bass Publishers.

———. (1998). *Consilience: the unit of knowledge*. New York: Alfred Knopf.

Zhang, B. (1987). Marxism and humanism: A comparative study from the perspective of modern socialist economic reforms. *Biology and Philosophy, 2*, 463–474.

———. (1994). *Marxism and human sociobiology: The perspective of economic reforms in China*. Albany: State University of New York Press.

9

Culture and the Evolution of the Human Mating System

Pouwel Slurink

To what extent is the human mating system affected by culture and vice versa? Whereas some traditional cultural anthropologists tend to believe in a kind of "superecological cultural variability" (e.g., Benedict, 1935), others, especially Marvin Harris, have been looking for the ecological factors that cause specific local mating patterns and habits (Divale & Harris, 1976; Harris, 1985). From an evolutionary comparative perspective, however, it becomes important to start identifying cross-cultural universals (e.g., Buss, 1989, 1994)—to locate the human life form or "ethogram" in the evolutionary tree of possible life forms— and *only then* to try to explain local variations on these universal themes (e.g., Flinn & Low, 1986, and some of the studies compiled in Betzig, 1997). If we place the variety of human societies within the framework of the variety of all animal societies, it becomes clear that all human societies share specific resemblances, despite the variations on which cultural anthropologists tend to focus. It can even be claimed that local variations in types of groupings in humans are not as dramatic as those of, for example, gorillas (Rodseth, Wrangham, Harrigan, & Smuts, 1991). In any case, it is clear that the relation between ecological factors and cultural variability is always mediated by an amount of "phylogenetic inertia" (Wilson, 1975), that is, by an underlying human nature that has been formed by past selective forces.

Should culture be interpreted as something that is merely superimposed on this underlying human nature? Sometimes this seems to be thought. For example, in an interesting article on the mating system of bee-eaters Emlen, Wrege, and Demong (1995) justify their choice of this species by noting that it has a mating system that is "largely unaffected by culture!" This can be read as implying that culture is merely a distorting factor that makes it ever more difficult to grasp the "original" human ethogram. If one sees it this way, however, one ignores the possibility that human culture is somehow more intimately linked to the human mating system. On the one hand, the original (pre)human mating system could have already been unique in particular aspects, and culture could be a result thereof. On the other hand, particular characteristics of evolved human sexual psychology and of the human mating system may only have emerged as a *result* of "gene-culture coevolution" (Lumsden & Wilson, 1981)—they may have evolved relatively recently in an era in which a cultural accumulation of nonhereditary information already formed an essential component of hominid life. To disentangle these possibilities, we have to reconstruct the original mating system of our prehominid ancestors, to list the properties in which modern humans have changed, and to compare, select, and integrate the different models that claim to explain the transition from the original prehominid mating system into the mating system that underlies modern human behavior.

A POSSIBLE LINK BETWEEN SEXUAL SELECTION AND NEOTENY

Recently, there has been a revival of interest in sexual-selection-based explanations of culture (e.g., Parker, 1987). Geoffrey Miller has refined a model in which the threefold brain enlargement during human evolution is explained as a result of the bilateral sexual selection of the sexes or of sexual selection in which the selected properties of one sex happen to be inherited by offspring of the other sex as well (Ridley, 1994, pp. 326–330; Mensel, 1995). Normally one would expect properties that evolve as a result of sexual selection to be represented especially in one sex, but as both sexes share most chromosomes, it is at least possible to imagine the sexual selection of properties that are highly advantageous to one sex and neutral to the other sex. Miller proposes that the most important trait that has been selected during human evolution is simply the ability to produce impressive courtship displays in the form of music, dance, poetry, rhetoric, and the like. Male humans would create art and culture just "to impress the girls," thus for the same reason that male peacocks display their feathers and ruffs defend their leks. Females also would need at least some creativity to be able to bind the males and lure them into investing in their offspring. This is called the "Scheherezade strategy" by Miller after the heroine of the *Arabian Nights* who had to tell the sultan a story every night to seduce him not to kill her after having slept with her. Miller claims that most artists have their peak at a relatively young age, just when they are most sexually active.

In his popular book *The Red Queen: Sex and the Evolution of Human Nature* (1994), Matt Ridley connects the idea of sexual-selected creativity with the already somewhat outmoded idea of neoteny, the idea that many human characteristics can be explained simply by the persistence of youthful characteristics in adult life, caused by the workings of genes that slow the maturation process. He reasons that in a situation with a certain degree of monogamous pair bonding and paternal assistance in childrearing, males should be particularly interested in females with a lot of residual reproductive capacity. If mating is just a transitory, noncommittal activity for males, there is no reason to be selective about female partners, but the more time it takes to concentrate on one particular female and the more the road to polygyny is blocked, the more important it becomes to have as many children with one female as possible. As a result, it would become adaptive for females to look as young as possible, and neoteny genes in women would continually be selected and even be inherited by their sons. As neoteny genes are supposed not only to cause someone to look younger, but also to influence brain/body ratio and overall behavioral flexibility and inclination to play and to learn, this would mean that they could cause an increase in general intelligence as well.

Neoteny is often too easily used as an explanation for human uniqueness, however. Brian Shea (1992) warns that theories that refer to neoteny are often too simple to account for uniquely human properties. Almost none of the morphological features associated with bipedal locomotion can be related to neoteny, for example, and while it is true that an adult human looks like a juvenile ape in that she or he has a relatively big brain and little prognathism, this resemblance is caused by completely different patterns of bone distribution. In particular, the construction of the pharynx of an adult human does not look like that of an juvenile ape, and the evolution of speech therefore cannot be attributed simply to neoteny. All in all, neoteny theory suffers from an overdose of explanatory monism, and it is not advisable to invoke neoteny too much as an explanatory *deus ex machina*.

Furthermore, Ridley himself notes that there is a general problem with sexual-selection-based theories of human evolution in that they are circular. As Van der Dennen notes, "prime-mover" theories of human evolution often are unable to reply to the question "What moved the prime mover?" (van der Dennen, 1995). Ridley himself answers that evolution often is circular and works by bootstrapping. There need not be a single cause-and-effect relation, because "effects can reinforce causes." "If a bird finds itself to be good at cracking seeds, then it specializes in cracking seeds, which puts further pressure on its seed-cracking ability to evolve" (Ridley, 1994, p. 332).

Ridley forgets here, however, that birds do not "find themselves good at cracking seeds" on any given day of their evolutionary history and do not specialize apart from the rest of an ecosystem. If they change their food habits, the most likely cause is a slight disturbance within the ecosystem because of geological or climatological factors (e.g., Grant, 1991). His argument that evolution is circular fails because evolution is driven by many external factors, such as the amount of solar energy, the composition of the atmosphere, and geological and climatological factors. If something like bootstrapping happens in evolution, there are always forces that set this bootstrapping process in motion.

Thus we have to conclude that if something like the sexual-selected neoteny mechanism has worked during specific periods of human evolution, we still have to look for a series of environmental pressures that drove it in the first place. To be more specific, we shall have to know why the unique combination of paternal investment and long-term "sex contracts" (Fisher, 1982) between males and females evolved, as these are absent in chimpanzees and bonobos. At the moment that these sex contracts were in place and males had to invest in particular females for a relatively long time, they also had good reasons to look especially for young (neotenous?) females with a lot of residual reproductive capacity. At the same time, females would have good reasons not only to look for "good genes," but for "good fathers" as well.

A POSSIBLE LINK BETWEEN PROLONGED CHILDREARING, PATERNAL INVESTMENT, AND CULTURAL ABILITIES

As we have seen, the neoteny hypothesis itself, although it has famous proponents (Gould, 1977), cannot account for the evolution of all human characteristics. (Shea, 1992, notes that it can account for the resemblance between the skull and face of a juvenile common chimpanzee and those of an adult bonobo, so even he does not exclude the possibility of the mechanism in some evolutionary trajectories.) It is easy, however, to think of other possible relations between characteristics of the human mating system and our cultural abilities. The most striking example is, of course, the extremely prolonged period of childhood dependence, which seems a condition for the cultural inheritance of large amounts of skills, practices, rituals, words, and knowledge. This prolonged childhood seems to result from a reduced rate of physical growth compared with apes and monkeys: for example, the permanent molars erupt at about the ages of six, eleven (or twelve), and eighteen in humans, whereas they erupt at the ages of three, six, and eleven in apes and at the ages of one and a half, three, and six in macaques (Leakey, 1994; Holly Smith, cited in Walker & Shipman, 1996). Following the anatomist Adolf Schultz, these ages are taken to represent the end of infancy, the beginning of adolescence, and the beginning of adulthood, respectively, as shown in figure 9.1 (based on Walker & Shipman, 1996).

These ages represent a revolution that has occurred in the mating system of our ancestors. It is difficult to imagine that a chimpanzee mother, without the aid of a father, could give her child so long a period of carelessness with relation to subsistence that the child could go on learning for decades, as children in our culture often do. Of course, it is true that in many cultures the periods in which children are dependent and the amount of paternal investment are limited. Nevertheless, it is reasonable to assume that there is a link between the unique property of our mating system—paternal investment coupled to an obsession with female fidelity (e.g., Daly & Wilson, 1988)—and the prolonged period of parental investment that might be a sine qua non for the acquisition of complex

Figure 9.1
Age of Eruptions of Permanent Molars

	Macaque	Chimpanzee	Modern human
End of infancy, 1st permanent molar	1 year, 5 months	3 years, 4 months	6 years
Beginning of adolescence, 2nd permanent molar	3 years, 3 months	6 years, 5 months	11/12 years
Beginning of adulthood, 3rd permanent molar	5 years, 10 months	11 years, 5 months	18 years

culture. Even in modern cultures, children in father-absent households have significantly less time to stay at home and absorb culture (Chisholm, 1993). Children from unstable families tend to start their sexual and reproductive career at an earlier age (Kim, Smith, & Palermiti, 1997) and therefore have less time for education. Children from small families, in which parents have relatively much time to invest, have more chance in getting jobs and becoming socially successful (Terhune, 1974, cited in Boyd & Richerson, 1993). At the other end of our evolutionary spectrum, it has been shown that female chimpanzees at Gombe that receive generous shares of meat produce more offspring that survive (McGrew, 1992, p. 110, combining data from Goodall, 1986, pp. 62, 310; Stanford, 1995).

But such a link between prolonged childrearing and the evolution of culture does not yet give an explanation for either of them. We still have to explain why some ancestral males started to invest in children and their mothers in exchange for a certain degree of paternal certainty (partly achieved by female fidelity, partly by male possessiveness, which was at some later stage reinforced by the cultural practice of marriage). We have to assume that there was a period in hominid evolution in which mothers simply could not do without the help of fathers, as a result of which children for which the mother was not able to obtain paternal investment were seriously at a disadvantage. If this was the case, however, what caused this increased dependence?

THE ORIGINAL HUCHIBO SOCIETY AND *AUSTRALOPITHECUS*

One possible explanation for an increase of paternal investment could be the increased dependence on meat. Foley (1987) speculated that 2.5 million years ago the genus *Australopithecus* was split into two as a result of the dry circumstances during the first ice ages: *Paranthropus* specialized on nuts, dry fruits, and roots, while *Homo* increased its level of meat intake to cope with the shortage of proteins. We now know that the chimpanzees at Gombe also hunt on a regular basis, especially during the dry season (Stanford, 1995). As I already indicated, females that are successful at obtaining meat are also the most successful mothers (McGrew, 1992, p. 110; Goodall, 1986, pp. 62, 310). Also, an alpha male at Kasoje was reported to distribute meat mainly to females with whom he consorted and to his mother (McGrew, 1992). So the necessary preadaptations may have existed to push the common ancestor of man and chimpanzee on the road toward "sex contract" in which paternal investment and paternity, or even paternal certainty, are exchanged.

What were the preadaptations that may have made such a shift possible? It has been recently stressed that the behavioral patterns of our ancestors were extremely diverse if we go back to the period in which New World monkeys and Old World monkeys were not separated (Small, 1995). In the New World monkeys we find a couple of characteristics that are typical for some hominids, for example, female dispersal, the existence of groups within groups (spider monkeys, Small, Robinson, & Janson, 1987), monogamy, and paternal investment (marmosets and tamarins, Hrdy, 1981; Kinzey, 1987). Thus the behavioral potential of our ancestors was already rich from the beginning, as is also proved by the variety of hominoid lifestyles: from monogamy (gibbon) to polygamy (gorilla), from almost solitary (big males in the orangutan) to extremely social (bonobo). Such lifestyles are, of course, a product of both phylogenetic inertia and ecological factors, like the presence of predators, the threat of conspecifics, and the dispersion and variety of food items.

To understand the origin of a mating system in which fathers invest in their offspring, it is probably most useful to start with a reconstruction of the behavioral patterns that we share with the species that are most related to us, the African apes. If this would allow us, for example, to reconstruct the type of society in which the common HUCHIBO ancestors lived, we could try to explain the divergence of humans, chimpanzees, and bonobos (hence HUCHIBOs; Slurink, 1993) as a result of specific local selection factors. Of course, we are particularly interested in factors like social life, male versus female dominance, sexual dimorphism, the different types of mating relationships, and the amount of paternal investment.

CHARACTERISTICS OF THE COMMON HUCHIBO ANCESTOR

Social Life

There has been much speculation and research on the differences in group size between bonobos and chimpanzees. Bonobos aggregate almost continually in relatively big parties. Bonobos are thought to experience less food competition as a result of their heavy reliance on terrestrial herbaceous vegetation (Wrangham, 1986), which is more evenly spread in the environment. Wrangham even speculates that they evolved from chimpanzees in the only corner of Africa's rainforest, the upper Congo, where chimpanzees did not live together with gorillas and where they could conquer the gorilla niche (Wrangham & Peterson, 1996). As a result of the omnipresence of their food, compared to the fruits that chimpanzees need, bonobos are more rarely forced to travel in small groups (Chapman, White, & Wrangham, 1994). Chimpanzees live in "fission-fusion" societies in which subgroups are continually formed, for example, at a particular fruit tree of which the fruits are ripe at that moment. As bonobos are a relatively recent species that probably split apart from the chimpanzees at about the same time that *Homo* was splitting apart from *Australopithecus*, their group sizes and other characteristics in which they resemble humans should be considered as products of independent evolution. Groups of the common ancestor are most likely to have been similar to those of chimpanzees. Perhaps Foley and Lee (1989) are right when they point out that the patchy grassland/bushland habitat in which they suppose that *Australopithecus afarensis* lived would promote larger group sizes because of predator avoidance. At some later stage hunting or intergroup competition may have forced groups to become even bigger. These groups must have formed subgroups continually, however, to constitute hunting parties or to patrol along the borders of the group territory. (If the idea of Aiello & Dunbar, 1993, is right and relative brain size correlates with group size, groups may have become gradually bigger during the evolution of the genus *Homo*.)

Male versus Female Dominance

Differences in group and party size are thought to explain some of the behavioral differences between chimpanzees and bonobos. While chimpanzee societies seem to be relatively male dominated, female bonds seem to be much stronger in bonobos (Parish, 1994). In chimpanzees, coalitions of males seem to form a center of power in the midst of the group; in bonobos, males are thought to be more markedly linked to particular females (De Waal, 1995). Male bonobos often need the support of their mothers to become powerful, and males can only become dominant if they have the support of equally dominant females in the group (Kano, 1992).

These far-reaching social differences are thought to have arisen as a result of banal ecological causes. Female bonobos are thought to be much more powerful as a result of the omnipresence of terrestrial herbaceous vegetation: groups do not have to split at any moment, which enables females to stay together and form relatively strong coalitions. Adolescent females can become part of the female social network of a group by starting an emotional and sexual relationship with more adult females. As a result of the power of these "lesbian matriarchies," bonobo males have become much less aggressive and "demonic" than chimpanzees. Whereas gorillas and chimpanzees are both very aggressive toward females and children, this strategy does not seem to work in bonobos (Wrangham & Peterson, 1996).

It is at least imaginable that there is a kind of continuum between more female-dominated and more male-dominated social systems that could explain both the differences between bonobos and chimpanzees and particular oscillations during human history. The more female-dominated social systems probably are promoted by a lack of predators (e.g., on Madagascar; Richard, 1987), relaxed food competition, and a low dependence on meat; the more male-dominated systems are promoted by a situation in which males can take advantage of the competition between females, by a dependence on meat, or by an increased level of intergroup competition. On the whole, humans seem to be more similar to chimpanzees than to bonobos in this respect. It would be interesting to investigate whether human societies that live under the threat of war become relatively male dominated.

Feminists have often speculated about an original human society in which females were more powerful than males and sometimes have referred in this context to Bachofen's work *Das Mutterrecht* (1861), which had considerable influence on Engels and thus on the Marxist tradition. Given the fact that in apes—in contrast to most monkeys—females disperse, that chimpanzee societies are mostly dominated by a small coalition of often related males, and that there are no human societies that are really female dominated, such an original matriarchal society is extremely unlikely. Even Hatshepsut could only become pharaoh by wearing an artificial beard. Whether we like it or not, there are good reasons to suppose that patriarchies are at least as old as gorillas and chimpanzees (Hrdy, 1997).

Sexual Dimorphism

Sexual dimorphism in all HUCHIBOs is relatively mild compared to more distantly related hominoids like gorillas and orangutans. Sexual dimorphism may have been relatively great in the common ancestor of all anthropoids, because a candidate for this genus, *Aegyptopithecus*, which has been found in the Fayum Depression in Egypt and is dated approximately 27 million years ago, shows considerable dimorphism (Frayer & Wolpoff, 1985). In orangutans and gorillas it is relatively great. The measures of the sexual dimorphism of *Australopithecus* differ: if one uses the canine teeth, it seems relatively great (Frayer & Wolpoff, 1985); if one uses the length of the hindlimb joints, it is somewhat above that of chimpanzees and bonobos, but below the sexual dimorphism of gorillas and orangutans (McHenry, 1991). On the basis of mandibular canines Frayer and Wolpoff (1985) have calculated a gradual decline of the sex differences from *Homo habilis* to *Homo erectus* and *Homo sapiens*, with a somewhat bigger difference in the European Neanderthals. Other authors, however, postulate that the sexual differences within the genus *Homo* were small from the very beginning (Stanley, 1996, pp. 178–179).

Sex differences in size are often thought to correlate with the amount of polygyny. They also result in different food habits. In chimpanzees females spend much more time in fishing for termites than males (McGrew, 1992, p. 91), and females also eat more insects generally. On the other hand, it is the males who do most of the hunting, especially on prey that is relatively difficult to get, like monkeys. Females sometimes catch ungulates, but generally they are more gatherers than hunters (McGrew, 1992, p. 103). McGrew notes cautiously that "it is tempting to interpret this difference as a possible pre-adaptation for the evolution of a system of sexual division of labor" (McGrew, 1992, p. 105).

Unimale or Multimale Groups?

If sex differences in size correlate with the amount of polygyny, australopithecines may have been relatively polygynous, even compared to chimpanzees. According to some theorists, the fact that chimpanzees live in multimale groups does not prove anything about the common HUCHIBO ancestor: it can have been like the gorilla in this respect. Schröder (1993) gives three arguments for a more gorillalike social system:

1. The remarkable sexual dimorphism in *Australopithecus*, "more likely indicating an intense competition between males to control access to females than gametic competition"
2. The fact that modern humans exhibit moderate polygyny, but not promiscuity
3. The fact that female gorillas do not show sexual swellings and that the sexual swellings of chimpanzees and bonobos could be a derived trait.

However, the idea that early hominids had a social structure somewhat more similar to that of gorillas than to that of chimpanzees remains implausible. Not only are we much more related to chimpanzees, as indicated by most molecular analyses (especially those of Sibley and Ahlquist; see the list in Tanner, 1987), but one of our oldest recently discovered hominid ancestors, *Ardipithecus ramidus* (White, Suwa, & Asfaw, 1994, 1995), displays many similarities to chimpanzees as well. Given the fact that the environments in which chimpanzees have lived have shrunk and expanded several times, but never completely vanished, it is not unreasonable to assume that the chimpanzee is still similar to the common HUCHIBO ancestor (Wrangham & Peterson, 1996). In that case bonobos and hominids are the product of isolated populations that have drifted apart into regions that were more deeply affected by the climatological and ecological events of the last five or six million years (Boaz, 1997).

As I noted already, Foley and Lee claim that the patchy grassland/bushland habitat in which they suppose that *Australopithecus afarensis* lived would promote larger group sizes because of predator avoidance. Larger group size implies that adult males must have associated together. Even in gorillas a dominant silverback male often tolerates one or more silverbacks—one extraordinary group in Rwanda even includes seven silverbacks (Wrangham & Peterson, 1996, p. 147)—so even gorillas cannot be said to live in unimale groups. Further, the discovery of the "First Family," a place where at least thirteen individuals of *Australopithecus afarensis* were found together (Afar Locality 333; e.g., Johanson & Edgar, 1996, p. 126), may give us a real hint of the group composition of that species. This group consisted of at least three large individuals who probably were males and at least two small-bodied individuals who may have been females.

Finally, Schröder's suggestion that the human mating system could have evolved directly from a more gorillalike polygynous system is implausible given the behavior of human females. It is probably a universal rule that the degree of female promiscuity correlates with the amount of sperm competition and, therefore, testes size in males (Martin & May, 1981; Harcourt, Harvey, Larson, & Short, 1981; Hrdy, 1997). If the human mating system was really characterized only by moderate polygyny, but not by promiscuity, the size of the human testes would be smaller. Given the fact that the human testes are halfway between those of gorillas (small) and chimpanzees (big), it is much more plausible to assume that the human mating system has evolved out of a more chimpanzeelike, partly promiscuous system as a result of a process of reproductive monopolization of females, which started as a result of some kind of ecological crisis.

Mating Relationships

All in all, although we do not know anything with certainty about the mating system of the original HUCHIBO ancestor and that of *Australopithecus*, we have good reason to use the chimpanzee as a model. In chimpanzees there exist three different types of mating relationships (Nishida & Hiraiwa-Hasegawa, 1987, p. 169, order changed):

1. Possessive matings of alpha males who may prevent other males from mating (and may occasionally use force or threats).
2. Opportunistic matings in which males copulate freely in the presence of other males.
3. Consortships in which a male and a female seclude themselves from the rest of society to have an exclusive relationship for a few days or even weeks. Often these consortships are initiated by males, and sometimes a male aggressively forces a female to follow him (Wrangham & Peterson, 1996).

Because the tendency to monopolize females in an aggressive way is shared with the gorilla, this probably has to be seen as the oldest mating pattern. It is interesting to speculate about the circumstances that would promote a specialization in one of these mating strategies:

1. Possessiveness is probably favored in situations in which males are not mutually dependent and are able to monopolize as many females as possible, and in which females are unable to form strong coalitions.

2. Promiscuity is probably favored in circumstances in which males are related or mutually dependent, in which aggressive possessiveness does not work as a result of female coalitions, or in which females may promote some competition to ensure fertilization by the strongest males.

3. Consortships are promoted by a situation in which females have an interest in having special relationships with particular males, perhaps because they need some extra support for their childrearing activities.

While the first strategy reminds one of gorillas, bonobos seem to have dropped this strategy altogether and to have evolved in the direction of promiscuity (Kano, 1992; Waal, 1995). The human mating system can be seen as descending from the third mating strategy. If this is true, the human mating system may have been promoted by a situation in which females had an interest in having special relationships with particular males. This may have been the situation in which our ancestors became increasingly dependent on meat.

Paternal Investment

In both chimpanzees and bonobos there does not seem to exist a special father-offspring bond, as it is unknown who has fathered a particular child. It may actually be in the interest of females to leave the question open as to who the father is as an anti-infanticide strategy (Hrdy, 1981). Perhaps this can explain why infanticide in chimpanzees occurs much more seldom than it does in gorillas, in which about one out of every seven children is killed and in which "it looks as though most infants unprotected by a silverback are killed" (Wrangham & Peterson, 1996, p. 148). However, even in a situation in which paternity is not certain, males may behave in accordance with an (unconscious) calculation of probabilities. In baboons there is no paternal certainty either, but males do sometimes help the children of their female "friends," partly to please their mothers, partly because they might be the fathers themselves (Strum, 1987). As I noted before, in chimpanzees there is a positive relationship between survival of offspring and the amount of meat that their mothers get at kills. Sometimes alpha males share their meat exclusively with females with which they have consorted. This is especially revealing if we realize that consortships do often result in successful conception (Goodall, 1986, pp. 471–477). Thus, although chimpanzee behavior gives us no indication of the existence of a father-child bond in the common HUCHIBO ancestor, sex contracts could have evolved as a result of an increased dependence on meat, and paternal investment could have increased gradually parallel to an increased paternal certainty.

SOME CHARACTERISTICS OF THE HUMAN MATING SYSTEM

In order to be able to make a reconstruction of the human evolutionary trajectory, we first have to list some of the typical features of the human mating system that seem to have been the object of selective forces during human evolution and that an adequate model should explain.

Altriciality: Helpless Infants, Dependent Mothers

It is currently thought that the increased encephalization during the evolution of *Homo*, together with the limits posed to a broadening of the hominid pelvis, necessitated a revolution in which babies were born relatively premature (an idea Portman defended in 1941; see Gould, 1977, p. 369; also defended by Waters, 1996). In fact, in comparison with other primates, a species with the general retardation in growth rate, with the brain size and longevity of humans, would need a gestation length of twenty-one months (Leakey, 1994). The early birth of the human baby has created a situation in which it lives as a kind of extrauterine embryo for more than one year, during which it needs much attention and care by the parents—which defines us as a clearly altricial species. Even in our modern, extremely egalitarian and efficient industrialized societies, many women stop working temporarily after childbirth. In most cases fathers are sorely needed in the raising of children, and some extra assistance by grandparents is very welcome as well. It is clear that this creates a social situation that is completely different from that which we see in bonobos and chimpanzees and that at best shows a dim resemblance to the behavior of a couple of New World primates.

Paternal Care

Probably about 80–90 percent of all children in all cultures have been fathered by their purported father. (Russel & Wells, 1986, estimate that *p* or paternity certainty is 87 percent and compare this with the *p* of 91 percent in Yanamamös, obtained via genetic research; Gangestad & Thornhill, 1997, point to figures from 1957 in which *p* was 93 percent, and Bellis & Baker, 1990, found that 6 percent of a sample of British women with one main partner reported their last act of sexual intercourse to be outside this relationship). As "extra-pair copulations" have simply to be considered part of monogamous breeding systems (this even goes for gibbons, as shown by Reichard, 1995), such figures show that it pays for human males to exchange paternal care for paternal certainty. In that respect Murdock could claim that the nuclear family was universal and that polygyny simply means that one man has more than one family (cited in Kinzey, 1987). Many psychological theories exist proposing effects of the presence or absence of the father at home (e.g., Chisholm, 1993), and there is reason to assume that the presence of a father of relatively high rank may have profound influences on the future rank and possibilities of a human child.

Menopause

Parental care in humans often continues well beyond the age at which children are able to reproduce themselves. Different authors have hypothesized that menopause is an adaptive phenomenon enabling older women to invest in their grandchildren rather than in their own children (Williams, 1957; Alexander, 1979, 1990; Hill & Hurtado, 1991; Pavelka & Fedigan, 1991). This may have been especially functional if the mother was high in rank and had many grandchildren. Apparently a mother who gradually has lost the advantage of being young and attractive can better use her acquired wisdom and power to assist several children at significant moments in the raising of grandchildren rather than simply to continue exhausting her own body and having children of her own. The evolution of menopause can probably be best explained within the context of the need of an increased period of dependence of young individuals on their family and especially within the context of the increased helplessness of the babies (Peccei, 1995).

Sex Differences

Although the sex differences in size are only moderate in our species, there are a couple of important physical and psychological differences between the sexes. The physical differences can be explained as a result of encephalization (width of the pelvis) and of sexual selection for neotenous mothers (relatively light complexion of the skin, hair, breasts). A couple of profound psychological differences between the sexes are attributed to a long stage of hunting and gathering during human prehistory. Females are better at remembering spatial configurations and objects and are very good in incidental, nondirectional learning of such configurations. Males are better at performing mental rotations and (as a result of that) at reading maps (Silverman & Eals, 1992). Women do better on precision manual tasks, too, and on mathematical calculation tests. Men, however, are more accurate in target-directed motor skills, such as aimed throwing, and do better on tests on mathematical reasoning (Kimura, 1992). In the use of speech, studies of aphasia suggest that women use their hemispheres more equally than men do (Kimura, 1992), which is also supported by the fact that their corpus callosum is bigger (e.g., Moir & Jessel, 1991). One of the effects seems to be that women have less difficulty in "finding the right word" to express their feelings and are generally more close to their feelings and to their body. Several other female psychological characteristics suggest that women are somewhat more inclined to stay at the home base and embellish it. It may be argued that throughout a large part of human evolution, females were somewhat more linked to the home base and relatively more involved in the raising of children, which would also explain their linguistic superiority (Dunbar, 1996).

Concealment of Ovulation and Sexual Privacy

It is generally agreed that the loss of estrus and the concealment of ovulation constitute a major difference between chimpanzees, on the one hand, and humans, on the other hand. Without calendars, many women themselves do not have even the slightest idea when they are ovulating, let alone their potential partners. Several hypotheses have been proposed to explain this difference; some of these are compared by Alexander (1990), who has given them eloquent names. The "prostitution hypothesis" explains concealment of ovulation in human females as a result of the necessity for females to obtain meat in exchange for sex. Females could obtain more meat by increasing their period of sexual attractivity (Symons, 1979, scenario A). The "cuckoldry hypothesis" sees concealment of ovulation essentially as a female reaction to a more monogamous lifestyle. By not advertising the exact moment of ovulation, females may have made it, in some situations, difficult for their partners and easy for their lovers to fertilize them, enabling them to get just the genes that they need most (Benshoof & Thornhill, 1979; Symons, 1979, scenario B; see also Schröder, 1993). Alexander's own favorite is the "paternal-care hypothesis," which stresses the ability of women to conceal the exact timing of ovulation in order to force a specific male partner to a more continuing investment (Alexander & Noonan, 1979).

An ingenious explanation of both estrus and its loss is offered by Hrdy (1981). Hrdy argues that the promiscuity of many female primates is a very effective way of confusing the issue of paternity and reducing the possibility of infanticide. By mating with a whole series of males, a female forces all these males to consider her children as possibly their own. In a situation in which females are monitored by harem leaders or husbands, the best way of continuing to confuse both these partners and extra-pair males about their possible paternity would be to conceal the moment of fertility. This would provide females the flexibility they need to spread illusions or at least confusions about the paternity of their children. Probably we should call this hypothesis the "confusion hypothesis."

Fortunately, some new empirical discoveries have gradually been made that may help us to choose among such hypotheses. The Austrian ethologist Karl Grammer discovered, for example, that the behavior of women may change around the time that they are ovulating as a result of a changed perception of androstenone: most of the time this odor repels them, but not so around the time of ovulation (Grammer, 1993). Grammer himself interprets this as proof for an explanation for concealed ovulation that stresses the female's chances of obtaining good genes outside the pair bond by mating quickly and at the right moment. Other researchers have shown that women can to some extent regulate the effectiveness of an insemination by having an orgasm or not (Baker & Bellis, 1993). Both discoveries can be cited as evidence in favor of a version of the cuckoldry hypothesis in which even females themselves are ignorant about their own intentions.

There is also evidence that can be used in support of other models, however. For example, if one compares the sexual behavior of chimpanzees and bonobos, it is striking that the duration of the maximum swelling in estrus is much longer in bonobos (20 days compared to 9.6 days; Kano, 1992). Whereas chimpanzee males compete intensely for copulations at the time that ovulation approaches, bonobos are much more indifferent and seldom fight. It can be argued that female bonobos conceal their ovulation (Wrangham & Peterson, 1996) in order to be able to protect their choice of the right father, which in their society need not be the most aggressive male. The advertising of ovulation in chimpanzees could be interpreted, then, as an adaptation to a male-dominated society that ensures both confusion about paternity and fertilization by the most dominant males. In bonobos the most aggressive males are no longer the most desirable fathers, and females no longer need to stimulate aggression between males; they only need to confuse. This would strengthen the confusion hypothesis, especially for bonobos.

One can argue that humans have evolved in an opposite direction, however. As noted, humans differ from both bonobos and chimpanzees in that females need some assistance of the father in the raising of offspring. If they would advertise their exact moment of ovulation, those males would not be interested anymore at other moments. Human females are therefore both attractive to males at each stage of the monthly cycle and cryptic about their exact moment of ovulation. Originally this system may have evolved out of the habit of male chimpanzees of sharing meat preferentially with females with which they have consorted. For *Australopithecus* the prostitution hypothesis may have been right. During the period of encephalization (*Homo*) such ephemeral exchanges would have become insufficient for the sustainment of the dependent mother and the helpless baby, however. Instead of an exchange of one copulation and one piece of meat, an exchange between an enlarged possibility of paternity and a lasting favoritism must have evolved, with a matching psychological motivation system (falling in love). For that period, the paternal-care hypothesis could well be right.

Perhaps the paternal-care hypothesis also needs to be supplemented by both the cuckoldry and confusion hypothesis. At the moment that societies started increasingly to consist of pair-bonded couples, females could still feel that they needed the protection of the most dominant males, which were not necessarily their own providers. The same concealment that helped them to bind their permanent partners may also have helped them to get the support of these dominant males and allowed them to swap partners at any moment that they found favorable.

We can conclude, therefore, that the different explanations for the concealment of ovulation do not exclude each other. If the original HUCHIBO ancestor

exhibited a mating system similar to that of chimpanzees, an increased dependence on meat may well have made it more attractive to females to join males
in consortships and to exchange sex for meat. These consortships may have
changed into somewhat longer periods at the time that more paternal investment
was needed. Concealment of ovulation in such a situation may have helped
females to keep their special friends or partners sexually interested while at the
same time enabling them to collect a set of superior genes occasionally.

THE HUNTING HYPOTHESIS AS AN EXPLANATION OF
THE ORIGINS OF *HOMO*

All this suggests that the increased period of helplessness of human infants
and, simultaneously, the increase in male investment have been the crucial factors that changed the mating system of the common HUCHIBO ancestor and
Australopithecus into the human lifestyle. As I said already, this increase in
paternal care could be explained by assuming a period of increasing dependence
on meat. The ice ages started 2.5 million years ago, and Africa became drier
and drier; to assume that one line of australopithecines became increasingly
dependent on meat is by no means unreasonable. There is much other evidence
as well that could point to an increased dependence on hunting. In an analysis
of the changes one would expect in a vegetarian species that is becoming carnivorous, Shipman and Walker (1989) enumerate the following:

1. An increase in either speed or sociality (adaptations required to catch prey)
2. A change in dentition or the appearance of a meat-processing industry
3. An increase in "free" time
4. Changes in the digestive tract
5. Either a decrease in body size or an increase in geographic range as a result of the
   availability of less food per square kilometer
6. A change to a more altricial pattern

One could argue that one can find at least two-thirds of these changes in the
transition from *Australopithecus* to *Homo* (the numbers correspond to those in
the preceding list):

1. The increase in brain size could point to a social life of increasing complexity (Aiello
   & Dunbar, 1993). This increased brain size may have been possible only as a result
   of an availability of more proteins (Aiello & Wheeler, 1995).
2. The Oldowan stone technology featuring sharp edges capable of slicing meat appears
   at about the same time that *Homo* appears. Compared to the molars of *Australopithecus*, the molars of early *Homo* were small, while the incisors were larger, which seems
   to point to a diet in which coarse plant foods were less important.
5. As Shipman and Walker noticed already, geographical expansion is characteristic of

Homo erectus. Since 1989, when they wrote their article, it has appeared that the geographical expansion of *Homo erectus* happened much earlier than originally thought, which strengthens their argument that it resulted from changing food habits necessitated by the first ice age.

6. Shipman and Walker argue that the relative brain size of early *Homo* was only possible as a result of an increase in gestation length, which they see as the most unambiguous sign that *Homo* is a "herbivore-turned-carnivore!"

Shipman and Walker also provide other evidence of both increased sociality and carnivorism in *Homo erectus* about 1.7 million years ago. They mention a female skeleton of *Homo erectus* from this period, KNM-ER 1808, with a large amount of ossified blood on her bones, which proves that she suffered from acute hypervitaminosis A and yet survived for several weeks prior to her death. They claim that the only way in which this would have been possible is if this unlucky female was supplied with water and possibly food and protection from predators during this period. At the same time, hypervitaminosis A is best explained by the consumption of meat: one can get it by either eating something like one hundred pounds of carrots or by eating one pound of carnivore liver. It seems likely that KNM-ER 1808 happened to eat somewhat too much liver, as is also suggested by the microwear of her teeth, which is comparable only to the microwear patterns that show up on the teeth of meat-and-bone-eating carnivores, like hyenas (Shipman & Walker, 1989; Walker & Shipman, 1996).

Another change that may have been the ultimate result of a change to a more carnivorous lifestyle is the increased dependence on a home base for the exchange of meat and other goods (Tooby & DeVore, 1986). The increased helplessness of the babies also may have necessitated such a change. The amount of offspring that a female could raise could increase by no longer bearing them individually, as in chimpanzees, but simply "storing" and feeding them at home base (Lovejoy, 1981). This tendency would reinforce the necessity of reliable paternal aid, which could only be obtained by giving the male an increased sense of paternal certainty. If *Homo* lived in a fission-fusion society centered at a home base, this may also have created the desirability of a communication system of increased complexity, either to report on the environment to the "home front" (Bickerton, 1990) or to form complex coalitions at the home front (Dunbar, 1996). If *Homo* was an efficient hunter, there may also have been more "free" time, which could be used for "cultural" displays. Of course, at the moment it is unclear whether we should project all these adaptations back as far as *Homo habilis* or *Homo rudolfensis*. They may have only emerged gradually, or as a result of additional crises.

In an analysis that still is close to that of Lovejoy (1981), Hill (1982) speculates that the transition to hunting would lead to male provisioning, which would allow females a greater freedom to concentrate on parental care.

This change would probably reduce infant mortality considerably, and thus, the *average* life span would increase. More importantly, with a greater number of organisms living to older ages, the advantages that could be obtained from averting causes of death later in life (aging) would increase greatly and thus provide the selection pressure for greater longevity. Organisms with a longer juvenile developing period might then be more able to outcompete others in adulthood (through learning, etc.), but such a longer period of development would necessitate an *increase* in the birth interval. This long period of juvenile dependency would, however, have an even more important consequence. If juvenile offspring had a very low probability of surviving their mothers' death at, for example, under ten years of age, it would be an unwise strategy for a female to continue to bear offspring when the probability of her death within the next ten-year period was quite high. Old females with a low probability of surviving another ten years should shift their reproductive strategy. The optimal strategy for a female under these conditions is to assist in the parental care of her own daughters' offspring, and to cease reproductive effort herself. (p. 539)

Thus a whole set of human characteristics seems to be explained by applying a version of the hunting hypothesis. Above that, it is strengthened by the analysis of fossilized bones and stone artifacts from several sites along the African Rift Valley (e.g., Bunn & Kroll, 1986).

Of course, as is well known, these same bones and artifacts are sometimes used to defend the hypothesis that early man was a scavenger, but several writers have pointed to the fact that this would bring our ancestors into serious competition with a list of other scavengers (Tooby & DeVore, 1986; Walker & Shipman, 1996). Also, scavenging and hunting are completely compatible, and both chimpanzees (Hasegawa et al., 1983) and Hazda hunter-gatherers in northern Tanzania (O'Connell et al., 1988) use both techniques at the same time, although scavenging in the Hazda accounts for only 20 percent of the carcasses and scavenging in chimpanzees is only rarely observed. The same pattern is found at the middle Pleistocene site at Aridos (Spain), where undisputed proof of elephant butchery was found that differs fundamentally from marginal scavenging (Villa, 1990). It should also be noted that many predators, from buzzards to lions, occasionally indulge in scavenging.

Another question is whether meat has ever been the exclusive nourishment of our ancestors (Tanner, 1987). This is unlikely, as we have a maximum sustained protein intake below about 50 percent of calories. It is even speculated that the ability of Eskimos to live on a diet with a protein intake of about 45–50 percent is due to a unique genetic capacity not seen in other populations (Speth, 1989). To discover the difference between the diet of a hominid and a real carnivore, one only has to compare one's dinner plate with the bowl of one's cat. Of course, as KNM-ER 1808 and the modern race of hamburger eaters demonstrate, meat is sometimes eaten more than is healthy and often is venerated as a supreme source of energy. As both chimpanzees and hunter-gatherers are predominantly vegetarian, the safest conclusion is that our ancestors have always been opportunists. Meat, however, may have enabled them to survive during periods of the ice ages in which the dry season became relatively long and exacting and, at a later stage, during the long winters on the Eurasian continent.

DOES THE HUNTING HYPOTHESIS EXPLAIN THE ORIGINS OF CULTURE?

This brings us to the question of whether the hunting hypothesis can explain the origins of culture. To some degree, it can. Hunting may have afforded the extra proteins needed to grow a big brain; it may have necessitated a more complex stone industry; it may have encouraged increasing cooperation and the need to pass on skills and techniques from generation to generation; and it may have encouraged the use of complex communication. Indirectly, it may have brought together individuals from three generations, thus encouraging cultural transmission. Finally, it may have eventually stimulated the occupation of home bases at which individuals from different generations could pass on their skills.

Hunting may not have been the only occupation of early *Homo*, however. Several writers have stressed that gathering may have been as important and that chimpanzees use tools predominantly in the context of nut cracking and insect collecting (Tanner, 1987). The first step that may have enabled *Australopithecus*, used to living along the border of tropical forests, to survive in a relatively dry environment may have been the opening up of new food sources below the ground: roots. In fact, in Tongo, a forest in eastern Congo with almost no rivers and lakes, a small population of chimpanzees lives with a tradition of digging and eating roots as a local adaptation to water shortage (Wrangham & Peterson, 1996). It seems that the skill of root digging is complex enough to stimulate a new dependence on the acquisition of skills through social learning, mainly between mother and offspring, as envisioned by Parker and Gibson (1979) and King (1994). From this perspective, the fact that *Australopithecus* has a brain that is slightly larger than that of the chimpanzee can be explained. Australopithecines may have been dependent on foods that were already hard enough to get to force them to relatively intelligent behavior, which may have preadapted particular populations for the even more demanding task of hunting and the processing of meat and bone marrow.

All in all, we can conclude that an increased dependence on difficult food sources, necessitating "extractive foraging" (Parker & Gibson, 1979), may have forced our ancestors to become smarter. Parker and Gibson even postulate that such a transition may have furthered their linguistic proficiency:

The prehistoric ecological transition to extractive foraging on foods that were both difficult to obtain and process would have resulted in mandatory parental provisioning of postweaning children. Abortive attempts by children to open tough nuts, dig deep tubers from the ground and engage in other complex activities would have resulted in need for parental aid. Many parents would have anticipated their children's difficulties in accomplishing these tasks and would have come to their aid as soon as interest was evidenced by the child by pointing, vocalizing, reaching, etc. The probable result would have been that certain vocal or manual gestures would have acquired specific meaning within individual mother-infant pairs. (p. 374)

A couple of questions remain, however. The first is why big brains and culture did not arise during the evolution of *Australopithecus* already. This question was given a first sketchy and speculative answer by Stanley (1996): their brain sizes may have been limited because they were unable to care for the helpless infants that need to be born if babies with big brains have to be born. As australopithecines probably were still partly adapted to a life in the trees, to which they had to flee from predators, they did not have their hands free to carry such infants. Only as a result of a climatic change that created an environment with fewer trees and with less food generally was a small population of australopithecines forced to start specializing increasingly on meat during the dry season, while they were unable to climb back into the trees for safety. The same skills that may have allowed them to hunt in groups may have enabled them to defend themselves from predators.

The second question is why our culture is so complex if it only evolved to enable us to hunt and why we tend to live in groups that are much bigger than would be efficient for group hunting. Obviously, living in relatively big groups has many disadvantages, especially for hunters. Also, as is shown by a variety of carnivores, one certainly needs to be clever to be able to hunt, but one does not need to be able to write poetry. Why would humans have started to live in groups of increasing size, and why would their brains have become bigger and their culture much more elaborate than would be required for mere hunting?

THE INTERGROUP-COMPETITION HYPOTHESIS

Alexander, who, together with Noonan, developed the paternal-care hypothesis to explain concealed ovulation, thought the hunting hypothesis insufficient to explain the "uniquely unique" characteristics of our species and developed the theory of intergroup competition or balances of power (e.g., Alexander, 1990; see also the chapter by van der Dennen in this volume). The idea is that our ancestors at some time in the past became "their own hostile force of nature" and that an arms race resulted between neighboring cultures that forced them to progress both in within-group cooperation and in the development of weapons, new ways of food production, the creation of cooperation-stimulating myths, and the like. The model depends to some extent on the notion of "ecological dominance" (e.g., Alexander, 1990) because it presupposes an uncorrected population pressure.

As I have shown elsewhere (Slurink, 1993, 1994), there are few signs that the necessary ecological dominance was really achieved in early *Homo*. Also, one would expect a transitory stage in which an ecologically dominant predator first became gradually more dangerous. Therefore, I do not see the hunting hypothesis and the intergroup-competition hypothesis as incompatible, but only as referring to different interpretations of the evolution of the genus *Homo*. An increase in parental care and a home base to exchange food and to protect increasingly helpless juveniles can already have been a characteristic of a carnivorous primate like *Homo ergaster/erectus*. A situation in which there were no longer other predators powerful enough to cope with *Homo ergaster/erectus* and its protected home bases could have given rise to increased competition for favorable sites and to the necessity to join relatively big and strong groups.

Thus the home base may have played a crucial role during human evolution. During the transition to a lifestyle as hunter it enabled our ancestors to find a place to exchange food; gradually it became more important as a place where children could be reared and protected; but finally, it became itself a scarce resource and the object of competition among different groups. Only this last factor can explain adequately why group sizes in our species seem to be above the optimum with relation to cooperative hunting (Alexander, 1979). Also, only this last factor can explain a tendency to socially respected monogamy.

PAIR BONDS AND REPRODUCTIVE-OPPORTUNITY REWARDS IN A MULTIMALE SOCIETY

All in all, the pattern that suggests itself is that man started as a predator, but at some relatively late point in prehistory increasingly turned into a warrior. Probably this can help us to explain the typical paradox of a species in which males to some extent mutually respect each other's relationships to particular females. The external pressure of a threat from foreign groups created a situation in which group members became mutually dependent and were forced to extinguish sources of intragroup conflict like conflict over females. There was a need for rules that would curtail an escalation of intragroup conflicts, and a tendency toward "reproductive-opportunity leveling" (Alexander, 1987) would do so.

At first sight the concept of reproductive-opportunity leveling might seem an artificial deus ex machina. However, one can find many examples in the anthropological literature that show how it might work. For example, the Mehinaku of Brazil have very outspoken ideas about what it is to be a real man: a real man is someone who is not lazy, who regularly provides food for the people, and who shares it altruistically. A real man is also a good wrestler and a strong personality. Anyone who does not fulfill this image is looked down upon by both men and women. The important point is that the women of these disrespected men, as a result of this lack of respect, also deceive them. To cite David Gilmore (1993), who uses studies of Thomas Gregor: "The sexual norms of the Mehinaku allow tacitly that a women deceives a bad wrestler. Knowing this, most of these women have adulterous relationships while their husbands are sulking helplessly" (p. 124). The important point is, of course, that a bad wrestler also makes a bad warrior and that the norms of manhood refer to some extent to cooperativeness and potential heroism.

This is even more clear in the Yanomamö. Chagnon (1988) has shown that men who have made the most victims in intergroup conflicts, that is, those who are the best killers, also have the most women. Of course, it would be important to show that cowardice and desertion are also punished and thus that satisfying the norm of the society is the only way to be reproductively successful.

Perhaps these examples show that Alexander's concept of reproductive-opportunity leveling is not entirely correct. Probably it should be replaced by reproductive-opportunity trading or even by the idea of a reproductive-opportunity rewards system. The point is that not just anyone in a society gets reproductive opportunities, but rather, the coalition of dominant individuals rewards those men whom they find helpful or indispensable. We should not forget that the balance-of-power model is not a model of Wynne-Edwardsian group selection, but a model explaining why human societies are characterized by so much moralistic aggression toward noncooperators and why such a relatively high level of cooperation can be achieved among nonrelated individuals.

The idea behind the model of Alexander is that the only way in which a multimale society in which paternity was totally uncertain could turn into a society in which paternity was certain, but in which children were nevertheless safe from other males, was by introducing an extra motivation for males to cooperate. To cite Alexander:

Prevention of infanticide . . . would be a massively important way that a male might help his female and the offspring he sires. Suppose a female begins to restrict her copulations, excluding certain males or excluding all but a single male. In a primate resembling chimpanzees we are justified in assuming that such a female would place her offspring in jeopardy of infanticide by the disenfranchised males within her own group. Because of her loyalty to the male who mated with her, it would profit him to defend her offspring against attack, at least under circumstances where this would not have been the case before, and assuming that his loyalty had some chance of being effective in preventing infanticide. If unity among males is sufficiently important, then rudimentary social reciprocity among males in connection with defense of the group or the "exporting" of aggression . . . could cause a male's importance to the group, and the importance of overall amicability among males, to prevent males who could not copulate with a particular female from attacks on her offspring or on the male who undertakes to defend them. Obviously respecting the right of the offspring of other individuals or families to exist and go about their business is also part of the social cooperativeness—the moral system—of humans today. (1990, p. 32)

CONCLUSION

As might be expected from an evolutionary perspective, human culture seems to be the coincidental product of a series of cumulative adaptive changes. These may have started as a result of ecological instability (Potts, 1996), which forced particular groups of chimpanzee-like HUCHIB ancestors to open up new food sources, which required new cognitive abilities. Probably the genus *Homo* resulted from a group of australopithecines that no longer was able to retreat into the trees and that became increasingly dependent on meat at the beginning of the ice ages. The birth of helpless children stimulated the origin of more or less exclusive pair bonds within the multimale societies of these early humans. At some later stage these pair bonds may have been one of the most important requirements of more complex societies because they enabled the origin of a reproductive-opportunity rewards system that allowed the evolution of a complex division of labor. The increased internal complexity of societies was probably driven by arms races between such societies, which resulted from the ecological dominance that was achieved by *Homo* at some point during prehistory.

ACKNOWLEDGMENT

Research on this chapter was supported by the Foundation for Philosophy and Theology (SFT), which is subsidized by the Netherlands Organization for Scientific Research (NWO).

REFERENCES

Aiello, L., & Dunbar, R. I. M. (1993). Neocortex size, group size, and the evolution of language. *Current Anthropology, 34*, 184–93.

Aiello, L., & Wheeler, P. (1995). The expensive-tissue hypothesis. *Current Anthropology, 36*, 199–221.

Alexander, R. D. (1979). *Darwinism and human affairs*, Seattle: University of Washington Press.

———. (1981). *The biology of moral systems*. New York: Aldine de Gruyter.

———. (1990). *How did humans evolve? Reflections on the uniquely unique species*. Special Publication No. 1. Ann Arbor: Museum of Zoology, University of Michigan.

Alexander, R. D., & Noonan, K. (1979). Concealment of ovulation, parental care, and human social evolution. In N. Chagnon & W. Irons (Eds.), *Evolutionary biology and human social behavior*. North Scituate, MA: Duxbury Press.

Bachofen, J. (1867). *Das Mutterrecht*. Basel: Beno Schwabe.

Baker, R. R., & Bellis, M. A. (1993). Human sperm competition: Ejaculate manipulation by females and a function for the female orgasm. *Animal Behaviour, 461*, 887–909.

Bellis, M. A., & Baker, R. R. (1990). Do females promote sperm competition? Data for humans. *Animal Behaviour, 40*, 997–999.

Benedict, R. (1935). *Patterns of culture*. London: Routledge.

Benshoof, L., & Thornhill, R. (1979). The evolution of monogamy and concealed ovulation in humans. *Journal of Social and Biological Structures, 2*, 95–106.

Betzig, L. (Ed.). (1997). *Human nature: A critical reader*. New York and Oxford: Oxford University Press.

Bickerton, D. (1990). *Language and species.* Chicago: University of Chicago Press.

Boaz, N. T. (1997). *Eco Homo: how the human being emerged from the cataclysmic history of the earth.* New York: Basic Books.

Boyd, R., & Richerson, P. J. (1993). Culture and human evolution. In D. T. Rasmussen (Ed.), *The origin and evolution of humans and humanness.* Boston and London: Jones & Bartlett.

Bunn, H. T., & Kroll, E. M. (1986). Systematic butchery by Plio-Pleistocene hominids at Olduvai Gorge, Tanzania. *Current Anthropology, 27,* 5 431–452.

Buss, D. M. (1989). Sex differences in human mate preferences: Evolutionary hypotheses tested in 37 cultures. *Behavioral and Brain Sciences, 21,* 1–49.

———. (1994). *The evolution of desire: Strategies of human mating.* New York: HarperCollins.

Chagnon, N. (1988). Life histories, blood revenge, and warfare in a tribal population. *Science, 239,* 985–992.

Chapman, C. A., White, F. J., & Wrangham, R. W. (1994). Party size in chimpanzees and bonobos. In R. W. Wrangham, W. C. McGraw, F. B. M. de Waal, & P. G. Haltne (Eds.), *Chimpanzee cultures.* Cambridge, MA: Harvard University Press.

Chisholm, J. S. (1993). Death, hope, and sex: Life-history theory and the development of reproductive strategies. *Current Anthropology, 34,* 1–24.

Daly, M., & Wilson, M. (1988). *Homicide.* New York: Aldine de Gruyter.

Darwin, C. R. (1871). *The descent of man, and selection in relation to sex.* London: Murray.

Dawkins, R. (1976). *The selfish gene.* Oxford: Oxford University Press.

Dennen, J. M. G. van der. (1995). *The origin of war: The evolution of a male-coalitional reproductive strategy.* Groningen: Origin Press.

De Waal, F. B. M. (1995). Bonobo sex and society. *Scientific American, 272* (3), 58–64.

Divale, W., & Harris, M. (1976). Population, warfare, and the male supremacist complex. *American Anthropologist, 78,* 521–538.

Dunbar, R. (1996). *Grooming, gossip, and the evolution of language.* London: Faber & Faber.

Emlen, S. T., Wrege, P. H., & Demong, N. J. (1995). Making decisions in the family: An evolutionary perspective. *American Scientist, 83* (2), 148–157.

Fisher, H. E. (1982). *The sex contract.* New York: William Morrow & Co.

Flinn, M. V., & Low, B. S. (1986). Resource distribution, social competition, and mating patterns in human societies. In D. I. Rubenstein & R. W. Wrangham (Eds.), *Ecological aspects of social evolution* (pp. 217–243). Princeton NJ: Princeton University Press.

Foley, R. A. (1987). *Another unique species: patterns in human evolutionary ecology.* Harlow: Longman.

Foley, R. A., & Lee, P. C. (1989). Finite social space, evolutionary pathways, and reconstructing hominid behavior. *Science, 243,* 901–906.

Frayer, D., & Wolpoff, M. H. (1985). Sexual dimorphism. *Annual Review of Anthropology, 14,* 429–73.

Gilmore, D. (1990). *Manhood in the making: Cultural concepts of masculinity.* New Haven: Yale University Press.

Goodall, J. (1986). *The chimpanzees of Gombe.* Cambridge, MA: Belknap Press of Harvard University Press.

Gould, S. J. (1977). *Ontogeny and phylogeny.* Cambridge, MA: Belknap Press of Harvard University Press.

Grammer, K. (1993). 5-α-androst-16en-3α-on: A male pheromone? A brief report. *Ethology and Sociobiology, 14,* 201–208.

Grant, P. R. (1991). Natural selection and Darwin's finches. *Scientific American, 265*(4), 60–65.

Harcourt, A. H., Harvey, P. H., Larson, S. G., & Short, R. V. (1981). Testis weight, body weight, and breeding systems in primates. *Nature, 293,* 55–57.

Harris, M. (1985). *Culture, people, nature.* New York: Harper & Row.

Hasegawa, T., Hiraiwa, M., Nishida, T., & Takasaki, H. (1983). New evidence on scavenging behavior in wild chimpanzees. *Current Anthropology, 24,* 231–232.

Hill, K. (1982). Hunting and human evolution. *Journal of Human Evolution, 11,* 521–544.

Hill, K., & Hurtado, A. M. (1991). The evolution of premature reproductive senescence and menopause in human females: An evaluation of the "grandmother hypotheses." *Human Nature, 2*(4), 313–350.

Hrdy, S. B. (1981). *The woman that never evolved.* Cambridge, MA: Cambridge University Press.

———. (1997). Raising Darwin's consciousness: Female sexuality and the prehominid origins of patriarchy. *Human Nature, 8*(1), 1–49.

Kano, T. (1992). *The last ape: Pygmy chimpanzee behavior and ecology.* Stanford: Stanford University Press.

Kim, K., Smith, P. K., & Palermiti, A. L. (1997). Conflict in childhood and reproductive development. *Evolution and Human Behavior, 18,* 109–142.

Kimura, D. (1992). Sex differences in the brain. *Scientific American, 267* (3), 80–87.

King, B. J. (1994). *The information continuum: Evolution of social information transfer in monkeys, apes, and hominids.* Santa Fe: Sar Press.

Kinzey, W. G. (1987). A primate model for human mating systems. In W. G. Kinzey (Ed.), *The evolution of human behavior: Primate models.* Albany: State University of New York Press.

Leakey, R. (1994). *The origin of humankind.* London: Weidenfeld & Nicolson.

Lovejoy, C. O. (1981). The origin of man. *Science, 211,* 341–350.

Lumsden, C. J., & Wilson, E. O. (1981). *Genes, mind, and culture: The coevolutionary process.* Cambridge, MA: Harvard University Press.

Martin, R. D., & May, R. (1981). Outward signs of breeding. *Nature, 293,* 7–9.

McGrew, W. C. (1992). *Chimpanzee material culture: Implications for human evolution.* Cambridge: Cambridge University Press.

McHenry, H. M. (1991). Sexual dimorphism in *Australopithecus afarensis. Journal of Human Evolution, 20,* 21–32.

Mestel, R. (1995). Arts of seduction. *New Scientist, 148* (2009/2010), 28–31.

Miller, G. (1996). Sexual selection in human evolution: review and prospects. In C. Crawford & D. Krebs (Eds.), *Evolution and human behavior: Ideas, issues and applications.* New York: Lawrence Erlbaum.

Moir, A., & Jessel, D. (1991). *Brain sex: The real difference between men and women.* New York: Dell.

O'Connel, J. F., Hawkes, K., & Blurton Jones, N. (1988). Hadza scavenging: implications for Plio/Pleistocene hominid subsistence. *Current Anthropology, 29,* 356–363.

Nishida. Nishida, T., & Hiraiwa-Hasegawa, M. (1987). Chimpanzees and bonobos: Co-

operative relationships among males. In B. B. Smuts, D. L. Cheney, R. M. Seyfarth, R. W. Wrangham, & T. T. Struhsaker, *Primate societies* (pp. 165–177).
London & Chicago: University of Chicago Press.

Parish, A. R. (1994). Sex and food control in the "uncommon chimpanzee": How bonobo females overcome a phylogenetic legacy of male dominance. *Ethology and
Sociobiology, 15*, 157–179.

Parker, S. T. (1987). A sexual selection model for hominid evolution. *Human Evolution,
2*(3), 235–253.

Parker, S. T., & Gibson, K. R. (1979). A developmental model for the evolution of language and intelligence in early hominids. *Behavioral and Brain Sciences, 2*, 367–
408.

Pavelka, M. S. M., & Fedigan, L. M. (1991). Menopause: A comparative life history
perspective. *Yearbook of Physical Anthropology, 34*, 13–38.

Peccei, J. S. (1995). The origin and evolution of menopause: the altriciality-lifespan hypothesis. *Ethology and Sociobiology, 16*, 425–449.

Potts, R. B. (1996). *Humanity's descent: The consequences of ecological instability.* New
York: William Morrow & Co.

Reichard, U. (1995). Extra-pair copulations in a monogamous gibbon (*Hylobates lar*).
Ethology, 100, 99–112.

Richard, A. F. (1987). Malagasy prosimians: Female dominance. In B. Smuts et al. (Eds.),
Primate societies. Chicago: University of Chicago Press.

Ridley, M. (1994). *The red queen: sex and the evolution of human nature.* Harmondsworth: Penguin.

Robinson, J. G., & Janson, C. H. (1987). Capuchins, squirrel monkeys, and atelines:
socioecological convergence with old world primates. In B. B. Smuts, D. L. Cheney, R. M. Seyfarth, R. W. Wrangham, and T. T. Struhsaker (Eds.), *Primate societies.* Chicago & London: University of Chicago Press.

Rodseth, L., Wrangham, R. W., Harrigan, A. M., & Smuts, B. B. (1991). The human
community as a primate society. *Current Anthropology, 32* (3) 221–254.

Russell, R. J. H., & Wells, P. A. (1986). Estimating paternity confidence. *Ethology and
Sociobiology, 8*, 215–220.

Schröder, I. (1993). Concealed ovulation and clandestine copulation: A female contribution to human evolution. *Ethology and Sociobiology, 14*, 381–389.

Shea, B. T. (1992). Neoteny. In S. Jones, R. Martin, & D. Pilbeam (Eds.), *The Cambridge
encyclopedia of human evolution* (p. 104). Cambridge: Cambridge University
Press.

Shipman, P., & Walker A. (1989). The costs of becoming a predator. *Journal of Human
Evolution, 18*, 373–392.

Silverman, I., & Eals, M. (1992). Sex differences in spatial abilities: Evolutionary theory
and data. In J. H. Barkow, L. Cosmides, & J. Tooby (Eds.), *The adapted mind:
Evolutionary psychology and the generation of culture* (pp. 533–553). New York
and Oxford: Oxford University Press.

Slurink, P. (1993). Ecological dominance and the final sprint in hominid evolution. *Human Evolution, 8*, 265–273.

Slurink, P. (1994) Causes of our complete dependence on culture. In R. A. Gardner, B. T.
Gardner, B. Chiarelli, & F. X. Plooij, *The ethological roots of culture* (pp. 461–
494). NATO—ASI series, Series D: Behavioural and Social Sciences, Vol. 78.
Dordrecht; Boston, London: Kluwer Academic Publishers.

Small, M. (1995). Making a monkey of human nature. *New Scientist, 146* (1981), 30–33.

Smuts, B. B., Robinson, J. G., & Janson, C. H., et al. (Eds.). (1987). *Primate societies.* Chicago: University of Chicago Press.

Speth, J. D. (1989). Early hominid hunting and scavenging: The role of meat as an energy source. *Journal of Human Evolution, 18*, 329–343.

Stanford, C. B. (1995). Chimpanzee hunting behavior and human evolution. *American Scientist, 83*(3), 256–261.

Stanley, S. M. (1996). *Children of the ice age.* New York: Harmony Books.

Strum, S. C. (1987). *Almost human: A journey into the world of baboons.* London: Elm Tree Books.

Symons, Donald A. (1979). *The evolution of human sexuality.* New York: Oxford University Press.

Tanner, N. M. (1987). The chimpanzee model revisited and the gathering hypothesis. In W. G. Kinzey (Ed.), *The evolution of human behavior: Primate models* (pp. 3–27). Albany: State University of New York Press.

Thornhill, R., Gangestad, S. W., & Comer, R. (1995). Human female orgasm and mate fluctuating, asymmetry. *Animal Behavior, 50*, 1601–1615.

Tooby, J., & DeVore, I. (1987). The reconstruction of hominid behavioral evolution through strategic modeling. In W. G. Kinzey (Ed.), *The evolution of human behavior: primate models* (pp. 183–237). Albany: State University of New York Press.

Villa, P. (1990). Torralba and Aridos: Elephant exploitation in middle Pleistocene Spain. *Journal of Human Evolution, 19*, 299–309.

Walker, A., & Shipman, P. (1996). *The wisdom of the bones.* New York: Alfred A. A. Knopf.

Waters, J. (1996). Helpless as a baby. Internet version. http://www.dircon.co.uk/jdwaters/HAAB.htm.

White, T. D., Suwa, G., & Asfaw, B. (1994). *Australopithecus ramidus*, a new species of early hominid from Aramis, Ethiopia. *Nature, 371*, 306–312.

———. (1995). Corrigendum to *Australopithecus ramidus*, a new species of early hominid from Aramis, Ethiopia. *Nature, 375*, 88.

Williams, C. G. (1957). Pleiotropy, natural selection, and the evolution of senescence. *Evolution, 11*, 398–411.

Wilson, E. O. (1975). *Sociobiology: The new synthesis.* Cambridge, MA: Belknap Press of Harvard University Press.

Wrangham, R. W. (1986). Ecology and social relationships in two species of chimpanzee. In D. Rubenstein & R. Wrangham (Eds.), *Ecological aspects of social evolution: Birds and mammals* (pp. 352–378). Princeton, NJ: Princeton University Press.

Wrangham, R. W., McGrew, W. C., Waal, F. B. M. de, & Heltne, Paul G. (Eds.). (1994). *Chimpanzee cultures.* Cambridge, MA: Harvard University Press.

Wrangham, R. W., & Peterson D. (1996). *Demonic males: Apes and the origins of human violence.* Boston: Houghton Mifflin.

10

Human Evolution and the Origin of War:
A Darwinian Heritage

Johan M.G. van der Dennen

In several respects Darwin was a child of his time. His ideas about war, and especially the role of warfare in human evolution, were derived, at least in part, from Spencer and Bagehot, who, in turn, derived their ideas, at least in part, from Hobbes, Ferguson (though probably not directly), Malthus, and even Lamarck. Recurrent themes during Darwin's times were (1) group-selection arguments; (2) orthogenesis or ortholinear progress and directedness of evolution; (3) belief in the inheritance of acquired characteristics; (4) obedience, cohesiveness, and compactness as decisive factors in mankind's progress from barbarism to civilization; (5) in-group–out-group dual morality and ethnocentrism; (6) ubiquity and constancy of warfare from the very beginning of the hominids and a consequent (7) instinct of belligerence; (8) the intimate connection between hunting and warfare; and (9) the (mostly implicit) idea of "balance of power."

HOBBES AND FERGUSON

In 1967 the Scottish philosopher Adam Ferguson published *An Essay on the History of Civil Society*, probably the first attempt at an empirical investigation of the origins of war using ethnographic data. His analysis seemed to confirm Hobbes (1651): the primitive state was indeed a state of war (*status hostilis*): "We have had occasion to observe, that in every rude state the great business is war; and that in barbarous times, mankind, being generally divided into small parties, are engaged in almost perpetual hostilities" (*Essay*, 3.5).

Among the Hobbesian motives for war—competition, diffidence, and glory—

Ferguson clearly assigns priority to glory. Both cannibals and kings fight for honor more than for booty or any other material interest: "Mankind not only find in their condition the sources of variance and dissension: they appear to have in their minds the seeds of animosity, and they embrace the occasions of mutual opposition, with alacrity and pleasures" (quoted in Dawson, 1996, p. 4). The basic cause of war is rivalry, and Ferguson sees positive value in it, where Hobbes had seen only a necessary evil. Ferguson points out that warfare enforces civic unity, engenders civic virtue, promotes social organization, and in fact may be an essential condition for the very existence of civilization (Dawson, 1996).

In addition to maintaining the balance of power *between* societies, Ferguson ascribes to warfare the function of maintaining solidarity and morale *within* societies. In-group amity depends upon out-group enmity, and vice versa. This idea could also be found, in primordial form, in classical authors, especially the Roman historians (e.g., Sallust), but Ferguson probably offers the first analysis of the phenomenon of ethnocentrism in history.

MALTHUS

The essence of Malthus's (1798) doctrine—which catalyzed Darwin's selectional paradigm (Smillie, 1995)—is that a population tends to increase faster than the means of subsistence and that this increase is checked by wars, epidemics, and famines, to which he subsequently added "moral restraint." He regarded warfare in the earlier ages of the world as "the great business of mankind" and considered as one of the first causes and most powerful impulses of war "undoubtedly an insufficiency of room and food; and greatly as the circumstances of mankind have changed since it first began, the same cause still continues to operate and to produce, though in smaller degree, the same effects" (van der Dennen, 1990, p. 757). Before Malthus, many authors had indicated the demographic factor (i.e., overpopulation) as one of the principal causes of war. The ancient Stoics, for example, Plutarch, and Renaissance neo-Stoics had already speculated that warfare belongs to a providential scheme designed to keep populations from outgrowing their food supply (Dawson, 1996). Malthus, however, generalized the theories into a "law" in which war functions as one of the effective checks on population. Since that time, this idea has become quite common in various formulations (van der Dennen, 1975, 1983, 1990, 1995).

In his *Cours de philosophie positive* (1830–42), Comte dismissed the ancient notion of a peaceful golden age at the dawn of history. On the contrary, perennial and savage warfare forced social solidarity as a defense against enemy groups (see also the ideas of Spencer and Bagehot) discussed later in this chapter). After Ferguson and Malthus, it was increasingly believed that the ancient "savage" and the non-Western preliterate societies were belligerent and violent. John McLennan (1826), a Scottish sociologist and lawyer, put the argument emphatically: "Lay out the map of the world, and wherever you find populations unrestrained by the strong hand of government, there you will find perpetual feud, tribe against tribe, and family against family'' (quoted in Blainey, 1998, p. 58). Similarly, the world's authority on ancient law, Sir Henry Maine (1867), saw tribal war as a frequent occurrence. In Darwin's times, the once-noble savage was turning into an ignoble, barbarous, and belligerent blute (Blainey, 1988). Darwin himself had occasion to observe the "savage" Fuegians during his voyage on the *Beagle* and compared them with "wild animals."

SPENCER

Herbert Spencer, the founding father of social Darwinism, argued that the inheritance of acquired characteristics was the only possible evolutionary force responsible for the evolution of human morality. His vision was that the inheritance of acquired characteristics (a theory of evolution associated with the French naturalist Lamarck) would bridge the gap between biological and cultural evolution, forging them into one grand seamless process (Cronin, 1991). In his *Social Statics* (1851) Spencer preached the inexorable progress in the course of history from a violent and chaotic early human state to higher stages that led ultimately to civilization and peace. War, bloodshed, enmity, and cruelty—those "manifold evils"—were endemic and inevitable in early history, mandated by predatory instincts. The forces that were working out the ultimate "great scheme of perfect happiness" took no account of incidental suffering and exterminated "such sections of mankind as stand in their way, with the same sternness as they exterminate beasts of prey and herds of useless ruminants" (p. 454).

With a few exceptions, most primitives were unsociable and warlike. They were in the early "egoistic" stage. However, the general direction of social evolution, according to Spencer, was from egoism to altruism. Relentless Malthusian population pressure producing recurrent wars between societies was the triggering mechanism that, despite its antisocial character, helped impel humanity forward into higher civilization. Challenges like war and crowding fostered among conquering races qualities of social cohesion, mutual aid, inventiveness in artifacts and weapons, economic specialization, and human differentiation: "From the very beginning the conquest of one people over another has been, in the main, the conquest of social man over anti-social man; or, strictly speaking, of the more adapted over the less adapted" (1851, p. 455; compare the ideas of Bagehot discussed later).

Spencer soon coined the phrase "survival of the fittest" (later adopted and adapted by Darwin) to describe this process. In later writings he explored the several ways that warfare led to progress:

Warfare among men, like warfare among animals, has had a large share in raising their organizations to a higher stage. The following are some of the various ways in which it has worked. In the first place, it has had the effect of continually extirpating races which, for some reason or other, were least fitted to cope with the conditions of existence they were subject to. The killing-off of relatively feeble tribes, or tribes relatively wanting in endurance, or courage, or sagacity, or power of co-operation, must have tended ever to maintain, and occasionally to increase, the amounts of life-preserving powers possessed by men. . . . A no less important benefit bequeathed by war, has been the formation of large societies. By force alone were small nomadic hordes welded into large tribes; by force alone were large tribes welded into small nations; by force alone have small nations been welded into large nations. (Spencer, 1873, pp. 193–194)

War thus had played a vital role in emancipating humans from an unruly, savage state. War had brought social cohesion during the militant stage of social evolution, the basis for emerging nation-states and empires. In contemporary industrial societies, however, war and militarism, Spencer asserted, had become retrogressive and dysgenic, eliminating the "best elements" of the population on the battlefield.

Struggle and violent competition ("pugnacity" or "fighting instinct" were the contemporary terms), bloodshed, and cruelty were generally regarded by the social Darwinists as the crude filtering mechanisms by which species evolved and natural progress occurred. It was this "nasty" aspect of natural selection that allegedly struck the nineteenth-century imagination, the emphasis on differential mortality and the ideas of the "law of the jungle" and "Nature red in tooth and claw" as the harsh ruling principles governing not only animals in their habitats but also humans in their societies (Crook, 1994; van der Dennen, 1990, 1995). Unless they were living in "splendid isolation" human societies must be considered to be universally belligerent:

Excluding a few simple groups such as the Esquimaux, inhabiting places where they are safe from invasion, all societies, simple and compound, are occasionally or habitually in antagonism with other societies; and, as we have seen, tend to evolve structures for carrying on offensive and defensive actions. . . . Already we have ample proof that centralized control is the primary trait acquired by every body of fighting men, be it hordes of savages, groups of brigands, or mass of soldiers. And this centralized control, necessitated during war, characterizes the government during peace. (Spencer, 1876, vol. 1, pt. 2, p. 576)

Spencer was also one of the first, after Ferguson, to discuss what we call today "ethnocentrism," or the phenomenon of in-group-out-group differentiation. In his *Principles of Ethics* (1892–93) he wrote: "Rude tribes and . . . civilized societies . . . have had continually to carry on an external self-defence and internal co-operation—external antagonism and internal friendship. Hence their members have acquired two different sets of sentiments and ideas, adjusted to these two kinds of activity. . . . A life of constant external enmity generates a code in which aggression, conquest and revenge, are inculcated, while peaceful occupations are reprobated. Conversely a life of settled internal amity generates a code inculcating the virtues conducing to a harmonious co-operation" (Spencer, 1892, vol. 1, p. 322). These two different sets of sentiments and ideas he called the "code of amity" and the "code of enmity." The theme of ethnocentrism cum xenophobia was later elaborated by Sumner (1906, 1911), who also coined the term "ethnocentrism."

LAMARCK

The claim that conflict was necessary for progress was thus by no means of exclusively Darwinian origin. Most theories of evolution at the time were progressionist in tone, and few were free of the assumption that ortholinear progress was intended to occur by nature or its creator (Bowler, 1986). Also, the idea of group selection was widely prevalent, as was the idea of inheritance of acquired characteristics, associated with the name of the much-maligned French naturalist Lamarck.

It is a curious irony of history that even Lamarck (1873) himself noted the possibility that conflict might have played a role in human evolution. He thought (anticipating Kortlandt's ideas on this matter by more than a century) that the first race of apes to gain a prehuman level of intelligence would have driven all its rivals into remote corners of the earth and would have prevented their further advancement (Bowler, 1986).

BAGEHOT

In his ideas on obedience and discipline, to be discussed later, Darwin acknowledged his debt to the journalist Walter Bagehot (Crook, 1994; Dawson, 1996), a disciple of Spencer who has been called the first self-proclaimed social Darwinist because of his explicit attempts to link Darwinian biological evolution and Spencerian social evolution, obvious in the title of his famous work *Physics and Politics, or, Thoughts on the Application of the Principles of "Natural Selection" and "Inheritance" to Political Society* (1872). In spite of this title, the principles of Bagehot, like those of Spencer, were Lamarckian rather than Darwinian and Bagehot attempted to reconstruct the pattern of growth of political civilization in the manner of evolutionary ethnologists like Lubbock and Tylor, from whom he drew some of his data. Spencer and his followers assumed that the inheritance of acquired characteristics was quite common, and hence all cultural evolution must be linked to biological evolution. They perceived human evolution as a rapid, largely purposeful and directed Lamarckian process (Peel, 1972; Dawson, 1996; van der Dennen, 1990, 1995).

Like Spencer, Bagehot assumed that war had been a major agent in this process: "Progress is promoted by the competitive examination of constant war" (p. 64). Like Spencer, he emphasized that warfare succeeds not so much through the genocidal elimination of rivals as by promoting superior organization and obedience to leadership: the most obedient and the tamest tribes are the strongest. "The compact [probably meaning the same as Spencer's "cohesive"] tribes win, and the compact tribes are the tamest. Civilisation begins, because the beginning of civilization is a military advantage'' (p. 47). There was no doubt in his mind that the "strongest killed out the weakest as they could." Progress, habitually thought of as a normal fact in human society, is actually a rare occurrence among peoples. Of the existence of progress in the military art there can be no doubt, however, nor of its corollary that the most advanced will destroy the weaker, that the more compact will eliminate the scattered, and that the more civilized are the more compact (Hofstadter, 1995). As we have seen, Spencer (1851) had already voiced a similar conviction. Wars also encouraged innovation and variability. Darwin noted Bagehot's argument that warfare could result in racial mixtures that begat "beneficial variability."

DARWIN

In his first notebook on transmutation of species (1837) Darwin denied that humans had instinctive urges or "hereditary prejudices" to conquer each other (Darwin, *Beagle Diary*, December 22, 1835; *First Notebook on Transmutation of Species*, p. 89 [July 1837–February 1838]; cited in Gruber & Barrett, 1974, pp. 184–187). *The Descent of Man* (1871) gives Darwin's fullest account of war and human instincts, and it borrowed more frankly than *On the Origin of Species* from current social theory: the anthropology of Maine and Lubbock and the social psychology of Spencer, Galton, and Bagehot (Jones, 1980; Moore, 1986; Crook, 1994). Though he did not talk in terms of instinctive pugnacity in humans, he clearly acknowledged that endemic warfare and genetic usurpation had been important selective forces in human history. E. O. Wilson says that *The Descent of Man* was, in this respect, "a remarkable model that foreshadowed many of the elements of modern group selection theory" (1975, p. 573). In this magnificent two-volume book Darwin argued that any social animal would acquire a moral sense once its intellect had developed to a certain level (cf. de Waal, 1996). The crucial problem was to explain why human intelligence had developed further than that of any other social animal, so as to permit the emergence of the moral sense.

Darwin was also aware of the ethnocentrism and xenophobia in social organisms. In animals living in groups, he wrote, "sympathy is directed solely towards members of the same community, and therefore towards known, and more or less loved members, but not to all the individuals of the same species" (1871, vol. 1, p. 163). As regards humans, Darwin stated that "the confinement of sympathy to the same tribe" must have been the rule. This was for him one of the chief causes of the low morality of the savages. "Primeval man," he argued, "regarded actions as good or bad solely as they obviously affected the welfare of the tribe, not of the species." Among the living tribal peoples, he added, "the virtues are practised almost exclusively in relation to the men of the same tribe," and the corresponding vices "are not regarded as crimes" if practiced on other tribes (1871, vol. 1, pp. 182, 179).

Apparently Darwin had formed the opinion that natural selection acts to a great extent through intergroup competition. In his own words, "Natural selection, arising from the competition of tribe with tribe, . . . would, under favourable conditions, have sufficed to raise man to his high position" (1871, vol. 1, p. 97). This competition, in his opinion, could be carried out through direct conflict, even in bloody forms. "When of two adjoining tribes one becomes less numerous and less powerful than the other," he maintained, "the contest is settled by war, slaughter, cannibalism, slavery, and absorption" (1871, vol. 1, p. 202). He was quite aware, however, that competition between groups had to be combined with cooperation within them (Melotti, 1987).

In *The Descent of Man*, Darwin explicitly suggested that warfare had been at one time an agent in human evolution. He observed that prehistoric humans as well as contemporary "savage" societies were constantly at war with each other. There was a never-ending fight for survival. The nobler sides of humanity, the moral faculties, sociality, and social sympathies, had their darker side, for they were forged in war and were, in turn, used to improve fighting and warfare (Darwin was indebted to Bagehot on this issue) (Crook, 1994). Intelligence and skill in hunting and fighting had been a critical factor, he asserted, in man's cultural progression: "We can see, that in the rudest state of society, the individuals who were the most sagacious, who invented and used the best weapons or traps, and who were best able to defend themselves, would rear the greatest number of offspring. As a tribe increases and is victorious, it is often still further increased by the absorption of other tribes" (Darwin, 1871, vol. 1, p. 196). From the remotest times tribes that were not only robust, but socially cohesive and skilled in organization, technology, and weaponry (Darwin called this "superiority in the arts"), tribes that included "a great number of courageous, sympathetic and faithful members, who were always ready to warn each other of danger, to aid and defend each other," had genetically usurped other tribes (1871, vol. 1, p. 199). As peoples highly endowed with social, but also military, discipline triumphed over others, "the social and moral qualities would tend slowly to advance and be diffused throughout the world" (p. 200).

In the following passages I liberally paraphrase Cronin's (1991) account. Darwin starts by considering competition between groups. If a group that has a high proportion of unselfishly devoted members comes into conflict with a group that has a high proportion of selfish members, it is easy to see that the group of altruists will triumph. Their discipline, fidelity, courage, and other such qualities will soon ensure victory. "Let it be borne in mind how all-important in the never-ceasing wars of savages, fidelity and courage must be. The advantage which disciplined soldiers have over undisciplined hordes follows chiefly from the confidence which each man feels in his comrades. Obedience, as Mr. Bagehot has well shown, is of the highest value, for any form of government is better than none. Selfish and contentious people will not cohere, and without coherence nothing can be effected. A tribe rich in the above qualities would spread and be victorious over other tribes" (Darwin, 1891, vol. 1, pp .199–200).

But the problem is to explain how unselfishness ever got off the ground in the first place: "[H]ow within the limits of the same tribe did a large number of members first become endowed with these social and moral qualities, and how was the standard of excellence raised?" (Darwin, 1891, vol. 1, p. 200). Unselfish members would not have the most offspring, Darwin realized—quite the contrary: "It is extremely doubtful whether the offspring of the more sympathetic and benevolent parents, or of those which were the most faithful to their comrades, would be reared in greater number than the children of selfish and treacherous parents of the same tribe. He who was ready to sacrifice his life . . . rather than betray his comrades, would often leave no offspring to inherit his noble nature. The bravest men, who were always willing to come to the front in war, and who freely risked their lives for others, would on an average perish in larger number than other men" (Darwin, 1891, vol. 1, p. 200). He conceded that the problem looks almost intractable: "Therefore it seems scarcely possible . . . that the number of men gifted with such virtues, or that the standard of their excellence, could be increased through natural selection, that is, by the survival of the fittest" (1891, vol. 1, pp. 200–201).

Darwin saw two ways out of the difficulty. One is reciprocity: "[E]ach man would soon learn that if he aided his fellow-men, he would commonly receive aid in return." But when he turned to his other solution, he seemed to suggest that individual sacrifice for the sake of the group can evolve because it pays off in intergroup competition: "It must not be forgotten that although a high standard of morality gives but a slight or no advantage to each individual man and his children over the other men of the same tribe, yet that an advancement in the standard of morality and an increase in the number of well-endowed men will certainly give an immense advantage to one tribe over another. There can be no doubt that a tribe including many members who, from possessing in a high degree the spirit of patriotism, fidelity, obedience, courage, and sympathy, were always ready to give aid to each other and to sacrifice themselves for the common good, would be victorious over most other tribes; *and this would be natural selection*" (Darwin, 1871, vol. 1, p. 166; italics added).

This passage, as Cronin (1991) comments, is puzzling. Darwin specifically said that he was now tackling the problem of how altruism gets established *within* the group; he took care to remind his readers "that we are not here speaking of one tribe being victorious over another." Yet he seemed to be speaking of exactly that. The problem with Darwin's theory is a common problem with group-selectionist arguments: how could group selection spread some trait that genic selection or individual selection on its own would not favor; "it is hard to imagine natural selection resolving a direct conflict between group welfare and individual welfare in favor of the group" (Wright, 1994, p. 707). The problem is still unsolved.

Like most social theorists of the nineteenth century, Darwin assumed that history was the record of man's unilineal progress from savagery to civilization, and his theory of social progress was similar in many respects to Spencer's (Greene, 1959). He wrote to Lyell in 1859: "I can see no difficulty in the most intellectual individuals of a species being continually selected; and the intellect of the new species thus improved, aided probably by effects of inherited mental exercise. I look at this process as now going on with the races of Man; the less intellectual races being exterminated'' (*Life and Letters*, II, p. 7: quoted in Greene, 1959, p. 320). The acquisition of tools, the use of fire, and the ''half-art and half-instinct'' of language would have stimulated the development of the brain and of the ''social sentiments.'' These, in turn, brought about group progress through imitation of the inventions and discoveries of the most gifted members of the group. The practice of each new art of war ''must likewise to some degree strengthen the intellect. If the invention were an important one, the tribe would increase in number, spread, and supplant other tribes. In a tribe thus rendered more numerous there would always be a greater chance of the birth of other superior and inventive members. If such men left children to inherit their mental superiority, the chance of the birth of still more ingenious members would be somewhat better, and in very small tribes decidedly better'' (see also Lumsden & Wilson's comment on this quote, 1983, pp. 760–65).

Meanwhile, social solidarity and common morality developed by a similar process of natural selection. Every advance in morality and social solidarity would have survival value for the group in which it occurred, Darwin added: ''At all times throughout the world tribes have supplanted other tribes: and as morality is one important element in their success, the standard of morality and the number of well-endowed men will thus everywhere tend to rise and increase'' (Darwin, 1871, vol. 1, p. 166). Tribes in which noble behavior was high would come to dominate those in which it was low. Thus group selection would favor tribes of brave and self-sacrificing individuals over the selfish and cowardly, whose undiscipline and lack of moral fiber would result in their succumbing should the groups come into conflict (Richards, 1987; Cronin, 1991).

What Darwin had in mind was clearly a model of group selection: groups constantly being supplanted, conquered, incorporated, or exterminated by other groups, the whole process being driven by intergroup competition. As Alexander (1974) suggested, humans are an excellent model for the kind of group selection Darwin envisioned.

WALLACE

As we have seen, the human capacity for culture and all that it entails (intelligence, language, morality, altruism, justice, and so on posed a real and serious problem to the early evolutionists (Greene, 1959; Cronin, 1991). Alfred Russell Wallace, the cofounder of classical Darwinian evolutionary theory, for example, became more and more convinced that natural selection could not possibly account for man's advanced mental attributes and the distinctly human brain. What is worse, some of these refined capacities would even have been a downright nuisance and a danger ''in the severe struggle he [the savage] has to carry on against nature and his fellow-man'' (Wallace, 1891, p. 192). Wallace

(1864) spoke of the "decreasing combative and destructive propensities" as an early factor in the development of man's social and sympathetic traits. From the very beginning, the various races of man would henceforth continue with very little morphological modification. In the mental and moral sphere, however, there would be a severe competition resulting in the spread of the best-endowed races and the gradual extinction of the less gifted ones. In this competition some races would "advance and become improved merely by the harsh discipline of a sterile soil and inclement seasons" (Greene, 1959, p. 377), while others, inhabiting tropical regions, would stagnate from lack of environmental challenge. In Wallace's opinion, the true "grandeur and dignity of Man" lay in his unique ability to transcend the law of natural selection that ruled the fates of all lower animals. Looking at the future, Wallace painted a dithyrambic picture of progressive cultural advance issuing from the steady predominance of "the more intellectual and moral" races over the "lower and more degraded" races in the clash of cultures. These ideas were clearly derived from Spencer (Greene, 1959; Peel, 1972).

HUXLEY

Darwin strongly disagreed with Wallace's view, as did, initially, Thomas Huxley. Eventually, however, Huxley (whose epithet "Darwin's bulldog" fitted him) came to believe that human morality must have been the result of cultural evolution only. The struggle for existence in nature, he held, is so profoundly red in tooth and claw that it would smother a developing morality at birth because morality must necessarily work *against* nature: "[S]ince law and morals are restraints upon the struggle for existence, the ethical process is in opposition to the principle of the cosmic process [the Hobbesian war of each against all], and tends to the suppression of the qualities best fitted for success in that struggle" (1894, p. 37). Huxley characterized these doctrines as the "gladiatorial" theory of existence, embodying an ethic of "reasoned savagery," as the weak were perpetually eliminated by the strong, or the most ruthless, or the most "aggressive" individuals, groups, or nations. The "weakest and stupidest went to the wall" (Huxley, 1888, p. 204). The toughest and shrewdest survived. Not surprisingly, Huxley rejected the myth of the noble savage (Greene, 1959; Richards. 1987: Cronin, 1997).

GUMPLOWICZ

In his *Der Rassenkampf*, Gumplowicz (1883) claimed to have found the genesis of human society in the primal conflicts of primitive hordes bonded together by intense feelings of kinship and instinctive pugnacity against rival hordes and aliens. According to him, mankind has a polygenist origin: each race comes from a distinct stock. Consequently, fierce antagonism and hatred have always existed among the human races and will continue to divide them till the end of time. "The perpetual struggle of the races is the law of history," Gumplowicz concluded, "while perpetual peace is nothing but the dream of the idealists" (van der Dennen, 1995, p. 220).

SUMNER

Continuing Spencer's functionalist line of thought was the American sociologist William Graham Sumner. "It is the competition of life," Sumner asserted, "which makes war, and that is why war always has existed and always will. It is the condition of human existence" (van der Dennen, 1990, p. 153). The foundation of human society, said Sumner (1911; Sumner & Keller, 1927), is the man/land ratio. Conflict over the means of subsistence is the underlying fact that shapes the nature of human society. When population presses upon the land supply, earth hunger arises, races of men move across the face of the world, militarism and imperialism flourish, and conflict rages. Where men are few and soil is abundant, the struggle for existence is less savage: "Wherever there is no war, there we find that there is no crowding" (van der Dennen, 1990, p. 153). Sumner emphasized group factors (including the binding power of folkways and mores) more strongly than did Spencer, and he was considerably less optimistic about the direction of evolutionary change (Hofstadter, 1955; Schellenberg, 1982).

Sumner (1906, 1911), who coined the term "ethnocentrism" for the dual code of conduct identified by Spencer, heavily implicated ethnocentrism and its collateral, xenophobia, in the evolution of warfare. In his *Folkways*, Sumner, echoing Spencer and Bagehot, wrote: "The exigencies of war with outsiders are what make peace inside, lest internal discord should weaken the in-group for war. The exigencies also make government and law in the in-group, in order to prevent quarrels and enforce discipline. Thus war and peace have reacted on each other, and developed each other, one within the group, the other in the inter-group relations. The closer the neighbors, the stronger they are, the intenser the warfare, and then the intenser is the internal organization and discipline of each" (van der Dennen, 1987, p. 4).

MARSHALL, INGE, AND HOBSON

The special blend of militant nationalism, pugnacious patriotism, and expansionist imperialism is called jingoism. In his book *The Psychology of Jingoism*, Hobson (1901) attributed it to man's "ancient savage nature" lurking somewhere in "sub-conscious depths," under the superstructure or thin veneer of civilization. He spoke of the "animal hate, vindictiveness, and bloodthirstiness" that lurked in the mildest-mannered patriot. Inge (1915) also traced the "perverted patriotism" that according to him caused war to "the inborn pugnacity of the *bête humaine*." These are by now familiar variants of Plato's "beast within."

Marshall (1898), also writing in the fin de siécle instinct-psychology tradition, included among his "tribal instincts of a higher type" the patriotic instinct, which was aroused by aggressive threats from neighboring nations, or by opportunity for tribal aggrandizement. He explained the self-sacrificial behavior of warriors in terms of biological sacrifice, a form of extreme altruism that paid off in "tribal advantage" (Crook, 1994).

KROPOTKIN

Far from being an uncritical devote of the "noble savage myth," as he is often represented, Kropotkin (1902) argued that the life of "savages" was split between two sets of actions and ethics, that applying within the group and that applying to outsiders: "Therefore, when it comes to a war the most revolting cruelties may be considered as so many claims upon the admiration of the tribe. This double conception of morality passes through the whole evolution of mankind, and maintains itself until now" (quoted in Crook, 1994, p. 109).

MCDOUGALL

McDougall (1908) developed the by now familiar theme that social solidarity and altruism arose from the need to organize for war. As group combat superseded individual fighting in early human history, he contended, success came to depend more and more upon the capacity of individuals for "united action, good comradeship, personal trustworthiness," and especially the ability "to subordinate their impulsive tendencies and egotistic promptings to the end of the group" (Crook, 1994, p. 135). McDougall adopted the concept of an "instinct of pugnacity" that had been propagated by William James (1890), who had also developed a theory of the coevolution of human hunting and warfare (hunting will be discussed later in this chapter).

PATRICK, READ, AND DAVIE

According to Patrick (1915), "man the fighting animal" had evolved out of conditions of incessant conflict between races, with the continuous extermination of the unfit. Survival in this perpetual struggle had been the product of order and mutual aid within groups, but with fear, hatred, and the rule of might prevailing between groups.

Read (1920) contended that hominids and early humans formed hunting packs that were predisposed to be aggressive toward all outsiders. "Wars strengthened the internal sympathies and loyalties of the pack or tribe and its external antipathies, and extended the range and influence of the more virile and capable tribes."

The next author to elaborate the theme of ethnocentrism in relation to primitive warfare was Davie (1929), who sketched a truly Hobbesian picture of the "savage" world, pointing out that the relation of primitive groups to one another, where agreements or special conditions have not modified it, is one of isolation, suspicion, hostility, and war, a *status hostilis*, if not a regular *status belli*. Yet within the tribe the common interest against every other tribe compels its members to unite for self-preservation. "Thus a distinction arises between one's own tribe—the 'in-group'—and other tribes—the 'out-group'; and between the members of the first peace and cooperation are essential, whereas their inbred sentiment toward all outsiders is one of hatred and hostility. These two relations are correlative" (van der Dennen, 1987, p. 6). Thus, Davie did not add much to Sumner's arguments in terms of theoretical sophistication. He did, however, summarize the then available ethnological evidence of ethnocentrism in preindustrial societies from all over the world.

KEITH

In his *Darwinism and What It Implies* (1928), Sir Arthur Keith wrote: "Competition is not confined to human rivalries and struggles; it pervades the whole kingdom of life; it is the basis of Darwin's doctrine of evolution; it has been, and ever will be, the means of progressive evolution" (pp. 18–19). Already during World War I Keith had begun to feel that tribal warfare played a role in evolution. He drew attention to the gregarious instinct in mankind and postulated that at a very early stage in our evolution, human beings had joined into competing tribal groups, of which the modern nations were the civilized equivalent. What is now called patriotism is the modern manifestation of this "herd instinct" that had once been essential for evolution (Bowler, 1986).

In his *Essays on Human Evolution* (1946) and *A New Theory of Human Evolution* (1948), he expounded the final version of his thesis. Nature had conditioned human instincts to ensure the maintenance of barriers between tribes, so that each could develop its full potential in isolation. Keith now openly criticized Elliot Smith's theory of a prehistoric golden age, insisting that the early tribes of mankind had always been in conflict with one another. Tribal competition had been vital both to the original differentiation and to the later development of the races.

Both cooperation and competition are deeply rooted in the evolutionary history of all social species. However, their highest development is found in man, the species that has risen highest in the scale of beings. For Keith, this was hardly a coincidence: indeed, the success of the human species had been secured by cooperation within groups and competition between them. This combination also favored the rise of both the "good" and the "bad" qualities of human beings. Both of these sets of opposite qualities, he stressed, have survival values, and only their balance can assure progressive, evolutionary change (Melotti, 1987; cf. the idea of Ferguson).

HUNTING

In *New Theory* Keith also pointed to another possible source of human belligerence. This was the hunting instinct, the importance of which had been noted by a number of earlier writers. Aristotle (*Politics*, 1.8) had already suggested that warfare originated as an offshoot of hunting, but later writers apparently paid little attention to this idea (Dawson, 1996). William James (1890), George Crile (1916), Harry Campbell (1917), and Carveth Read (1917, 1920) all anticipated Raymond Dart's and Robert Ardrey's "hunting hypothesis"—revived later by Washburn and his carnivorous psychology school—by emphasizing how man, the hunting animal, hunted in packs that intensified human combativeness. In war they simply hunted each other (some other predecessors of the hunting hypothesis are discussed in Cartmill [1994]). Hunting and warfare, both almost exclusively male activities in primitive groups, were explained as products of the sexual division of labor and associated with the ubiquitous human social patterns of male bonding and male supremacy (e.g., Washburn & Lancaster, 1968; cf. Tiger & Fox, 1971).

No one seems to have paid much attention to the predecessors of the hunting hypothesis. Cartmill (1994) notes that even Raymond Dart was not aware of their work. According to Bowler (1986), Dart assumed from the start that *Australopithecus* engaged in the hunting of at least small animals, although it was only later in his career that he came to see hunting as a major force in the shaping of human nature. In his *Adventures with the Missing Link* of 1959, Dart now spoke openly, in a convoluted prose with high protein content, of modern humanity's loathsome cruelty as a continuation of the blood lust of our carnivorous and cannibalistic ancestors (p. 201). Ardrey drove home this theme, becoming a leading exponent of what was sometimes known as the anthropology of aggression (Bowler, 1986).

Besides the hunting hypothesis, which still attracts many scholars, the two main themes regarding the origin and evolution of human warfare, that is, ethnocentrism and balance-of-power thinking, have found advocates after the emergence of (human) ethology, sociobiology, and human ecology. I shall briefly discuss the recent developments.

ETHNOCENTRISM

Spencer's, Sumner's, and Davie's views exerted considerable influence on subsequent social thought. For example, Murdock (1949), a prominent American anthropologist, repeated that "intergroup antagonism is the inevitable concomitant and counterpart of in-group solidarity" (1949, p. 83). Ethnocentrism is a major explanatory category in contemporary theories of primitive warfare. The founding father of modern sociobiology, E. O. Wilson (1978), drawing heavily from Leach's (1965) split-universe imagery, regards it as a culturally hypertrophied biological predisposition (which has evolved via kin selection): "Primitive men cleaved their universe into friends and enemies and responded with quick, deep emotion to even the mildest threats emanating from outside the arbitrary boundary. . . . The force behind most warlike policies is ethnocentrism, the irrationally exaggerated allegiance of individuals to their kin and fellow tribesmen'' (quoted in van der Dennen, 1987, p. 6).

Meyer (1977, 1990) also regards ethnocentrism and xenophobia as cultural hypertrophies. He argues that the extreme ethnocentrism on the primitive level sets preconditions for violent interaction, while specific conditions serve as triggers. He suggests that the basic motivation in violent encounters between members of distinct groups is not ''aggression'' impelled by some sort of drive, instinct, or appetite, but ''fear,'' fear generated by the position of the cultural ''we-group'' in a threatening universe made up of ''they-groups'' that endanger the social cosmos by their very existence. While any social system requires boundary maintenance and mutual identification of actors, man's condition as a psychocultural animal brings about hypertrophications of these needs.

According to Eibl-Eibesfeldt (1970, 1975, 1982), destructive intergroup aggression in humans depends, to a large extent, on ''cultural pseudospeciation.'' Owing to this process, first analyzed by Erikson (1966), ethnic groups tend to perceive one another as different species and to behave accordingly. Therefore, war appears to be the result of our innate repulsion for outsiders, not a simple effect of aggressive drive. By instigating intergroup aggression, pseudospeciation favors intragroup solidarity, friendship, and altruism.

The group, according to Eibl-Eibesfeldt (1982), is an important level of selection in humans, and many traits that are disadvantageous to the individuals are stabilized by selection at this level. He dwells upon indoctrinability and the inclination to polarize values. According to him, these traits, which create a readiness for self-sacrifice for the group, are difficult to explain by sheer individual selection and probably initially evolved for the defense of the family. The theme of ethnocentrism is an evolutionary context has been elaborated more recently by van den Berghe (1981); Reynolds, Falger, and Vine (1987); Shaw and Wong (1989); Peres and Hopp (1990); Vanhanen (1992); and Flohr (1994); see also van der Dennen (1995) for review. These authors all agree with the basic tenets of van der Berghe's theory of ethnocentrism as extended kin selection. Ethnic sentiments are extensions of kinship sentiments. Ethnocentrism is thus an extended form of nepotism—the propensity to favor kin over nonkin. Both cooperation and conflict in human societies follow a calculus of inclusive fitness.

BALANCE OF POWER

Lorenz (1966) and Monod (1971), discussing the evolutionary origins of human destructive intraspecific aggression, found its roots in the ''malignant'' intergroup intraspecific selection that had occurred, in their opinion, in the late Stone Age. In fact, they believed that when man had more or less mastered the inimical forces of his extraspecific environment (such as hunger, cold, and predatory animals), war became the main selective factor in human evolution (cf. Alexander's ideas to be discussed later on).

This rationale for Paleolithic intergroup conflict was particularly emphasized by Bigelow (1969, 1972, 1975), who, in contrast to Lorenz and Monod, maintained that intergroup intraspecific selection had beneficial effects on human evolution. In particular, it brought about the relatively rapid tripling in size of the emergent human brain (a theme also elaborated by Pitt, 1978). Bigelow's main thesis is that intergroup competition conferred a strong selective advantage on cooperation within and between groups. He points out that cooperation for conflict was imperative for sheer survival and that it also had important spinoffs in peacetime. Throughout man's evolutionary history the more successful groups were those that were better organized, and especially those whose members had the intelligence and foresight to cooperate with other groups. Even the highest human virtues were brought forth in response to the dangers threatened by the lowest of human qualities. These apparent opposites, for him, "were not even two different sides of the same coin, but were as intimately interdependent as our brains and hearts are" (1969, p. 7). Though the aggressive competition and cooperative behavior seem to be opposing tendencies, he emphasizes, they are highly interdependent parts of a single system and have evolved together (cf. the views of Ferguson, Sumner, Davie, and Keith). Remarkably, Bigelow did not deduce that humans have some kind of innate propensity (such as an aggressive "instinct" or disposition) to intergroup violence. On the contrary, he suggested that exactly because war had selected for cooperation within and between groups, and hence for communication and intelligent self-control, humans are now in a favorable position to prevent and abolish war (Melotti, 1987).

At some time in the course of human evolution, other human groups replaced the big feline predators as the major threat to survival, and Bigelow believes that this transition marked the beginning of distinctively human evolution. The transition was probably very gradual, but at the dawn of history other humans had clearly become man's most dangerous enemy.

In the case of human evolution, a very powerful and relentless selective force was acting against the smallest-brained humans, even after their brains were twice as large as those of any other primate. Had this not been so, the average size of the human brain would not have been doubled within such a relatively brief span of evolutionary time. This force was acting on the human line alone, for in no other primate species were the smaller-brained individuals placed at such a severe selective disadvantage. The selective force that produced distinctively human evolution seems to have been contained within the human species itself.

Bigelow believes that it is unlikely that our ancestors would have lived in peace throughout the Pleistocene, only to begin, very suddenly, to fight savage wars at the dawn of history. Therefore, it is reasonable to assume that intergroup competition was a selective factor before our ancestors became human, and long before they became big-game hunters. It is important to note, Bigelow asserts, that intergroup competition can be an important selective force without high levels of violence and killing.

Wilson suggested that warfare might have continued to serve this adaptive function well into historic times: "By current theory, genocide or genosorption strongly favoring the aggressor need take place only once every few generations to direct evolution. This alone could push truly altruistic genes to a high frequency within the bands. . . . The turnover of tribes and chiefdoms estimated from atlases of early European and Mideastern history suggests a sufficient magnitude of differential group fitness to have achieved this effect" (1975, pp. 573–574).

Alexander (1979, 1987, 1990; Alexander & Borgia, 1978), another ancestor of modern evolutionary biology, even reasoned that especially *human* groups would be expected to have been amenable to powerful group selection: "For two reasons human social groups represent an almost ideal model for potent selection at the group level. First, the human species is (and possibly always has been) composed of competing and essentially hostile groups that frequently have not only behaved toward one another in the manner of different species, but also have been able quickly to develop enormous differences in reproductive and competitive ability because of cultural innovation and its cumulative effects. Second, human groups are uniquely able to plan and act as units, to look ahead and purposely carry out actions designed to sustain the group and improve its competitive position" (Alexander and Borgia, 1970, 1978, p. 470).

Alexander (1979) agrees with Bigelow that intergroup aggressive competition was a prime mover in human evolution and that it selected for intragroup altruism as well as for other forms of complex behavior. He argues that warfare must have had a long period of preadaptation: hominids first formed small bands for defense against predators, then increasingly turned to the hunting of game; at some point the primary purpose of group organization became defense against other hominid bands. At the last stage escalation became necessary to achieve a margin of safety, setting up the self-perpetuating chain reaction known as the balance of power. Alexander revives Spencer's notion that warfare explains why large groups are advantageous and why human groups have tended to increase in size throughout history.

The attribute that could explain human uniqueness, he argues, was an increasing prominence of direct intergroup competition—when the hominids or humans became "ecologically dominant" (Slurink, 1993, 1994; van der Dennen, 1995)—leading to an overriding significance in balances of power among competing social groups, in which social cooperativeness and eventually culture became the chief vehicle of competition. Alexander and Borgia (1978) had already underlined that human individuals must have begun to win the reproductive race by cooperating to compete. Culture continually rebuilds the differences between neighboring human populations. Culture is the great unbalancer (or "pseudospeciator") that reinforces human tendencies to live and compete in groups and to engage in an unusual (and unusually ferocious) group-against-group competition (Melotti, 1987).

Alexander also endorses the war-makes-states theory: the rise of the nation-state "occurred as a result of the interactions of neighboring competitive and hostile groups as they expanded their alliances and cemented unities in a balance of power race" (Alexander, 1979, p. 249). According to Tiger and Fox (1971), wars arrived on the scene when tools were perfected into weapons, and the expanding and convoluting human brain was able to cope with organization, on the one hand, and the development of the categories of friend and foe, on the other. Baer and McEachron (1982) also include weapons technology in their account of the evolution of intergroup conflict. With the development of in-creasingly dangerous weapons, the groups became more closed and aggressive. Intergroup conflict rose to higher levels, and this led to a positive biocultural feedback system. Intergroup conflict selected for greater intelligence (greater ability to learn, communicate, plan, have foresight, and the like), and such en-hanced mental capacity in turn created better weapons and made the group itself a better fighting unit. Thus the selective pressures for conflict increased in force, and the cycle began again, at a higher level. This process probably operated in other ways as well. Closing the hominid groups enhanced the genetic relatedness within the groups and decreased it between them. This must have fostered ag-gressive xenophobia.

In Corning's (1983) evolutionary scenario, an ever-widening system of syn-ergistically interacting biological and social traits enabled the evolving hominids to cooperate and compete with one another in meeting survival and reproductive needs. Thus the key to human successful adaptation was competition through cooperation or cooperation for competition, a behavioral system that must have been effective long before the recent millennia for which we have evidence of overt, direct conflict between human groups. According to Corning, egoistic cooperation, kin selection, and functional group selection concurred to reinforce individual selection rather than working against it.

FINAL COMMENTS

The themes of hunting, ethnocentrism, and balance-of-power thinking do not, of course, exhaust the Darwinian legacy in relation to the evolutionary origin of war. Other themes, also tangentially touched upon by the authors dealt with above, are hominid/human bipedalism, exponential brain expansion, and mating system; (hyper)sociality and altruism; intergroup agonistic behavior in group-territorial organisms (especially social carnivores and primates), lethal male raid-ing in chimpanzees (*Pan troglodytes*)—and the distinct possibility that the chimpanzee–human common ancestor already had this lethal male raiding pattern in its behavioral repertoire. These and similar considerations have driven Slurink (1993, 1994) and van der Dennen (1995) to develop a more or less integral scenario of the evolution of hominid/human warfare which emphasizes phylogenetic continuity between humans and nonhuman primates, and which does not stipulate that war is a one-time cultural invention.

Van der Dennen's (1995) investigation of the evolutionary origins of intergroup conflict in social carnivores and primates identified (a) the capability to form polyadic coalitions (selfish and opportunistic cooperation with more than one conspecific) as the necessary precondition, which in turn required (b) sociality; (c) Machiavellian intelligence; and (d) proto-ethnocentrism. Proto-ethnocentrism is supposed to imply some kind of group identity, that is, the ability to recognize ingroup versus outgroup members, to discriminate between these categories, and to preferentially treat ingroup members to positive reciprocal (altruistic) interactions such as protection, nepotism, and sharing of resources. It also outlines the phylogenetic and socio-ecological principles governing group formation, ingroup altruism, outgroup antagonism, and intergroup agonistic behavior.

In this enterprise the authors may feel like true and proud heirs of Darwin.

REFERENCES

Alexander, R. D. (1974). The evolution of social behavior. *Annual Review of Ecology and Systematics, 5*, 325–383.

———. (1979). *Darwinism and human affairs*. Seattle: University of Washington Press.

———. (1987). *The biology of moral systems*. New York: Aldine de Gruyter.

———. (1990). *How did humans evolve? Reflections on the uniquely unique species*. Special Publication No. 1. Ann Arbor: Museum of Zoolooy, University of Michigan.

Alexander, R. D., & Borgia, G. (1978). Group selection, altruism, and the levels of hierarchical organization of life. *Annual Review of Ecology and Systematics, 9*, 449–474.

Ardrey, R. (1976). *The hunting hypothesis*. New York: Atheneum.

Baer, D., & McEachron, D. L. (1982). A review of selected sociobiological principles: Application to hominid evolution I: The development of group social structure. *Journal of Social and Biological Structures, 5* (1), 69–90.

Bagehot, W. (1872). *Physics and politics: or, Thoughts on the application of the principles of "natural selection" and "inheritance" to political society*. London: Henry S. King & Co.; New York: Appleton.

Bigelow, R. (1969). *The dawn warriors: Man's evolution toward peace*. Boston: Little, Brown.

———. (1972). The evolution of cooperation, aggression, and self-control. *Nebraska Symposium on Motivation, 20*, 1–57.

———. (1975). The role of competition and cooperation in human aggression. In M. A. Nettleship, R. D. Givens, & A. Nettleship (Eds.), *War, its causes and correlates* (pp. 235–261). The Hugue: Mouton.

Blainey, G. (1988). *The great sea-saw: A new view of the western world, 1750–2000*. London: Macmillan.

———. (1986). *Theories of human evolution: A century of debate, 1844–1944*. Oxford: Blackwell.

Campbell, H. (1917). *The biological aspects of warfare.* London: Balliere.

Cartmill, M. (1994). *A view to a death in the morning: Hunting and nature through history.* Cambridge, MA: Harvard University Press.

Comte, A. (1830–42). *Cours de philosophie positive.* Paris: Bachelier.

Corning, P. A. (1983). *The synergism hypothesis: A theory of progressive evolution.* New York: McGraw-Hill.

Crile, G. M. (1916). *A mechanistic view of peace and war.* New York: Macmillan.

Cronin, H. (1991). *The ant and the peacock: Altruism and sexual selection from Darwin to today.* Cambridge: Cambridge University Press.

Crook, P. (1994). *Darwinism, war, and history: The debate over the biology of war from the "Origin of Species" to the First World War.* New York: Cambridge University Press.

Dart, R. A. (1959). *Adventures with the missing link.* New York: Harper & Brothers.

Darwin, C. R. (1871). *The descent of man, and selection in relation to sex.* London: Murray.

————. (1873). *The expression of the emotions in man and animals.* London: Murray.

Darwin, C. (1891). *The descent of man, and selection in relation to sex* (second revised and augmented edition). London: John Murray.

Darwin, F. (Ed.) (1887). *The life and letters of Charles Darwin.* London: John Murray.

Davie, M. R. (1929). *The evolution of war: A study of its role in early societies.* New Haven: Yale University Press.

Dawson, D. (1996). The origins of war: Biological and anthropological theories. *History and Theory, 35* (1), 1–28.

De Waal, F. B. M. (1996). *Good natured: The origins of right and wrong in humans and other animals.* Cambridge, MA: Harvard University Press.

Eibl-Eibesfeldt, I. (1970). *Liebe und HaB: Zur Naturgeschichte elementarer Verhaltensweisen.* Munich: Piper Verlag.

————. (1975). *Krieg und Frieden aus der Sicht der Varhaltensforschung.* Munich: Piper Verlag.

————. (1982). Warfare, man's indoctrinability, and group selection. *Zeitschrift film Tierpsychologie, 60* (3), 177–198.

Erikson, E. H. (1966). Ontogeny of ritualisation in man. *Philosophical Transactions of the Royal Society of London, 251B,* 337–349.

Ferguson, A. (1767). *An essay on the history of civil society.* Reprint. Chicago: Aldine, 1966.

Flohr, A. K. (1994). *Fremdenfeindlichkeit: Biosoziale Grundlagen von Ethnozentrismus.* Opladen: Westdeutscher Verlag.

Greene, J. C. (1959). *The death of Adam: Evolution and its impact on Western thought.* Ames: Iowa State University Press.

Gruber, H. E., & Barrett, P. H. (1974/1987). *Darwin on man: A psychological study of scientific creativity, together with Darwin's early and unpublished notebooks.* London: Wildwood House.

Gumplowicz, L. (1883). *Der Rassenkampf.* Innsbruck: Wagner Verlag.

Hobbes, T. (1651). *Leviathan; or, The matter, form, and power of a common-wealth, ecclesiastical and civil.* London: Crooke.

Hobson, J. A. (1901). *The psychology of jingoism.* London: G. Richards.

Hofstadter, R. (1955). *Social Darwinism in American thought, 1860–1915.* Boston: Beacon Press.

Huxley, T. H. (1863). *Evidence as to man's place in nature.* London: Williams & Norgate.

Huxley, T. H. (1888). The struggle for existence in human society. *Nineteenth Century, 23,* 161–80 (reprinted in *Collected Essays.* London: Macmillan, 1893–1894).

———. (1894). *Evolution and ethics, and other essays.* New York: D. Appleton & Co.

Inge, W. R. (1915, July). Patriotism. *Quarterly Review, 224,* 73–75, 83.

James, W. (1890). *The principles of psychology.* New York: Holt.

Jones, G. (1980). *Social Darwinism and English thought.* Brighton: Harvester Press.

Keith, A. (1928). *Darwinism and what it implies.* London: Watts.

Keith, A. (1946). *Essays on human evolution.* London: Watts.

———. (1948). *A new theory of human evolution.* London: Watts.

Kortlandt, A. (1972). *New perspectives on ape and human evolution.* Amsterdam: Stichting voor Psychobiologie.

Kropotkin, P. (1902). *Mutual aid: A factor of evolution.* London.

Lamarck, J. B., de Monet, Chevalier de. (1873). *Philosophie zoologique* (new ed.). Paris: Savy.

Lankester, E. R. (1905). *Nature and man.* Oxford: Clarendon Press.

Leach, E. R. (1965). The nature of war. *Disarmament and Arms Control, 3,* 165–183.

Lorenz, K. (1966/1967). *On aggression.* London: Methuen; New York: Bantam Books (originally, *Das sogenannte Böse: Zur Naturgeschichte der Aggression!* Wien: Borotha Schoeler Verlag, 1963).

Lumsden, C. J., & Wilson, E. O. (1983). *Promethean fire: Reflections on the origin of mind.* Cambridge, MA: Harvard University Press.

Maine, H. J. S. (1861). *Ancient law, its connections with the early history of society and its relation to modern ideas.* London: John Murray.

Malthus, T. R. (1798). *An essay on the principle of population.* London: Johnson.

Marshall, H. R. (1898). *Instinct and reason.* New York: Macmillan.

———. (1915). *War and the ideal of peace.* New York: Diffield.

McDougall, W. (1908). *An introduction to social psychology.* London: Methuen; Boston: Luce.

———. (1927). *Janus: The conquest of war.* London: Kegan Paul, Trench, Trubner & Co.

———. (1964). The instinct of pugnacity. In L. Bramson & G. Goethals (Eds.), *War: Studies from psychology sociology, anthropology* (pp. 33–43). New York: Basic Books.

McLennan, J. F. (1886). *Studies in ancient history.* London: Bernard Quaritsch.

Melotti, U. (1985). Competition and cooperation in human evolution. *Mankind Quarterly, 25,* 323–351.

———. (1987). Ingroup/outgroup relations and the issue of group selection. In V. Reynolds, V. S. E. Falger, & I. Vine (Eds.), *The sociobiology of ethnocentrism* (pp. 94–111).

———. (1990). War and peace in primitive human societies. In J. M. G. van der Dennen & V. S. E. Falger (Eds.), *Sociobiology and conflict* (pp. 241–246).

Meyer, P. (1977). *Kriegs- und Militärsoziologie.* Munich: Goldmann.

Meyer, P. (1990). Human nature and the function of war in social evolution: A critical review of the naturalistic fallacy. In J. M. G. van der Dennen & V. S. E. Falger (Eds.), *Sociobiology and conflict: Evolutionary perspectives on competition, cooperation, violence and warfare* (pp. 227–240). London: Chapman & Hall.

Monod, J. (1971/1975). *Chance and necessity: An essay on the natural philosophy of modern biology*. New York: Knopf.

Moore, J. (1986). Socializing Darwinism. *Radical Science, 20*, 47–48.

Murdock, G. P. (1949). *Social structure*. New York: Macmillan.

Patrick, G. T. W. (1915). The psychology of war. *Popular Science Monthly, 87*, 155–168.

Peel, J. D. Y. (Ed.).(1972). *Herbert Spencer on social evolution*. Chicago: University of Chicago Press.

Peres, Y., & Hopp, M. (1990). Loyalty and aggression in human groups. In J. M. G. van der Dennen & V. S. E. Falger (Eds.), *Sociobiology and conflict* (pp. 123–130).

Pitt, R. (1978). Warfare and hominid brain evolution. *Journal of Theoretical Biology, 72*, 551–575.

Read, C. (1905). *The metaphysics of nature*. London: A. and C. Black.

Read, C. (1917). On the differentiation of the human from the anthropoid mind. *British Journal of Psychology, 8*, 395–422.

McLothi, P. S., *The sociobiology of ethnocentrism: Evolutionary dimensions of xenophopia, discrimination, racism and nationalism*. London: Croom Helm.

McLothi, P. S. & Hopp, M. *Sociobiology and conflict: Evolutionary perspectives on compatision, cooperation, violence and warfare*. London: Chapman and Hall.

———. (1920). *The origin of man and his superstitions*. Cambridge: Cambridge University Press.

Reynolds, V., Falger, V. S. E., & Vine, I. (Eds.). (1987). *The sociobiology of ethnocentrism: Evolutionary dimensions of xenophobia, discrimination, racism, and nationalism*. London: Croom Helm.

Richards, G. (1987). *Human evolution: An introduction for the behavioral sciences*. London: Routledge & Kegan Paul.

Schellenberg, J. A. (1982). *The science of conflict*. Oxford: Oxford University Press.

Shaw, R. P., & Wong, Y. (1989). *Genetic seeds of warfare: Evolution, nationalism, and patriotism*. London: Unwin Hyman.

Slurink, P. (1993). Ecological dominance and the final sprint in hominid evolution. *Human Evolution, 8* (4), 265–273.

———. (1994). Causes of our complete dependence on culture. In R. A. Gardner, P. T. Gardner, B. Chairelli, and F. X. Puooij. (Eds.), *The ethological roots of culture* (pp. 461–474). Dordrecht: Kluwer Academic Publishers.

Smillie, D. (1995). Darwin's two paradigms: An "opportunistic" approach to natural selection theory. *Journal of Social and Evolutionary Systems, 18*, 231–256.

Smith, G. Elliot. (1933). *The diffusion of culture*. London: Watts.

Spencer, H. (1851). *Social statics*. London: John Chapman.

———. (1873–81). *Descriptive sociology* (8 vols.). New York: Appleton.

———. (1876). *The principles of sociology* (3 Vols.). London: Williams & Norgate.

———. (1892–93). *The principles of ethics*. London: Williams & Norgate.

———. (1902). *Social statics, abridged and revised, together with "the man versus the state."* London: Williams & Norgate.

Sumner, W. G. (1906). *Folkways: A study of the sociological importance of usages, manners, customs, mores, and morals*. Boston: Ginn.

———. (1911). *War and other essays*. New Haven: Yale University Press.

———. (1913). *Earth-hunger and other essays*. New Haven: Yale University Press.

———. (1963). *Social Darwinism: Selected essays of William Graham Sumner.* Englewood Cliffs, NJ: Prentice-Hall.

———. (1964). War. In L. Bramson & G. Goethals (Eds.), *War: Studies from psychology, sociology, anthropology* (pp. 205–228). New York: Basic Books.

Sumner, W. G., & Keller, A. G. (1927). *The science of society* (4 Vols.). New Haven: Yale University Press.

Tiger, L., & Fox, R. (1971). *The imperial animal.* New York: Holt, Rinehart & Winston.

van der Berghe, P. L. (1981). *The ethnic phenomenon.* New York: Elsevier.

van der Dennen, J. M. G. (1975). Population dynamics and violence. Unpublished manuscript, Polemological Institute, Groningen University.

———. (1983). *Theories of war causation: Vol. 6. The demographic-ecological school.* Unpublished manuscript, Polemological Institute, Groningen University.

———. (1990). Origin and evolution of "primitive" warfare. In J. M. G. van der Dennen & V. S. E. Falger (Eds.), *Sociobiology and conflict* (pp. 149–188). *Sociobiology and conflict: Evolutionary perspectives on competition, cooperation, violence and warfare.* London: Chapman and Hall.

———. (1995). *The origin of war: The evolution of a male-coalitional reproductive strategy.* Groningen: Origin Press.

van der Dennen, J. M. G. (1987). Ethnocentrism and in-group/out-group differentiation: A review and interpretation of the literature. In V. Reynolds, V. Falger, & I. Vine (Eds.), *The sociobiology of ethnocentrism: The evolutionary dimensions of xenophobia, discrimination, racism and nationalism* (pp. 1–47). London: Croom Helm.

Vanhanen, T. (1992). *On the evolutionary roots of politics.* New Delhi: Sterling Publishers.

Wallace, A. R. (1864). Origin of the human races and the antiquity of man deduced from the theory of "natural selection." *Journal of the Anthropological Society, 2,* 158–170.

———. (1870). *Contributions to the theory of natural selection.* London: Macmillan.

———. (1891). *Natural selection and tropical nature: Essays on descriptive and theoretical biology.* London: Macmillan.

Washburn, S. L., & Lancaster, C. S. (1968). The evolution of hunting. In R. B. Lee & I. DeVore (Eds.), *Man the hunter* (pp. 293–303). Chicago: Aldine Atherton.

Wilson, E. O. (1975). *Sociobiology: The new synthesis.* Cambridge, MA: Belknap Press of Harvard University Press.

———. (1978). *On human nature.* Cambridge, MA: Harvard University Press.

Wright, R. (1994). *The moral animal.* New York: Pantheon.

11

Ethnic Conflicts and Ethnic Nepotism

Tatu Vanhanen

Ethnic conflicts seem to be common in all ethnically plural societies and across all cultural boundaries, although such conflicts are not always termed "ethnic." Conflicting ethnic groups may be national, tribal, racial, religious, linguistic, cultural, or communal groups. Conflicts between castes and with indigenous people belong to the same category. Experience shows that political and economic interest conflicts become easily organized along ethnic lines in ethnically divided societies. In extreme cases, ethnic conflicts paralyze societies and culminate in violent confrontations and civil wars. Social scientists have sought theoretical explanations for ethnic conflicts. Why are ethnic conflicts so common? Why has it been so difficult for ethnic groups to live in harmony with each other and to agree on the sharing of scarce resources? Further, how are we to accommodate ethnic conflicts? Social scientists have not yet agreed on a satisfactory theoretical explanation for ethnic conflicts (see, for example, Rabushka & Shepsle, 1972; Horowitz, 1985; Smith, 1987; Rupesinghe, 1988; Gurr, 1993; Gurr & Harff, 1994; McGarry & O'Leary, 1994; Glickman, 1995; Hutchinson & Smith, 1996). Many researchers argue that ethnic conflicts are cultural conflicts that need a separate explanation in each case. Consequently, there cannot be any general theoretical explanation for them. Other researchers have attempted to formulate more general explanations for ethnic conflicts. My intention is to derive a theoretical explanation for the persistence of ethnic conflicts from a sociobiological kin-selection theory and to test it by empirical evidence.

ETHNIC NEPOTISM

I have argued in my books *Politics of Ethnic Nepotism: India as an Example* (Vanhanen, 1991) and *On the Evolutionary Roots of Politics* (1992) that it is possible to deduce a cross-culturally valid ultimate explanation for ethnic conflicts from a Darwinian interpretation of politics and from the sociobiological theory of kin selection. According to my Darwinian interpretation of politics, the struggle for scarce resources is the central theme of politics everywhere. This central and universal theme of politics can be derived from the Darwinian theory of evolution by natural selection. According to that theory, all organisms have to struggle for survival because we live in a world of scarcity in which all species are able to produce much more progeny than can be supported by the available resources. The permanent discrepancy between the number of individuals and the means of existence makes the struggle for survival inevitable and omnipresent (see Darwin, 1859, pp. 114–172; Dobzhansky, Ayala, Stebbing, Valentine, 1977, pp. 96–99; Mayr, 1982, pp. 479–480). Politics is one of the forums of this struggle. The evolutionary roots of politics lie in the necessity to solve conflicts for scarce resources by some means. We should understand that universal competition and struggle in human societies are an inevitable consequence of the fact that we live in a world of scarcity and that we are programmed to further our own survival by all available means. The Darwinian theory explains why this must be so. Thus it provides an ultimate evolutionary explanation for the necessity and universality of conflicts in all human societies (Vanhanen, 1992, pp. 24–27).

But why are so many conflicts taking place along ethnic lines? I think that we can deduce an answer to this question from the assumption that different kinds of behavioral predispositions evolved in the struggle for scarce resources. One of the politically relevant behavior strategies that evolved to help their users in the struggle for scarce resources seems to be our universal tendency to nepotism. The term ''nepotism'' refers to favoritism toward kin and can be explained by the sociobiological theory of kin selection or inclusive fitness, which was originally formulated by W. D. Hamilton (1964). He looked at evolution from the gene's point of view and realized that natural selection tends to maximize inclusive fitness, the survival of one's genes through one's own offspring and through relatives who have the same genes. It means that the logic of evolution presupposes individual selfishness together with favoritism toward relatives. Animals that behave in such a way have the best chances to reproduce their genes (see Wilson, 1975, pp. 117–118; Dawkins, 1976; Maynard Smith, 1979, pp. 180–182; Alexander, 1980, pp. 45–56; Barash, 1982, pp. 67–74; Daly & Wilson, 1983, pp. 45–50). Pierre L. van den Berghe (1981, p. 7) says that ''whenever cooperation increases individual fitness, organisms are genetically selected to be nepotistic, in the sense of favoring kin over nonkin, and close kin over distant kin.''

Ethnic groups can be perceived as extended kin groups. The members of an ethnic group tend to favor their group members over nonmembers because they are more related to their group members than to the remainder of the population. The members of the same ethnic group tend to support each other in conflict situations. Van den Berghe extended kin selection to ethnic groups. "My basic argument is quite simple: ethnic and racial sentiments are extension of kinship sentiments. Ethnocentrism and racism are thus extended forms of nepotism— the propensity to favor kin over nonkin" (van den Berghe, 1981, p. 18; see also Dunbar, 1987; Meyer, 1987; Silverman, 1987). This is an important insight. Our tendency to favor kin over nonkin has extended to include large linguistic, national, racial, religious, and other ethnic groups. Van den Berghe used the term "ethnic nepotism" to describe this phenomenon (van den Berghe 1981, p. 222; see also van den Berghe, 1996; Rushton, 1986, pp. 145–146). I have used "ethnic nepotism" to refer to this kind of nepotism at the level of extended kin groups (Vanhanen, 1991).

From the perspective of ethnic nepotism, it does not matter what kinds of kin groups are in question. The crucial characteristic of an ethnic group is that its members are genetically more closely related to each other than to the members of other groups. Therefore, in my study "ethnic group" refers not only to racial, tribal, and national groups, but also to linguistic groups, castes, and old religious communities (cf. Horowitz, 1985, pp. 41–54; Riggs, 1995). A problem with this definition is that people are related to each other at many levels, from the level of nuclear family to the level of *Homo sapiens*. Consequently, ethnic groups are never absolutely distinct and exclusive. Any level can provide a basis for ethnic nepotism. It depends on the situation as to what level of ethnic cleavages becomes politically relevant.

I assume that our behavioral predisposition to ethnic nepotism, because of its evolutionary roots, is shared by all human populations. This assumption makes it reasonable to present universal hypotheses on political consequences of ethnic nepotism. I hypothesize that (1) significant ethnic divisions tend to lead to ethnic interest conflicts in all societies, and (2) the more a society is ethnically divided, the more political and other interest conflicts tend to become channeled along ethnic lines.

In principle, these hypotheses are testable, which means that it is possible to falsify them by empirical evidence. The first hypothesis claims that because ethnic nepotism is shared by all human populations, ethnic tensions and conflicts can be expected in all ethnically divided societies. It would be possible to falsify this hypothesis by empirical evidence that shows that there is no systematic relationship between the extent of ethnic divisions and the emergence of ethnic conflicts. The second hypothesis complements the first one by claiming that the degree of ethnic conflicts depends on the degree of ethnic cleavages. In other words, the more ethnic groups differ from each other genetically, the higher the probability and intensity of conflicts between them. It would be possible to falsify this hypothesis by empirical evidence that shows that there is no clear relationship between the degree of ethnic divisions and the degree of ethnic conflicts.

However, I have to complement these hypotheses by a reservation. The relationship between ethnic divisions and ethnic conflicts does not need to be automatic and uniform. There are intervening factors that may increase or decrease the intensity of conflicts. I assume that political and social institutions constitute important intervening factors. Depending on their nature, they can help to accommodate ethnic interest conflicts, or they can deepen them. It can be further assumed that political institutions based on equality and reciprocity are better adapted to accommodate ethnic interest conflicts than institutions based on hegemonic and unequal relations between ethnic groups. This is the case because we may have an evolved tendency to reciprocity (for reciprocity, see Trivers, 1981, 1985). Reciprocity evolved in the struggle for existence in the same way as nepotism and other politically relevant behavioral predispositions.

In politics, reciprocity appears in numerous forms from the level of individual relationships to the relations between social groups, nations, and international alliances. The basic rules remain the same in all variations: we are ready to perform services and to help others if we can expect that the receiver will reciprocate. We are disposed to accept balanced reciprocal relationships and to resist unequal reciprocity. I assume that these basic rules of reciprocity are cross-culturally similar in all human societies.

It is also reasonable to assume that the same rules of reciprocity apply to the relationships between ethnic groups. Consequently, political institutions based on balanced reciprocity and equality between ethnic groups may provide a better institutional framework for harmonious relations between ethnic groups than institutions biased to favor some groups and discriminate against some others. This assumption leads me to hypothesize that (3) at the same level of ethnic cleavages, the degree of ethnic conflicts tends to be higher in societies in which political institutions are biased to favor some ethnic groups and discriminate against some others than in societies in which political institutions are based on balanced reciprocity and equality between ethnic groups. In principle, this hypothesis is testable, but it presupposes detailed information on the nature of institutional arrangements. Such information is not easily available, and I have not yet gathered enough of it to test this hypothesis.

The evolutionary theory of ethnic conflicts just outlined differs from other theories of ethnicity and ethnic conflicts in some important respects. First, according to this theory, ethnic groups are basically kinship groups, although the level of kinship and, consequently, the size of a kinship group can vary greatly depending on the situation. In this respect, I agree more with scholars who emphasize the primordial origins of ethnic groups than with scholars who regard ethnic groups as easily changeable and malleable cultural groups. However, the assumption about the primordial origins of ethnic groups does not mean that they are assumed to be unchangeable and eternal. What type of ethnic groups become politically relevant depends on the political and social situation. I emphasize the crucial importance of genetic distance between ethnic groups. From this perspective, there are great differences in the degree of ethnic cleavages. The cleavage between blacks and whites, for example, is much deeper than the cleavage between white Catholics and Protestants, although the latter groups can also be regarded as ethnic groups if people have belonged to these different religious communities over several centuries and if intermarriages between Catholics and Protestants have remained relatively rare. Blacks and whites constitute clearly different ethnic groups, whereas it is not always self-evident whether we should regard Catholics and Protestants as different ethnic groups, although they certainly can be regarded as different cultural groups.

Second, my theoretical explanation of ethnic conflicts differs from many other theoretical explanations because it is derived from the principles of evolutionary theory and from a Darwinian interpretation of politics. According to my theory, interest conflicts between ethnic groups are inevitable because ethnic groups are genetic kinship groups and because the struggle for existence concerns the survival of our genes through our own and our relatives' descendants. Therefore, it has been rational for relatives to ally with each other in political and other struggles for scarce resources and survival. Our behavioral predisposition to ethnic nepotism evolved in the struggle for existence because it was rational and useful. It is reasonable to assume that ethnic nepotism is equally shared by all human populations. Consequently, all human populations and ethnic groups have approximately equal potential to resort to ethnic nepotism in interest conflicts. This explains the otherwise strange fact that ethnic interest conflicts seem to appear in all countries where people belong to clearly different ethnic groups and that ethnic interest conflicts have appeared within all cultural regions and at all levels of socioeconomic development. It is difficult to imagine any cultural explanation of ethnic conflicts that could account for the appearance of these conflicts across all cultural and civilizational boundaries.

Finally, I would like to emphasize that the same evolutionary argumentation provides a theoretical basis for the accommodation of ethnic conflicts. Because we have a behavioral predisposition to accept balanced reciprocity in our relations with others, it is reasonable to assume that institutional arrangements based on balanced reciprocity and equality between ethnic groups make it easier to mitigate ethnic interest conflicts than institutions based on equality of ethnic groups and on unequal reciprocity.

A RESEARCH PROJECT ON ETHNIC CONFLICTS

My intention is to make a comparative study of ethnic conflicts in which I test the three hypotheses by empirical evidence covering contemporary states. A difficult problem is how to measure the degree of ethnic divisions and the degree of ethnic conflicts and how to find reliable and comparable empirical data on indicators intended to measure hypothetical concepts of "ethnic divisions" and "ethnic conflicts."

Indicators of Ethnic Divisions

I have measured major ethnic divisions by taking into account three aspects of ethnic cleavages: (1) cleavages based on race or color, (2) cleavages based on nationality, language, or tribe, and (3) cleavages based on differences between old religious communities. All of these cleavages divide the population into separate ethnic groups, although their deepness varies. I assume that ethnic divisions based on race or color are genetically the deepest ones because they may be tens of thousands of years old, whereas most religious cleavages are usually not older than one or two thousand years and often only a few hundred years or even less.

Thus we have three operationally defined indicators to measure three dimensions of ethnic divisions. In each dimension, the level of ethnic divisions will be measured by the percentage of the largest ethnic group of the country's total population. Together the three percentages measure the relative degree of ethnic homogeneity, and the inverse percentages indicate the degree of ethnic heterogeneity. There are different ways to combine the three percentages. I have combined them into an index of ethnic heterogeneity (EH) by summing up the three percentages of ethnic heterogeneity. The possible range of the index of ethnic heterogeneity varies from 0 to 297, but the actual range varies from 0 to 177. Unfortunately, space does not allow me to further detail these the three components of EH.

Measures of Ethnic Conflicts

It is easier to get data on the existence of ethnic interest conflicts than to measure their relative significance. My intention is to measure their relative significance by taking into account two principal dimensions of ethnic conflicts: (1) institutional ethnic conflicts and (2) conflicts in which participants resort to various forms of coercion (violent ethnic conflicts).

I use the relative significance of ethnic parties and organizations and significant ethnic inequalities in governmental institutions to measure the institutional dimension of ethnic conflicts. The criteria of estimations are given in the following scale of institutionalized ethnic conflicts. The scale extends from 0 to 100. Each country gets an estimated score that is intended to indicate the relative significance of institutionalized ethnic conflicts during the period 1990–96. The higher the score is, the higher the estimated degree of institutionalized ethnic conflicts.

0 = no significant ethnic organizations; no significant ethnic inequalities in governmental institutions

5 = the share of ethnic parties, less than 10 percent of the votes cast in parliamentary or presidential elections; some ethnic parties or other organizations; some ethnic inequalities in governmental institutions

10 = the share of ethnic parties, 10–14 percent; some prominent ethnic organizations; clear ethnic inequalities in governmental institutions.

20 = the share of ethnic parties, 15–29 percent; significant ethnic organizations; permanent ethnic interest conflicts; significant ethnic inequalities in governmental institutions

40 = the share of ethnic parties, 30–49 percent; ethnic organizations cover a significant part of the population; extensive ethnic interest conflicts; large ethnic inequalities in governmental institutions

60 = the share of ethnic parties, 50–69 percent; most interest organizations are ethnic ones; ethnic interest conflicts are more important than other types of interest conflicts; serious ethnic inequalities in governmental institutions

80 = the share of ethnic parties, 70–89 percent; nearly all interest organizations are ethnically based; ethnic interest conflicts or inequalities in governmental institutions dominate national politics

100 = the share of ethnic parties, 90–100 percent; all significant interest organizations are ethnic by nature; practically all interest conflicts between groups take place along ethnic lines

It should be noted that the different types of criteria given for each level of the scale are intended to complement each other, but they can also be used alternatively. Consequently, the scores of some countries may be principally based on the significance of ethnic parties, whereas the scores of other countries are based on the significance of other types of ethnic organizations or on the significance of institutionalized ethnic inequalities. The electoral data on the share of ethnic parties of the votes cast in elections can be used to indicate the significance of ethnic organizations in many democracies, whereas some other kind of information is needed to estimate the relative significance of ethnic organizations in nondemocracies and also in many democracies in which institutionalized ethnic conflicts take place outside party politics. Ethnically based military, guerrilla, or underground organizations are important in many nondemocracies. In these criteria, "ethnic interest conflicts" refer to the existence of permanent ethnic tension, to repeated incidents of ethnic discrimination, and to other possible forms of nonviolent ethnic conflicts.

I have not found any direct indicator to measure the relative significance of violent ethnic conflicts. The extent and relative significance of such conflicts has to be estimated on the basis of available information. The scope of ethnic conflicts characterized by the use of coercion extends from demonstrations, riots, strikes, destruction of property, and sabotage to attacks on persons, violent clashes between groups, arrests, killing of people, rebellions, terrorism, forceful deportation of people, ethnic guerrilla wars, separatist wars, ethnic civil wars, ethnic cleansings, and genocides. The problem is how to estimate and compare the significance of different types of coercive ethnic conflicts. My estimations on the significance of violent ethnic conflicts are based on a scale whose scores vary between 0 and 100. The higher the score is, the higher the estimated significance of coercive ethnic conflicts. Estimations on the extent and significance of coercive or violent ethnic conflicts are based on the following criteria:

0 = no information on significant ethnic demonstrations, riots, or violence, although there may be tension between ethnic groups

5 = occasional ethnic demonstrations or riots; attacks on single persons; destruction of property

10 = sporadic ethnic demonstrations, riots, or violence involving single persons or small groups; destruction of property in ethnic riots

20 = serious ethnic conflicts involving ethnic communities; demonstrations, riots, and attacks on persons; local ethnic guerrilla movements; occasional ethnic terrorist acts

40 = repeated violent conflicts between ethnic groups; suppression of particular ethnic groups; ethnic rebellions; significant ethnic terrorism

60 = violent ethnic conflicts dominate politics; separatist ethnic wars in some parts of the country

80 = prolonged violent ethnic conflicts involving significant parts of the country; ethnic civil wars or separatist wars; forced deportation of people; ethnic cleansings

100 = ethnic wars dominate politics completely; genocidal ethnic violence

Because the two scales on instutionalized and violent ethnic conflicts are intended to measure the same phenomenon from two different perspectives, it is reasonable to combine the scores of the two dimensions into a combined index of ethnic conflicts (EC). This is done simply by adding the scores. This means that the values of the index of ethnic conflicts can vary between 0 and 200. This index will be used as the principal empirical indicator of ethnic conflicts in this study. Unfortunately, space does not allow me to give and document the data on which these estimations are based.

Research Hypotheses

The original hypotheses on the relationship between ethnic nepotism and ethnic conflicts can now be reformulated into research hypotheses using operationally defined variables in the place of original hypothetical concepts. It is possible to test the first hypothesis by separating the countries with "significant ethnic divisions" from countries without significant ethnic cleavages and, second, the countries with "ethnic interest conflicts" from countries without ethnic interest conflicts. This can be done by determining threshold values for EH and EC. It is not self-evident what the threshold values should be. Let us assume that countries for which the index of ethnic heterogeneity is 10 or less are without significant ethnic divisions and that countries for which the index of ethnic conflicts is 10 or less are without significant ethnic conflicts. The first hypothesis can now be reformulated as follows: (1) the value of EC tends to be 10 or less for countries whose EH value is 10 or less, and vice versa.

The first research hypothesis can be tested by a simple method of cross-tabulation. The countries should cluster to the same diagonal. Deviating cases will be in the opposing diagonal. The strength of the hypothesized relationship can be measured by Yule's Q (for Yule's Q, see Buchanan, 1980, pp. 83–84). The hypothesis allows some deviating cases because my empirical variables are rough and errors of measurement and judgment are possible. However, if the number of deviating cases is large and the strength of the hypothesized relationship is not at least moderate, the hypothesis should be regarded as falsified.

The second hypothesis presupposes a clear positive correlation between the composite index of ethnic heterogeneity (independent variable) and the index of ethnic conflicts (dependent variable). The second hypothesis can now be given in the following form: (2) the values of the index of ethnic heterogeneity (EH) are positively correlated with the values of the index of ethnic conflicts (EC).

The second research hypothesis can be tested by correlation and regression analyses. The correlation coefficient measures the strength of the relationship. Because the hypothesized relationship between the level of ethnic divisions and the level of ethnic conflicts is positive, correlations should be clearly positive. Negative or weak positive correlations would falsify the hypothesis. Regression analysis can be used to disclose how well single countries are adapted to the average relationship between the two variables and which countries deviate most from the regression line. Detailed analysis could then be focused on deviating cases.

SUMMARY OF PRELIMINARY RESULTS

The first research hypothesis was tested by cross-tabulating the 183 countries of this study by the significance levels of ethnic heterogeneity (EH higher than 10) and of ethnic conflicts (EC higher than 10). The results are given in table 11.1.

The first research hypothesis claims that significant ethnic divisions tend to lead to ethnic conflicts in all societies. It does not exclude the possibility of ethnic conflicts at lower levels of EH, but it presupposes that ethnic conflicts emerge in all countries with EH higher than 10 index points. Table 11.1 shows that empirical evidence supports this hypothesis very strongly. There are not more than 21 contradicting countries (11.5 percent), and nearly all of these deviations are very small or completely insignificant.

The strength of the hypothesized relationship was measured by Yule's Q. It varies from +1.0 for perfect positive association to −1.0 for perfect negative association. A perfect positive association occurs if there are no cases at all in

Table 11.1
Cross-tabulation of 183 Countries according to the Significance Levels of Ethnic Divisions and Ethnic Conflicts in 1990–1996

Ethnic Conflicts (EC)	EH 10 or Less N	EH Higher Than 10 N	Total N
10 or less	36	15	51
Higher than 10	6	126	128
Total	42	141	183

the b cell, or none in either b or c cells (see Buchanan, 1980). In this tabulation, Yule's Q is 0.96, which indicates an extremely strong positive relationship.

The second research hypothesis tries to explain cross-cultural regularities in the intensity and extent of ethnic conflicts by ethnic nepotism. The more a society is ethnically divided (EH), the more ethnic nepotism tends to channel political and other interest conflicts along ethnic lines. The hypothesis was preliminarily tested by correlation and regression analyses. The correlation between EH and EC is 0.718, which means that the index of ethnic heterogeneity statistically explains 52 percent of the variation in the index of ethnic conflicts. The unexplained part of the variation is 48 percent. It is due to possible measurement errors and to all other explanatory variables, including the significance of institutional factors. Thus empirical evidence supports the second research hypothesis, although there seem to be many countries that deviate from the average relationship between EH and EC.

Regression analysis complements the results of correlation analysis. Its results disclose how well the average relationship between EH and EC applies to single countries. In this case, the regression equation is y est. $= 6.718 + 0.899x$. Figure 11.1 summarizes and illustrates the results graphically.

In figure 11.1, negative residuals (countries below the regression line) indicate that the level of ethnic conflicts is lower than expected on the basis of the regression equation of EC on EH. Positive residuals (countries above the regression line) indicate that the level of ethnic conflicts is higher than expected. Figure 11.1 shows that most countries are relatively near the regression line, which means that the level of ethnic conflicts is approximately at the expected level in these countries. Several countries, however, deviate from the regression line considerably. The more a country deviates from the regression line, the more deviating a case it is. The names of the most deviating countries are indicated in figure 11.1, as well as the ethnically most divided countries (EH higher than 100).

My evolutionary explanation for the universality of ethnic conflicts in ethnically divided societies is based on the assumption that ethnic nepotism belongs to human nature. We have evolved to favor our relatives in the struggle for existence because this has been an adaptive behavior pattern. Of course, there

Figure 11.1

Results of Regression Analysis in Which the Index of Ethnic Heterogeneity Is Used as the Independent Variable and the Index of Ethnic Conflicts Is Used as the Dependent Variable in the Comparison Group of 183 Countries in 1990–96

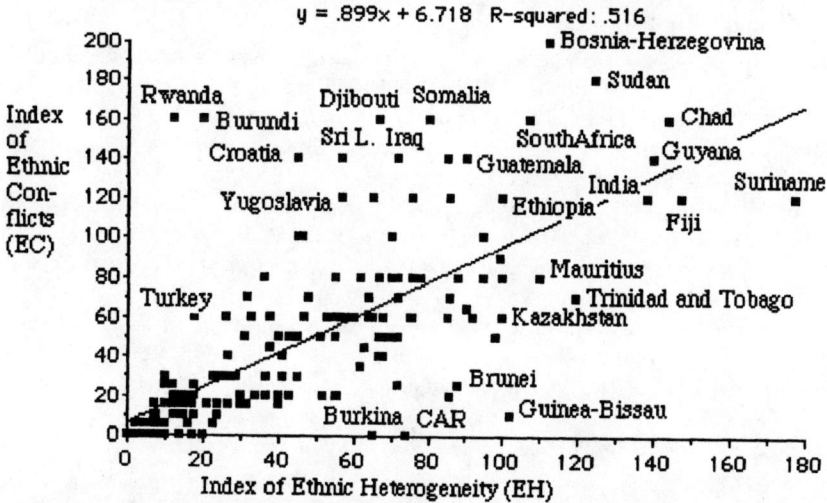

$y = .899x + 6.718$ R-squared: .516

Index of Ethnic Conflicts (EC) — vertical axis, Index of Ethnic Heterogeneity (EH) — horizontal axis

are good proximate explanations for every ethnic conflict, but the fact that ethnic cleavages have led to some kinds of ethnic conflicts in all societies implies the existence of a common causal factor. According to my interpretation, our shared disposition to ethnic nepotism is the common factor behind all ethnic conflicts. Our disposition to ethnic nepotism is bound to lead to ethnic interest conflicts in all ethnically divided societies. It does not seem possible to avoid the emergence of ethnic conflicts in such societies, but it may be possible to mitigate them by inventing social and political institutions that help to accommodate the interests of different ethnic groups. It is a challenge for social scientists to invent institutions that could make it easier for different ethnic groups to coexist in our increasingly crowded world.

REFERENCES

Alexander, Richard D. (1980). *Darwinism and human affairs*. London: Pitman Publishing.

Barash, David P. (1982). *Sociobiology and behavior* (2nd ed.). London: Hodder & Stoughton.

Buchanan, William. (1980). *Understanding political variables* (3rd ed.). New York: Charles Scribner's Sons.

Connor, Walker. (1992). *Ethnonationalism: The quest for understanding*. Princeton, NJ: Princeton University Press.

Daly, Martin, & Wilson, Margot. (1983). *Sex, evolution, and behavior* (2nd ed.). Boston: PWS Publishers.

Darwin, Charles. (1859/1981). *The origin of species by means of natural selection; or The preservation of favored races in the struggle for life.* Harmondsworth: Penguin Books.

Dawkins, Richard. (1976). *The selfish gene.* London: Granada Publishing.

Dobzhansky, Theodosius, Ayala, Francisco J., Stebbins, G. Ledyard, & Valentine, James W. (1977). *Evolution.* San Francisco: W. H. Freeman & Co.

Dunbar, Robin I. M. (1987). Sociobiological explanations and the evolution of ethnocentrism. In V. Reynolds, Vincent S. Falger, & Jan Vine (Eds.), *The sociobiology of ethnocentrism* (pp. 48–59). London: Croom Helm.

Glickman, Harvey. (Ed.). (1995). *Ethnic conflict and democratization in Africa.* Atlanta, GA: African Studies Association Press.

Gurr, Ted Robert. (1993). *Minorities at risk: A global view of ethnopolitical conflicts.* Washington, DC: United States Institute of Peace Press.

Gurr, Ted Robert, & Harff, Barbara. (1994). *Ethnic conflict in world politics.* Boulder, CO: Westview Press.

Hamilton, W. D. (1964). The genetic evolution of social behavior. In A. L. Caplan (Ed.), *The sociobiology debate* (pp. 191–209). New York: Harper & Row.

Horowitz, Donald L. (1985). *Ethnic groups in conflict.* Berkeley: University of California Press.

Hutchinson, John, & Smith, Anthony D. (Eds.). (1996). *Ethnicity.* Oxford: Oxford University Press.

Maynard Smith, John. (1979). *The theory of evolution* (3rd ed.). Harmondsworth: Penguin Books.

Mayr, Ernst. (1982). *The growth of biological thought: Diversity, evolution, and inheritance.* Cambridge MA: Berkeley Press of Harvard University Press.

McGary, John, & O'Leary, Brendan. (Eds.). (1994). *The politics of ethnic conflict regulation.* London and New York: Routledge.

Meyer, Peter. (1987). Ethnocentrism in human social behaviour: Some biosociological considerations. In V. Reynolds, Vincent S. Falger, & Jan Vine (Eds.), *The sociobiology of ethnocentrism* (pp. 81–93). London: Croom Helm.

Rabushka, Alvin, & Shepsle, Kenneth. (1972). *Politics in plural societies: A theory of democratic instability.* Columbus, OH: Charles E. Merrill.

Riggs, Fred. (1995). *Ethnonational rebellions and viable constitutionalism*: Preliminary draft.

Rupesinghe, Kumar. (Ed.). (1988). *Ethnic conflict and human rights.* Oslo: Norwegian University Press.

Rushton, J. Philippe. (1986). Gene-culture coevolution and genetic similarity theory: Implications for ideology, ethnic nepotism, and geopolitics. *Politics and the Life Sciences, 4* (2), 144–148.

Silverman, Irwin. (1987). Inclusive fitness and ethnocentrism. In V. Reynolds, Vincent S. Falger, & Jan Vine (Eds.), *The sociobiology of ethnocentrism* (pp. 112–117). London: Croom Helm.

Smith, Anthony D. (1987). *The ethnic origins of nations.* New York: Basil Blackwell.

Trivers, R. L. (1981). Sociobiology and politics. In E. White (Ed.), *Sociobiology and human politics* (pp. 1–43) Lexington, MA: Lexington Books.

———. (1985). *Social evolution.* Menlo Park, CA: Benjamin/Cummings.

van den Berghe, Pierre L. (1981). *The ethnic phenomenon.* New York: Praeger.

————. (1996). Does race matter? In John Hutchinson & Anthony D. Smith (Eds.), *Ethnicity* (pp. 57–63). Oxford: Oxford University Press.

Vanhanen, Tatu. (1991). *Politics of ethnic nepotism: India as an example*. New Delhi: Sterling Publishers.

————. (1992). *On the evolutionary roots of politics*. New Delhi: Sterling Publishers.

Wilson, Edward O. (1975). *Sociobiology: The new synthesis*. Cambridge, MA: Belknap Press of Harvard University Press.

12

Mating Patterns and Their Role in the Formation and Structure of the Abbad Tribe in Jordan

A. J. Nabulsi

Arab tribes are self-defined communities of divergent economies. They often include complex structures organized in a manner dictated by sociocultural, historical, and geographic factors. Relationships within the tribe are governed by their patrilateral relatedness, that is, genealogy, which is a description of tribal history (Oppenheim, 1939, 1943; Peters, 1960; Lancaster & Lancaster, 1992). One can differentiate between microgenealogy, which describes the actual paternal descent lines of the smaller subunits, and macrogenealogy, which provides the framework of relationships between the tribal subunits, (Lancaster, 1981). One may then suggest that mating behavior plays a major role in the structure and organization of Arab tribes, despite these being considered as loosely structured or even as representing political assemblies (e.g., Khoury & Kostiner, 1990). The definability of Arab tribes as biological entities, as breeding populations, thus becomes eminent. Investigating the mating patterns (random, assortative, or inbreeding) within these tribes would be the first step in this direction. Any association between the factors that may have influenced mating patterns and the makeup of the tribes can possibly provide valuable information on the long prevalence of this type of social organization, going back thousands of years, and on the maintenance of the regional population in a largely volatile environment.

In a case study on the biological structure of Arab tribes, the Abbad tribe in Jordan was the subject of an anthropological investigation. Empirical data and supplement any information on the tribe were compiled exclusively through house visits and interviews. Documents and assays that helped in the reconstruction of some family trees were provided by many genealogists from Abbad.

Figure 12.1
The Geographic Distribution of the Abbad Tribe in the Balqa Area in Jordan

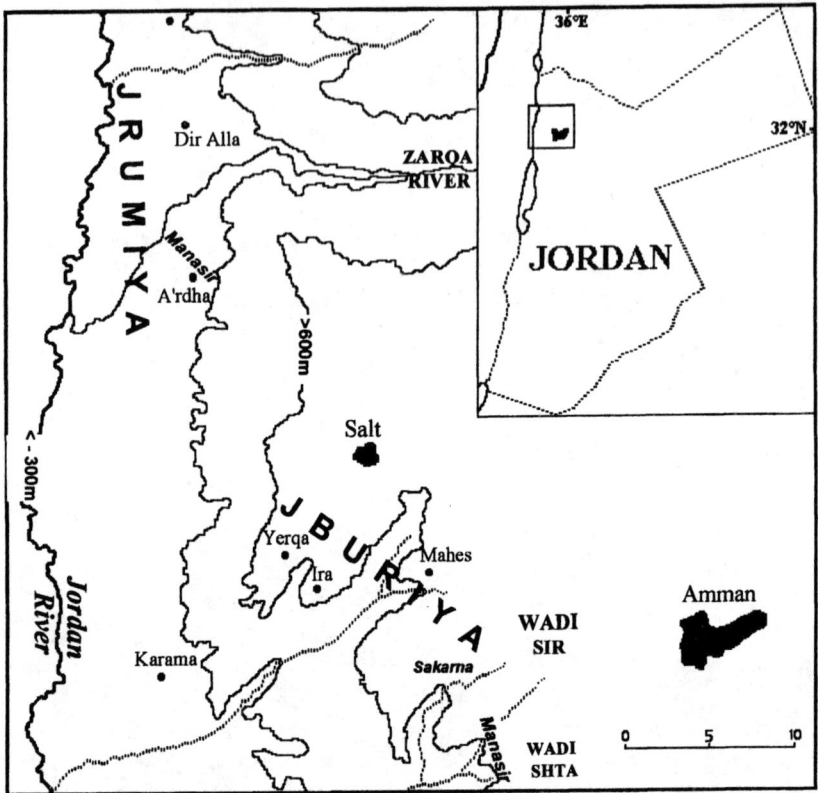

Abbad is one of the larger tribes in Jordan (about 120,000 individuals). It is primarily agricultural and resides mainly in the Balqa area (figure 12.1). The tribe consists of two divisions: the northern Abbad, Jrumiya, in the Jordan valley and the southern Abbad, Jburiya, on the Eastern Heights (figure 12.2). The Jrumiya division includes two subdivisions, the Jrum and the Za'anif, each consisting of six groups. Except for the Manasir group, each of the Jrumiya groups considers itself to be unilineal. The Manasir is the largest group in Abbad and consists of four inclusive subgroups. This group is also geographically divided into Manasir Wadi-Shta, at the Eastern Heights, and A'rdha Manasir in the Jordan valley. Six groups are allocated to the Jburiya division, in which all but two groups include various numbers of lineages.

In a previous study, the mating patterns were examined (Nabulsi, 1995). Four proportional male samples representing the different levels of Abbad's structure were gathered for empirical analysis. The samples were to represent the Jrumiya

Figure 12.2
Schematic Representation of the Structure of the Abbad Tribe

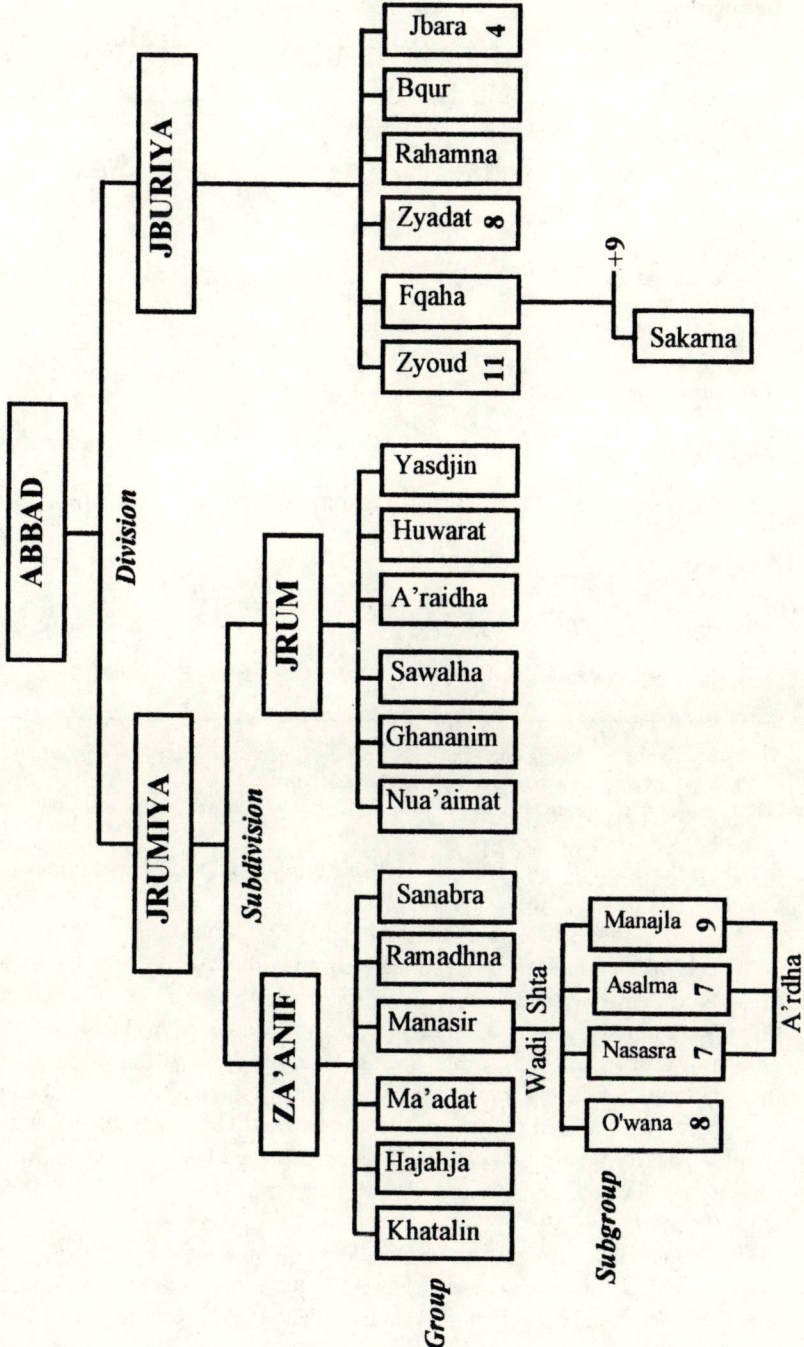

Note: Numbers refer to the number of lineages in the inclusive subunits.

Table 12.1
Demographic Parameters Estimated in the Abbad Tribe

	JRUMIYA	JBURIYA	MANASIR	SAKARNA	TOTAL
1ST cousin	14.8%	15.8%	14.4%	21.8%	17.3%
1ST cousin once removed	31.5%	19.7%	17.8%	25.5%	23.0%
within lineage	9.3%	10.9%	17.8%	19.1%	15.2%
cum. within lineage	55.6%	46.1%	50.0%	66.4%	55.5%
cum. within Abbad	83.3%	94.7%	93.3%	93.6%	91.2%
Exogamous regional	16.7%	3.9%	5.6%	2.7%	7.6%
Exogamous nonregional	0	1.3%	1.1%	3.6%	1.2%
Inbreeding coefficient F	0.0192	0.0171	0.0146	0.0216	0.0180
Sample size	54	76	90	120	340
Sample mean age (years)	29.3	26.4	28.6	29.0	28.3
Average family size/ males	9.1/4.4	10.2/5.4	10.0/4.9	9.1/4.9	9.7/5.0

Note: The relative inbreeding coefficients for autosomal traits (F) estimation were based only on first-degree-cousin marriages of the parental generation.
N.B. Cumulative values are not additive due to exclusion of non-Abbad elements.

division, the Jburiya division, the Manasir group, and the Sakarna lineage from Fqaha group of the Jburiya division.

The demographic data, summarized in table 12.1, reveal that in all tribal levels 90 percent of the matings involved mates from within Abbad, and 30–50 percent of the marriages were consanguineous, particularly within the paternal descent line. The estimated inbreeding coefficients (F) varied from 0.0146 in Manasir groups to 0.0216 in the Sakarna lineage. Matings beyond the paternal descent lines were influenced by maternal links, that is, in the matrilateral line and were more common between neighboring communities. Thus the probability for any couple to marry would decrease the less closely they were related, starting with first-degree cousins, or the more distant geographically and in their tribal affiliation they were.

First-cousin marriage categories were less common in Abbad than the reported average in Jordan (Khoury & Massad, 1992) but more frequent than Lancaster's (1981) observations among the Rwala tribe. The Rwala is a camel-herding tribe, and its members are widely dispersed in the Syrian desert just north of the

Arabian peninsula. Since mobility is vital, restrictions on intermarriages between the subunits can result in reducing intertribal solidarity and consequently the survival options in unsuitable conditions, such as lack of rain. Abbad is primarily agricultural and its members are sedentary. Though land was a commonwealth among Abbad's members, families belonging to single lineages or groups preferred to live together either dispersed or concentrated in one area in the vicinity of their cultivated land. As families grow larger, they tend to practice first-cousin marriages to prevent the partitioning of their share of land and become involved in a process of reciprocal matings with neighboring families. This leads in some cases to the emergence of new settlements in which all members claim to be related or even of common descent, that is, the typical "one village one 'tribe' or clan" (Gulick, 1953; Lutfiyya, 1966), regardless of their actual paternal descent, as is well demonstrated by the Jbara group in Mahes (figure 12.1).

Under "isolation," small populations often reveal high inbreeding coefficients. Yet if tribal affiliation is to be considered as an isolating or discriminating factor, consanguineous marriages in Arab tribes appear to increase with growing population size. With the rapid population increase in Jordan, proliferation of tribal subunits is increasing while tribal solidarity is decreasing. This was indicated by increasing first-cousin and exogamous marriages in Abbad and segmentation in the inclusive groups.

All the subunits of Abbad consider themselves to be Yeman, that is, of southern Arab origin. Although some of the present subunits were reported to have resided in Balqa around the thirteenth century, the actual emergence of the tribe is claimed to have occurred some 250 years ago from a multitude of tribal elements and regional villagers. Since then, the structure of the tribe has been in continuous change as a result of fusion and segmentation. By the end of the nineteenth century the Abbad tribe achieved its current boundaries and structure (Oppenheim, 1943; Nabulsi, 1995). The emergence of the Abbad tribe from a multitude of lineages and clans was similar to that reported for other Arab tribes (Oppenheim, 1943; Lancaster, 1981). In these cases the unity of the different subunits is based on a common denominator that provides for each individual to be equal among equals. This could be common descent, as in the Rwala tribe, religious, as in the Nua'aimat tribe, or some other factor of relevance.

Pedigree analysis revealed that matings between the different subunits were very common in the early stage of Abbad's history. For example, the Sakarna joined Abbad in the eighteenth century as part of the Muherat, a lineage of the Fqaha-Jburiya, they consisted of the nuclear family of the ancestral parents. They intermarried with the Fqaha group, to which they were allocated, as well as with subunits from both divisions in Abbad. The growth of the family was combined with an increase in consanguinity in the paternal line. Later they segmented from the Muherat as an independent lineage within the Fqaha group. It appears that in the absence of common ancestry, the mating behavior of the diverse elements that formed Abbad aimed to provide a common denominator or cause upon which their unity could rest. Geographic and historical relationships were considered in the arrangement and integration of the various subunits, which led to the geographical differentiation of Abbad into two divisions and to complex substructures, as of the Manasir group (Jrumiya subdivision), as well as the allocation of smaller elements of the identical ancestries to different groups in Abbad, for example, the A'lawin of the Zyadat group and the A'laiwat of the Jbara group from the Jburiya division. Furthermore, Abbad's military alliances with other tribes did not lead to intermarriages. Exogamous marriages were almost restricted to neighboring tribal populations (Nabulsi, 1995).

The mating behavior of the tribal elements at the early stages of Abbad's history can be interpreted as a method to increase the effective size (Ne) of the population, that is, the number of reproductive individuals. The tribe is the medium that provides endurance for each of its members. Its structure is the description of relationships between the substructures. Under hard natural or artificial conditions that have resulted in high mortality and low life-expectancy rates over a long period, the formation of a larger tribe would provide a better environment in which to survive for the smaller subunits, clans, or lineages.

The Abbad tribe may thus be seen as an endogamic population consisting of more or less inbred subunits whose unity is achieved by past and present relationships. Relatedness, maternal links, geography, and the type of economy were the factors that affected the mating patterns and influenced the makeup and organization of the Abbad tribe. Unity between the single subunits, distinctive through their descent lines and consanguineous marriages, is achieved through reciprocal matings. The formation of the larger tribal community could be considered as an adaptation providing continuity and more options within a larger frame for all involved. The tribe as a defined unit distinct from other similar communities (Lancaster & Lancaster, 1992) may evolve or disintegrate, but the population itself remains.

REFERENCES

Gulick, J. (1953). The Lebanese village: An introduction. *American Anthropologist 55* 367–372.

Khoury, P. S., & Kostiner, J. (Eds.). (1990). *Tribes and state formation in the Middle East.* Berkeley: University of California Press.

Khoury, S. A., & Massad D. (1992). Consanguineous marriage in Jordan. *American Journal of Genetics, 43,* 769–775.

Lancaster, W. (1981). *The Rwala Bedouin today.* Cambridge: Cambridge University Press.

Lancaster, W., & Lancaster F. (1992). Tribal formation in the Arabian peninsula. *Arabian Archaeology and Epigraphy, 3,* 145–172.

Lewis, N. N. (1987). *Nomads and settlers in Syria and Jordan, 1800–1980.* Cambridge: Cambridge University, Press.

Lutfiyya, A. M. (1966). *Baytin: A Jordanian village.* The Hague: Mouton.

Nabulsi, A. J. (1995). Mating patterns of the Abbad tribe in Jordan. *Social Biology, 42,* 162–174.

Oppenheim, M. von. (1939). *Die Beduinen* (Vol. 1). Leipzig: Otto Harrassowitz.
————. (1943). *Die Beduinen* (Vol. 2). Leipzig: Otto Harrassowitz.
Peake, F. G. (1958). *History and tribes of Jordan*. Miami: University of Miami Press.
Peters, E. (1960). The proliferation of segments in the lineage of the Bedouin of Cyrenaica. *Journal of the Royal Anthropological Institute, 90*, 29–53.

13

Darwin's *Really* Dangerous Idea—The Primacy of Variation

J. Philippe Rushton

Evolution is the science of variation and selection. As such, establishing the Darwinian perspective in the social sciences has been much impeded by "political correctness," an ideology about social equality. Although there is much agreement about slow progress in establishing the Darwinian perspective in social science, there is little consensus as to the cause. I argue that egalitarianism has been the root cause of opposition to evolutionary thinking in the 25 years since E. O. Wilson's monumental *Sociobiology* (1975), the 35 years since William Hamilton's (1964) seminal formulations, and the 125-plus years since Charles Darwin's *Origin of Species* (1859) and *Descent of Man* (1871).

It is yet to be recognized, however, that much of the politically inspired resistance comes from evolutionary scientists themselves. By overemphasizing the search for universals, that is, pan-human traits (partly to show people's commonalities), many evolutionists eschew the very comparative method that created the Darwinian Revolution in the first place. While developing my general thesis, I provide specific evidence from personal experience over the last 15 years researching genetic variation in traits such as altruism, brain size, and IQ, and race, evolution, and behavior (Rushton, 1995b).

Although Darwinians emerged victorious in their nineteenth-century battles against biblical theology in academia and educated opinion, subsequently they lost this ground to liberal egalitarians, Marxists, cultural-relativists, and literary deconstructionists. From Herbert Spencer (1851) to the world depressions of the late 1920s and 1930s, the political right gained the ascendancy in using evolutionary theory to support their arguments, while the political left came to believe, perhaps correctly, that "survival of the fittest" was incompatible with social equality. Darwinian has been marginalized ever since the mid-1920s when the Boasian school of anthropology succeeded in decoupling the biological from the social sciences (Degler, 1991).

Although William McDougall proposed an "instinct" theory of personality, and G. Stanley Hall and others advanced an evolutionary perspective for developmental psychology, Darwinian views were swept away in the 1920s by various environmentalist dogmas. Freud's Oedipal theories and Watson's behavioral molding of individuals were compatible with Marx's assumptions of the malleability of entire social groups through government interventions.

Hostility to Nazi racial theories tainted any attempt to restore whatever remained of Darwinism to the social sciences. From the 1930s onward, scarcely anyone outside of Germany and its Axis allies dared to suggest that any group of individuals might be genetically different from any other in respect to behavior lest it should appear that the author was endorsing or excusing the Nazi cause. Even today, consideration of ethnic differences in intelligence leads to accusations of Nazism, even if one is a Jewish scientist describing how Jews average higher IQs than non-Jews (e.g., Seligman, 1992, Chapter 10; Herrnstein & Murray, 1994, Chapter 13).

Those who believed in the biological identity of all people, on the other hand, remained free to write what they liked, without fear of vilification. In the intervening decades, the idea of a genetically based core of human nature on which individuals and groups might differ was consistently derogated. This intellectual movement has been politically fueled by successively coupling it to Third World decolonization, the U.S. civil rights movement, the struggle against apartheid in South Africa, and the renewed debates over immigration.

Opposition to the existing political order has long been a tradition in universities. Today most evolutionary scientists subscribe to the concept of political equality rather than hierarchy. They also posit far more malleability to human nature than may be the case. Whether out of ideological or prudential considerations, current evolutionary scientists focus on pan-human traits that all people share (with the notable exception of sex differences), and they also emphasize "facultative adjustment" (the noncontroversial view that people alter their behavior depending on their circumstances), which, taken to extremes, denigrates heritable traits as causes of behavior.

THE PRIMACY OF VARIATION

Focusing on pan-human traits has solidified our knowledge of human nature and emphasized the continuity between humans and other primates (see, for example, Barkow, Cosmides & Tooby, 1992; Tooby & Cosmides, 1990; Bailey, 1997; and Wilson, 1994). However, ignoring or minimizing the role of heritable variation goes against the two basic postulates of Darwinian theory: (1) that genetic variation exists within species and (2) that differential reproductive success favors some varieties over others. In both *Origins* (1859) and *Descent*

(1871), Darwin left no doubt about the importance he ascribed to both individual and racial variation. For example:

Hence I look at individual differences, though of small interest to the systematist, as of high importance for us, as being the first step towards such slight varieties as are barely thought worth recording in works on natural history. And I look at varieties which are in any degree more distinct and permanent, as a step leading to more strongly marked and more permanent varieties; and at these latter, as leading to sub-species, and to species. . . . Hence I believe a well marked variety may be justly called an incipient species. (1859: 107)

Sir Francis Galton (1865, 1869) immediately recognized the implications of his cousin Darwin's theory for understanding the importance of variation in humans. He gathered evidence for the existence and heritable nature of variation, thus anticipating the concept of heritability and other later work in behavioral genetics. Galton carried out surveys and found, for example, that good and bad temper and cognitive ability ran in families. He discovered the phenomenon of regression-to-the-mean and argued that it implied family variation was heritable.

Galton also reviewed accounts contrasting the taciturn reserve of American Indians with the talkative impulsivity of Africans. He noted that these temperamental differences persisted regardless of climate (from the frozen north through the equator), religion, language, or political system (whether self-ruled or governed by the Spanish, Portuguese, English, or French). Anticipating later work on transracial adoption, Galton pointed out that the majority of individuals adhered to their racial type, even if they were raised by white settlers. He also wrote that the average mental ability of Africans was low, whether they were observed in Africa or in the Americas. In *Descent*, Darwin acknowledged Galton's work and also accepted the brain-size differences between Africans and Europeans found by Paul Broca and other nineteenth-century scientists.

In this chapter, I argue that the Darwinian-Galtonian paradigm was abandoned for political rather than scientific reasons. I also assert that the most recent scientific data are more understandable from the Darwinian-Galtonian perspective than the egalitarian one that displaced it.

BEHAVIORAL GENETICS

Consider the results shown in table 13.1. It presents data from a twin study of the heritability of altruism and aggression that I carried out in the early 1980s (Rushton, Fulker, Neale, Nias, & Eysenck, 1986). Over 500 pairs of twins were sent self-report questionnaires measuring their empathy, altruism, and aggression. Recall that twins come in two kinds—monozygotic (MZ) or identical twins, who share 100 percent of genes in common, and dizygotic (DZ) or fraternal twins, who share 50 percent of genes in common (as do ordinary siblings). Both correlational and model-fitting analyses showed that identical twins were

Table 13.1
Variance Components for Altruism and Aggression Questionnaires from 573
Adult Twin Pairs

Trait	Additive Genetic Variance		Shared Environmental Variance		Nonshared Environmental Variance	
	%	(%)	%	(%)	%	(%)
Altruism	51	(60)	2	(2)	47	(38)
Empathy	51	(65)	0	(0)	49	(35)
Nurturance	43	(60)	1	(1)	56	(39)
Aggressiveness	39	(54)	0	(0)	61	(46)
Assertiveness	53	(69)	0	(0)	47	(31)

Source: Adapted from Rushton, Fulker, Neale, Nias, and Eysenck (1986, p. 1195, Table 4). Copyright 1986 by the American Psychological Association. Reprinted with permission. Estimates in parentheses are corrected for unreliability of measurement.

twice as similar in their responses to altruism questionnaires as were fraternal twins. Genetic and environmental factors each contributed about 50 percent to the total variance in individual differences.

The results in table 13.1 also highlight an important distinction between two sources of environmental influence. Shared environmental factors (those common to children reared together) cause similarities in their behavior, while nonshared environmental factors (those unique to children reared together) cause differences. For example, parents with two unrelated adopted children provide a common rearing environment—a shared environment that should make the unrelated siblings similar in some respects. But the children also have individual interactions with their parents and distinct perceptions of family encounters that may influence each sibling in a unique way.

Research shows that it is the nonshared environment that accounts for most of the environmental influence on children's personalities and IQ. The most psychologically important environmental influences turn out to be those that make children in a family different from, not similar to, one another (Plomin & Daniels, 1987). A good example is the birth order effect, first-borns being more achieving and more authoritarian, while later borns are consequently "nurtured to rebel" and possibly be more creative (Sulloway, 1996).

Heritabilities of about 50 percent are typically found for behavioral characteristics including intelligence, mental illness, criminality, political values, vocational interests, and even religiosity (Plomin, DeFries, McClearn, & Rutter, 1997). The heritability of intelligence is now well established from numerous adoption, twin, and family studies. Particularly noteworthy are the heritabilities of around 80 percent found for adult twins reared apart (Bouchard, Lykken, McGue, Segal, & Tellegen, 1990). Moderate to substantial heritabilities for IQ have also been found in studies of nonwhites, including African Americans (Osborne, 1980) and Japanese (Lynn & Hattori, 1990). Even the most critical of meta-analyses find IQ almost 50 percent heritable (Devlin, Daniels, & Roeder, 1997).

In a twin study of cranial size based on external head measurements among 13- to 17-year-olds (472 individuals, boys and girls, blacks and whites), Rushton and Osborne (1995) found heritabilities of about 50 percent. Age, sex, and race differences were also found; cranial size increased from age 13 to 17, from girls to boys, and from blacks to whites. The heritabilities did not vary significantly by sex or race, although there was a trend for them to be lower in blacks than in whites, making it likely that the black sample had been reared in a less benign environment.

Unfortunately for social science, academics all too often sneer at heritability studies, not only when they cover obviously sensitive topics like IQ and brain size, but even studies of altruism. Many, including some evolutionists, were outraged when I published my results on altruism and aggression in the mid-1980s. Some even raised highly dubious objections to the well-established twin methodology. For example, sociobiologist David Barash (1995), reviewing my book *Race, Evolution, and Behavior*, claimed that it was impossible to separate genetic from environmental contributions. He argued, sarcastically, that a five foot six inch tall woman could not be divided into a genetic height of two feet nine inches and an environmental height of 2'9". But Barash's comment speaks only to his own ignorance of or antipathy to behavioral-genetic methods, not the inadequacy of heritability estimates. When the sample size is only one person, as in his example, there is no variance to be explained; but when the samples consist of 500 to 1,000 people, as in my studies, several statistical procedures, routinely used in plant and animal research, exist for partitioning variance.

Genetic influences on behavior during development are well illustrated in R. S. Wilson's (1983) longitudinal study which tested the IQ scores of some 500 pairs of twins at the ages of 3, 6, 9, 12, 18, 24, and 30 months, then yearly from 3 through 9 years, with a final followup at 15 years (see figure 13.1). The results show that the differentiation between monozygotic twins and dizygotic twins is not very pronounced in the early years when the environment (through gestation and other influences) has its major effect. But genetic influences are continuously at work, and so, by 6 years of age, while the monozygotic twin correlations had reached the upper 0.80s, the dizygotic twin correlations had dropped to about 0.50 to which the ordinary sibling correlations had risen— exactly as the genetic hypothesis would predict.

BRAIN SIZE AND COGNITIVE ABILITY

Variation in brain size is related to variation in cognitive ability. Galton (1888) was one of the first to quantify this relationship. His subjects were 1,095

Figure 13.1
Correlations Proportionate with Shared Genes for Mental Development

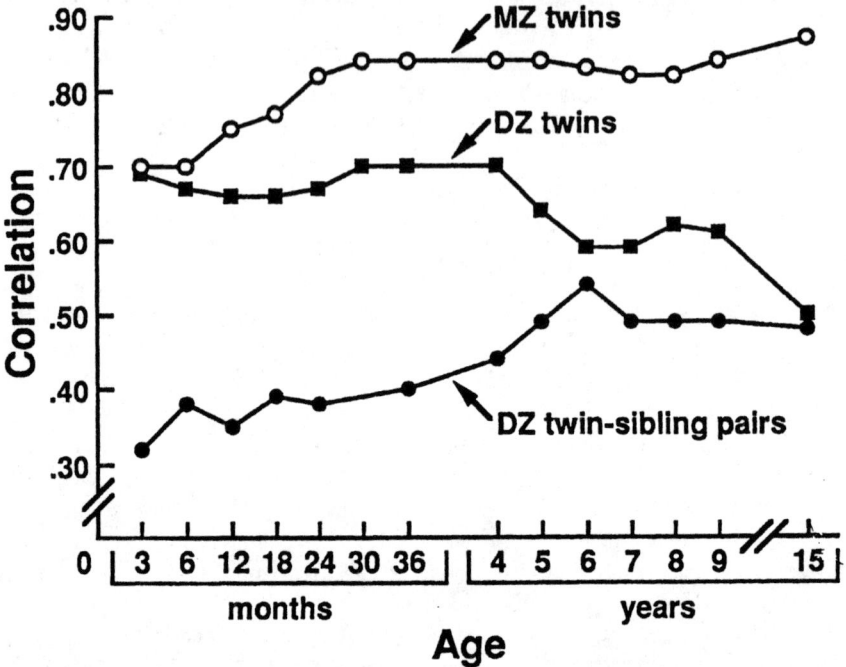

Note: Due to common and specific environmental influences, during the first months of life the differentiation between the two zygosity groups is not very pronounced, whereas that between DZ-twin–sibling sets is overpronounced. Genetic influences are continually at work, and by six years of age, while the MZ-twin correlations remained in the upper 0.80s, the DZ-twin correlations had dropped and the DZ-twin–sibling correlations had risen and were not significantly different from each other.

Source: Adopted from R. S. Wilson (1983, p. 311, figure 4). Copyright 1983 by the Society for Research in Child Development. Reprinted with permission.

Cambridge undergraduate men divided into those who had achieved first-class honors degrees and those who had not. Galton computed head volume by multiplying head length by breadth by height and plotting the results against age (19 to 25 years) and class of degree (A, B, C). He reported that: (1) cranial capacity continued to grow after age 19; and (2) men who obtained high honors degrees had a brain size from 2 to 5 percent greater than those who did not. In 1906, Sir Karl Pearson reexamined Galton's data using his newly developed correlation coefficient and found a small positive relationship between head size and university grade. This has remained the consensus scientific conclusion among those actually working in the field (see the review by Rushton & Ankney, 1996).

The published research that most clearly shows the correlation between brain size and intelligence employs state-of-the-art magnetic resonance imaging (MRI) technology, which creates a three-dimensional image of the brain of living subjects. In eight separate studies with a total sample size of 381 nonclinical adults, an overall correlation of 0.44 was found between MRI-measured brain size and IQ. This correlation is about as strong as the relationship between socioeconomic status of origin and IQ (which may itself be partly genetic). The relationship between a subtest's correlation with brain size and its loading on the general factor of intelligence (known as g), is even larger—0.60! (Jensen, 1994, 1998; Wickett, Vernon, & Lee, 1996).

RACIAL VARIATION

Brain Size

Modern studies confirm Darwin's view that races vary in average brain size— but not to the benefit of Caucasians. The racial gradient in decreasing mean brain size is now established as running from East Asians to Europeans to Africans. This three-way gradient has been found independently using three separate procedures: wet brain weight at autopsy, volume of empty skulls using filler, and volume estimated from external head sizes. The results from all these studies converge on the conclusion that the brains of East Asians and their descendants average about 17 cm³ (1 in³) larger than those of Europeans and their descendants, whose brains average about 80 cm³ (5 in³) larger than those of Africans and their descendants (see Rushton & Ankney, 1996, for review).

Because the modern data on racial variation in brain size are not as well known as they should be, I will summarize a few of the relevant studies. Using brain mass at autopsy, Ho, Roessmann, Straumfjord, and Monroe (1980) found that 811 European Americans averaged a brain weight of 1,323 grams, whereas 450 Africans Americans averaged 1,223 grams. Using endocranial volume, Beals, Smith, and Dudd (1984) found that East Asians averaged 1,415 cm³, Europeans averaged 1,362 cm³, and Africans 1,268 cm³. Using external head measurements from a stratified random sample of 6,325 U.S. Army personnel, Rushton (1992b) found that Asian Americans, European Americans, and African Americans averaged 1,416, 1,380, and 1,359 cm³, respectively. Using cranial measurements from tens of thousands of men and women aged 25–45 collated by the International Labour Office from around the world, Rushton (1994) found that Asians, Europeans, and Africans averaged 1,308, 1,297, and 1,241 cm³, respectively. Finally, an MRI study found that people of African and Caribbean background averaged a smaller brain volume than did those of European background (Harvey, Persaud, Ren, Baker, & Murray, 1994).

Racial differences in brain size show up early in life. Data from the National Collaborative Perinatal Project on 53,000 children show that Asian children averaged a larger head perimeter at birth than did white children, who averaged a larger head perimeter at birth than did black children. By age seven, the Asian children averaged highest in IQ scores, largest in head size, but smallest in body size; the black children averaged lowest in IQ, smallest in head size, but largest in body size; and the white children were intermediate on all three measures. In all three groups, head perimeter correlated with IQ (Broman, Nichols, Shaughnessy, and Kennedy, 1987; Rushton, 1997a).

IQ TEST SCORES

Paralleling the average differences in brain size are average differences in IQ. The race–IQ debate became international in scope when research demonstrated that East Asians average higher on tests of mental ability than do whites, whereas Africans and Caribbeans average lower (Lynn, 1982; Vernon, 1982). East Asians, measured in North America and in Pacific Rim countries, typically average IQs in the range of 101 to 111. Caucasoid populations in North America, Europe, and Australasia typically have average IQs from 85 to 115 with an overall mean of 100. African populations living south of the Sahara, in North America, in the Caribbean, and in Britain typically have mean IQs from 70 to 90 (Lynn, 1991, 1997).

Speed of decision making (reaction time) in 9- to 12-year-olds shows the same three-way racial pattern. In this task, children are asked to decide which of several lights stand out from others and move their preferred hand to press the corresponding button to turn off the light. All children can perform the task in less than one second, but higher-IQ children perform faster. Asian children from Hong Kong and Japan were, on average, faster than European children from Britain and Ireland, who in turn were faster than black children from South Africa (Lynn, 1991). This same three-way pattern of racial differences was independently replicated in a study of California school children using the same and similar elementary cognitive tasks (Jensen & Whang, 1993, 1994).

OTHER TRAITS

As shown in table 13.2, the East Asian-white-black racial matrix occurs on a surprisingly wide range of dimensions. For example, international police statistics from INTERPOL yearbooks averaged over several years show rates of murder, rape, and serious assault to be three times higher in African and Caribbean countries than in Pacific Rim countries, again with European countries intermediate (Rushton, 1990, 1995a). Similarly, the matrifocal family pattern disproportionately found among black Americans is also found in the Caribbean as well as in south-of-Sarah Africa (Draper, 1989). Whatever the causes of racial differences in crime and family structure turn out to be, they obviously apply well beyond American particulars.

Worldwide surveys show higher rates of sexual activity among blacks than among whites and especially among Asians. Racial differences in sexual activity

Table 13.2
Relative Ranking on Diverse Variables

Variable	Asians	Whites	Blacks
Brain size			
Autopsy data (cm³ equivalents)	1,351	1,356	1,223
Endocranial volume (cm³)	1,415	1,362	1,268
External head measure (cm³)	1,356	1,329	1,294
Cortical neurons (billions)	13.767	13.665	13.185
Intelligence			
IQ test scores	106	100	85
Decision times	Faster	Intermediate	Slower
Cultural achievements	Higher	Higher	Lower
Maturation rate			
Gestation time	?	Intermediate	Earlier
Skeletal development	Later	Intermediate	Earlier
Motor development	Later	Intermediate	Earlier
Dental development	Later	Intermediate	Earlier
Age of first intercourse	Later	Intermediate	Earlier
Age of first pregnancy	Later	Intermediate	Earlier
Life-span	Longer	Intermediate	Shorter
Personality			
Activity	Lower	Intermediate	Higher
Aggressiveness	Lower	Intermediate	Higher
Cautiousness	Higher	Intermediate	Lower
Dominance	Lower	Intermediate	Higher
Impulsivity	Lower	Intermediate	Higher
Self-concept	Lower	Intermediate	Higher
Sociability	Lower	Intermediate	Higher
Social organization			
Marital stability	Higher	Intermediate	Lower
Law abidingness	Higher	Intermediate	Lower
Mental health	Higher	Intermediate	Lower
Administrative capacity	Higher	Higher	Lower
Productive effort			
Two-egg twinning (per 1000 births)	4	8	16
Hormone levels	Lower	Intermediate	Higher
Secondary sex characteristics	Smaller	Intermediate	Larger
Intercourse frequencies	Lower	Intermediate	Higher
Permissive attitudes	Lower	Intermediate	Higher
Sexually transmitted diseases	Lower	Intermediate	Higher

Source: J. P. Rushton, *Race, evolution, and behavior* (New Brunswick, NJ: Transaction, 1995b), p. 5. Copyright 1995 by Transaction Publishers. Reprinted by permission.

have consequences, including elevated rates of AIDS and HIV-infection. As of January 1, 1998, statistics from the World Health Organization and Centers for Disease Control and Prevention revealed that 8 out of every 100 sub-Saharan Africans, 2 out of every 100 black Caribbeans, and 2 out of every 100 black Americans were living with HIV. The comparable figure for whites (either American or European) was less than 1 per 1,000; for East Asians, the figure was less than 1 per 10,000.

Controlling for IQ and social class substantially reduces but does not eliminate racial differences in rates of incarceration, illegitimate birthing, and sexual behavior (Herrnstein & Murray, 1994; cf. Gordon, 1997). More than IQ must be involved. One neurohormonal contributor to crime and sexual behavior is testosterone. Studies show 10 percent more testosterone in black college students and military veterans than in their white counterparts (Ellis & Nyborg, 1992), and more in whites than in East Asians. Sex hormone levels may also explain differential dizygotic twinning rates, which are less than 4 per 1,000 births among East Asians, 8 among Europeans, and 16 or greater among Africans. DZ twinning is known to be heritable through the race of the mother regardless of the race of the father, as found in East Asian/European crosses in Hawaii and European/African crosses in Brazil. (There is, however, no ethnic difference in MZ twinning rates.)

SELECTION AND THE EVOLUTION OF RACES

An evolutionary explanation for why Asians average the largest brains and the highest IQ is especially interesting because there has been a threefold increase in the relative size of the hominid brain over the last 3 million years. From an adaptationist perspective, unless large brains substantially contributed to evolutionary fitness (defined as increased survival of genes through successive generations), they simply could not have evolved. Metabolically, the brain is a very expensive organ. It uses about 5 percent of basal metabolic rate in rats, cats, and dogs, about 10 percent in rhesus monkeys and other primates, and about 20 percent in humans (Armstrong, 1990). Across species, large brains are related to other life history traits, such as longer gestation, slower rate of maturation, higher rate of offspring survival, lower reproductive output, and longer lifespan (Pagel & Harvey, 1988; Hofman, 1993).

The Out-of-Africa theory holds that *Homo sapiens* arose in Africa 200,000 years ago, expanded beyond Africa following an African/non-African split about 110,000 years ago, and then migrated east after a European/East Asian split about 40,000 years ago (Stringer & Andrews, 1988; Stringer & McKie, 1996). My extension of that theory (Rushton, 1995b) argues that since evolutionary selection pressures were different in the hot savanna where Africans evolved than in the cold Arctic where Mongoloids evolved, these ecological differences had not only morphological but also behavioral effects. The farther north the populations migrated "Out of Africa," the more they encountered the cognitively demanding problems of gathering and storing food, gaining shelter, making clothes, and raising children during prolonged winters. As these populations evolved into present-day Europeans and East Asians, they did so by shifting toward the K end of the $r–K$ dimension of reproductive strategies. That is, they underwent selective pressure for larger brains, slower rates of maturation, and lower levels of sex hormone with concomitant reductions in sexual potency and aggression and increases in family stability and longevity.

Why do so many variables correlate in so comprehensive a fashion? Why do East Asians average the largest brains and the lowest twinning rates, Africans the smallest brains and the highest twinning rate, and Europeans intermediate in both? The explanation I propose lies in $r–K$ life history theory. Following E. O Wilson (1975), a life history is a genetically organized suite of characters that evolved so as to allocate energy to survival, growth, and reproduction. Among r-strategies the emphasis is on egg production, and among K-strategies, parental care. As Johanson and Edey (1981, p. 236) succinctly summarized, quoting Owen Lovejoy: "More brains, fewer eggs, more 'K'."

HERITABILITY OF RACE DIFFERENCES

But is there any evidence to support my contention that the evolutionary scenario sketched above has produced genetic differences among the races? There is. Research has found that racial differences are more pronounced on subtests that are highly heritable than they are on less heritable IQ tests. This clearly supports the genetic hypothesis. So, too, do data showing that regression to the mean is greater for black children who have high IQ parents and siblings and less for black children who have low IQ parents and siblings than it is for their respective white counterparts.

Other supporting evidence comes from transracial adoption studies. Korean and Vietnamese children adopted into white American and white Belgian homes were examined by Clark and Hanisee (1982), Frydman and Lynn (1989), and Winick, Meyer, and Harris (1975). Although, prior to adoption, many had been hospitalized for malnutrition, they went on to develop IQs ten or more points higher than their adoptive national norms.

By contrast, black and mixed-race (black-white) children adopted into white middle-class families typically perform at a lower level than similarly adopted white children. In the well-known Minnesota Transracial Adoption Study, by age 17, adopted children with two white biological parents had an average IQ of 106, adopted children with one black and one white biological parent averaged an IQ of 99, and adopted children with two black biological parents had an average IQ of 89 (Weinberg, Scarr, & Waldman, 1992). It is sobering to realize how little the evolutionary perspective has taken hold in the social sciences when we consider that while they corroborate Francis Galton's observations made over 130 years ago (as mentioned earlier), the results of the

Minnesota Transracial Adoption Study are treated as highly controversial (when they are not simply ignored).

GENETIC SIMILARITY THEORY AND ETHNIC NEPOTISM

Even postulating an evolutionary basis for some *universal* traits has been anathema to the political left which fears that doing so may seem to justify social policies of exclusion. Ethnic nationalism, for example, is typically held to be an entirely cultural phenomenon, remediable through education and other intervention techniques. In "civilized" societies, notions of ethnic identity are considered archaic and "reactionary" (unless practiced by groups historically discriminated against, in which case they may be considered justifiable and even "liberationist"). Yet, ethnopolitical warfare, because it is a means by which genes can be replicated more efficiently, can best be understood within the evolutionary perspective.

Germane to this discussion is Genetic Similarity Theory and its implications for social assortment and ethnic nepotism (Rushton, 1989). Briefly, I have offered a new theory and presented empirical evidence that social identity, and especially ethnic identity, are constructed on the basis of genetic similarity in order to direct altruism toward those carrying similar genes (extended kin), thereby increasing their ability to replicate. From this perspective, xenophobia and genocide are seen as the "dark side" of altruism.

In-group-amity and out-group-enmity, group selection, and group replacement were put forth in general terms by nineteenth-century evolutionists such as Darwin, Spencer, and W. Graham Sumner. However, such early attempts fell very much out of favor with Marxist, sociological, behaviorist, and psychodynamic approaches. Recent theoretical and empirical advances and mathematical models have led to a reconsideration of these processes (see also Wilson & Sober, 1994).

THE MORALISTIC FALLACY AND BEHAVIORAL CREATIONISM

Although many high-profile members of evolutionary societies are notable for continuing to do battle against Christian fundamentalists and their creationist crusade (an argument some might think was settled in the nineteenth century), these same individuals are notably quiet when it comes to combating left-wing ideology and what anthropologist Vincent Sarich (1982, 1995) has called "behavioral creationism." The political left, I contend, poses a more serious and immediate threat to the advance of evolutionary science than does religious fundamentalism because fundamentalism has no clout in research universities whatsoever. The chilling effect of self-imposed censorship, euphemistically referred to as "heightened sensitivity," comes from within our own membership, our own academic institutions, and indeed even our own minds.

The deliberate withholding of evidence has become all too characteristic of evolutionary scientists when writing about race. Three highly praised recent books exemplify this trend: Gould's (1996) "revised and expanded edition" of *The Mismeasure of Man*, Diamond's (1997) *Guns, Germs, and Steel*, and Stringer and McKie's (1996) *African Exodus*. (I have reviewed the first two of these in detail—Rushton, 1997b; 1999).

Gould is the most well known of these three. In his 1981 edition of *Mismeasure*, he charged nineteenth-century scientists with "juggling" and "finagling" brain-size data in order to place Northern Europeans at the apex of civilization. Implausibly, he argued that Paul Broca, Francis Galton, and Samuel George Morton all "finagled" in the *same* direction and by *similar* magnitudes using *different* methods. Gould asks us to believe that Broca "leaned" on his autopsy scales when measuring wet brains by just enough to produce the same differences that Morton caused by "overpacking" empty skulls and that Galton caused with his "extra loose" grip on calipers while measuring heads!

Yet even before *Mismeasure*'s first edition (1981), new research was confirming the work of these nineteenth-century pioneers. Gould neglected to mention Van Valen's (1974) review which established a positive correlation between brain size and intelligence. As reviewed earlier in this chapter, the single most devastating development for Gould is the latest research on brain size. Was he asleep throughout the 1990s—called, with good reason, "The Decade of the Brain"?

Jared Diamond, another well-known evolutionary biologist and writer for *Discover* magazine, also joined the debate over racial differences in IQ. In a few *ex cathedra* pronouncements, Diamond branded the genetic argument "racist" (pp. 19–22), declared Herrnstein and Murray's (1994) *The Bell Curve* "notorious" (p. 431), and gave away his game when he pontificated: "The objection to such racist explanations is not just that they are loathsome but also that they are wrong" (p. 19). He summarizes his solution to one of philosophy and social science's most enduring questions in one credal sentence: "History followed different courses for different peoples because of differences among people's environments, not because of biological differences among peoples themselves" (p. 25).

Diamond's thesis is that the people of the Eurasian continent were environmentally, rather than biologically, advantaged. They had the good fortune to have lived in centrally located homelands that were oriented along an east-west axis, thereby allowing ready diffusion of their abundant supply of domesticable animals, plants, and cultural innovations. The north-south axis of Africa and of the Americas inhibited diffusion due to severe changes in climate. For example, the tropical jungle of Central America effectively stopped both the southward migration of domestic corn from Mexico and the northward migration of the domestic llama from Peru. Thus, the agriculturally wealthy Eurasians had a long head start in developing a surplus population with a division of labor that enabled civilization to arise.

It is sad to see an evolutionary biologist like Diamond failing to inform his readers that it is different environments that cause, via natural selection, biological differences among populations. Each of the Eurasian developments he describes created positive feedback loops, thereby selecting for increased intelligence and various personality traits (e.g., altruism, rule-following, ability to tolerate greater levels of population density). Subsequently, internecine tribal and ethnic warfare was a potent force in natural selection of human groups. Diamond omits to discuss how intergroup competition over scarce resources influences the human genotype, including why hominid brain size increased threefold over the last 3 million years.

A final example of political correctness by an important scholar who probably knew better is *African Exodus* by paleontologist Christopher Stringer at the British Museum of Natural History (written in collaboration with journalist Robin McKie). The parts of the book that review human origins are competent and very readable. Unfortunately, major errors appear in the book when it descends to the politically obligatory trashing of both *The Bell Curve* and my own work. In my case, instead of taking time to read, cite, and critique my 1995 book intelligently, the authors rely mainly on a 1994 account of it in the tabloid magazine *Rolling Stone*!

The basic political argument of *African Exodus* is as follows: "In any case, the story of our African Exodus makes it unlikely that there are significant structural or functional differences between the brains of the world's various peoples" (181). The logic here is especially odd given that other parts of the book present a fascinating discussion of how populations vary in jaw size and in number of teeth. For example, page 215 states: "Among Europeans, for example, it has been found that up to 15 percent of people have at least two wisdom teeth missing . . . while in east Asia, the figure can be as much as 30 percent in some areas." As an example of evolutionary pressure, the book describes how before modern medicine, impacted wisdom teeth often became infected and led to death.

The authors appear to find it plausible for evolution to act through differential death rates resulting from differences in the number of wisdom teeth and yet find it implausible that death rates could vary in different regions because of differential intelligence as an adaptation to extreme cold. While Stringer and McKie describe how noses and skin color have been shaped in different regions, they deny that there are any cognitive differences and they withhold from readers the modern literature on brain size and IQ. Perhaps least forthright in this regard is the citation (p. 177) of Beals, Smith, and Dodd's (1984) study of worldwide variation in cranial size (which I cited earlier) and their attribution of racial differences only to "climate," as though climate is not a likely potent source of natural selection for intelligence.

THE PERVASIVENESS OF THE EGALITARIAN DOGMA

In the United States the First Amendment protects the right of every citizen to free speech, and there is not much the government can do to silence unpopular ideas. In Canada and many Western European countries, however, "anti-hate" laws exist, as well as laws against spreading what is termed "false news." Governments can and do prohibit speech on topics they consider obnoxious. In Denmark, a woman wrote a letter to a newspaper calling national domestic partner laws "ungodly" and homosexuality "the ugliest kind of adultery." She and the editor who published her letter were targeted for prosecution. In Great Britain, the Race Relations Act forbids speech that expresses racial hatred, "not only when it is likely to lead to violence, but generally, on the grounds that members of minority races should be protected from racial insults."

In his book, *Kindly Inquisitors*, Rauch (1992) showed that even in the United States, despite the protections supposedly guaranteed by the First Amendment, nongovernmental institutions, including colleges and universities, have set up "anti-harassment" rules prohibiting, and establishing punishments for, "speech or other expression" that is intended to "insult or stigmatize an individual or a small number of individuals on the basis of their sex, race, color, handicap, religion, sexual orientation or national and ethnic origin." This decree, taken from Stanford's policy adopted in 1990, is more or less representative. One case at the University of Michigan became well known because it led a federal court to strike down the rule in question. A student claimed, in a classroom discussion, that he thought homosexuality was a disease treatable with therapy. He was formally disciplined by the university for violating the school's policy and victimizing people on the basis of sexual orientation.

A growing number of cases of intimidation and censorship are coming to light, especially of psychologists studying individual and group variation (see Hunt, 1999; Pearson, 1997). Noteworthy accounts have been provided by Arthur Jensen (1973) in the Preface to his *Educability and Group Differences*, by Richard Herrnstein (1973) in the Preface to his *IQ in the Meritocracy*, and by Hans Eysenck (1997) in his Introduction to Roger Pearson's *Race, Intelligence and Bias in Academe* and his (1997) autobiography *Rebel with a Cause*. Readers might also see Glayde Whitney's (1995) account of the reaction of his colleagues to his presidential address to the Behavior Genetics Association. Although the editors have persuaded me to delete much of my own personal account, interested readers can consult Hunt (1999), Pearson (1997), and Rushton (1998).

My book *Race, Evolution, and Behavior* was published by Transaction Publishers in 1995, at the same time as Herrnstein and Murray's (1994) *The Bell Curve*, and was soon caught up in that debate. Both books were reviewed in the *New York Times Book Review* (October 16, 1994) by Malcolm Browne, the *New York Times* science writer, along with a third book, Seymour Itzkoff's (1994) *The Decline of Intelligence in America*. Browne concluded his review with the statement that "the government or society that persists in sweeping this topic under the rug will do so at its peril."

Sweeping the topic under the carpet, however, is exactly what was attempted. One lurid article screaming "Professors of HATE" (in five-inch letters!) appeared in *Rolling Stone* magazine (October 20, 1994). Taking up the entire next page was a photograph of my face, hideously darkened, twisted into a ghoulish image, and superimposed on a Gothic university tower. In another long propaganda piece entitled "The Mentality Bunker" which appeared in *Gentleman's Quarterly* (November 1994), I was misrepresented as being an outmoded eugenicist and pseudoscientific racist. A photograph of me was published in brown tint reminiscent of vintage photos from the Hitler era.

It is difficult to disagree with Charles Murray's (1996, p. 575) conclusion in his analysis of the aftermath to *The Bell Curve* that in regard to heritable variation and race, science has "become self-censored and riddled with taboos—in a word, corrupt." I find the pervasiveness of the egalitarian orthodoxy in high places particularly worrying. In 1992, then editor-in-chief of *Nature*, John Maddox, attacked my work in a full-page lead editorial. Maddox likened the possibility of finding significant group differences in brain size to contradicting accepted views of an ellipsoid earth, continental drift, and relativity theory.

Another of the world's prestigious journals, *Science*, featured special issues documenting the underrepresentation of minority scientists (November 13, 1992, November 12, 1993, March 29, 1996). Unflinching statistics were accompanied by muddled analysis. First, the word *minority* is misleading. In truth, only blacks, Hispanics, and American Indians are underrepresented in science: Several other minorities are overrepresented. Adopting the criterion of being listed in *American Men and Women of Science* and using Weyl's (1989) ethnic classification of surnames, we find that Chinese are overrepresented relative to their numbers in the population by 620 percent, Japanese by 351 percent, and Jews by 424 percent. These figures cast doubt on prejudice as an explanation and, instead, suggest considering factors shown to be characteristic of the various groups.

CONCLUSION

The gene-based evolutionary models I have proposed to explain ethnocentrism and racial group differences may provide a catalyst for understanding individual differences and human nature. Such gene-based hypotheses, however, conflict with current orthodoxy in the social sciences which holds that behavioral differences are almost exclusively the result of social inequalities. Less well recognized is that they also conflict with current orthodoxy in evolutionary psychology which holds that there is a "universal human nature" (Tooby & Cosmides, 1990, p. 18). Michael Bailey searched the Medline database for articles published in the last decade that referenced the keywords—*evolution, genetics, behavior,* and *human*—and the combination of those words. Although he found that each word was referenced by several thousand articles, he found only one article that referenced all four (Bailey, 1997, p. 82). The r–K life-history theory that I have proposed unites the evolutionary psychology tradition begun by Darwin with the behavior genetic tradition begun by Galton. Building on the contemporary synthesis of E. O. Wilson (1975), it accounts for individual as well as racial variation by postulating that evolution selected a genetically organized suite of characters to optimize the allocation of metabolic energy to survival, growth, and reproduction.

Understanding the problems of the next millennium will require knowledge from biology as much as from the social and physical sciences. Effective public policies must be based on sound scientific conclusions rather than popular assumptions or misconceptions. As the world is made smaller by the global high-tech economy, competition and inequality among individuals and between groups might well increase rather than decrease.

Let us return to the problem we face—political correctness. Its central thesis is the environmental determinism of all important human traits. It stems from Marxism, and at worst it is a Marxist-Lysenkoist denial of genetics and a belief that social and economic oppression is at the root of all major individual and group differences (Pearson, 1997; Whitney, 1995). The Marxist invasion of liberal political sentiment has been so extensive that many of us think that way without realizing it. We censor ourselves lest we even dare to think the forbidden thoughts.

In an invited paper to the British Association for the Advancement of Science, Hans Eysenck wrote the encroachments on scholarship of what is now known as "political correctness":

It used to be taken for granted that it was not only ethically *right* for scientists to make public their discoveries; it was regarded as their *duty* to do so. Secrecy, the withholding of information, and the refusal to communicate knowledge were rightly regarded as cardinal sins against the scientific ethos. This is true no more. In recent years it has been argued, more and more vociferously, that scientists should have regard for the social consequences of their discoveries, and of their pronouncements; if these consequences are undesirable, the research in the area involved should be terminated, and the results already achieved should not be publicized. The area which has seen most of this kind of argumentation is of course that concerned with the inheritance of intelligence, and with racial differences in ability. (Eysenck, 1975, p. 1, emphasis in original)

Richard Lynn (1995) noted that many politically left-of-center scientists are currently in the same position as Christians were after the publication of *The Origin of Species*. He proposes that liberals do now what honest, intelligent Christians did then and still do today. Bite the bullet, and jettison those aspects of their world view (like egalitarianism) that are incompatible with the science of natural selection. "Political correctness" must be discarded if evolutionary theory is to achieve its full promise to become the unifying framework for the human sciences.

REFERENCES

Armstrong, E. (1990). Brains, bodies and metabolism. *Brain, Behavior and Evolution,* *36*, 166–176.

Bailey, J. M. (1997). Are genetically based individual differences compatible with species-wide adaptations? In N. L. Segal, G. Weisfeld, & C. Weisfeld (Eds.), *Uniting psychology and biology: Integrative perspectives on human development* (pp. 81–100). Washington, DC: American Psychological Association.

Barash, D. P. (1995). [Review of the book *Race, evolution and behavior.*] *Animal Behavior, 49,* 1131–1133.

Barkow, J. H., Cosmides, L., & Tooby, J. (1992). *The adapted mind: Evolutionary psychology and the generation of culture.* New York: Oxford University Press.

Beals, K. L., Smith, C. L., & Dodd, S. M. (1984). Brain size, cranial morphology, climate, and time machines. *Current Anthropology, 25,* 301–330.

Bouchard, T. J., Jr., Lykken, D. T., McGue, M., Segal, N. L., & Tellegen, A. (1990). Sources of human psychological differences: The Minnesota study of twins reared apart. *Science, 250,* 223–228.

Broman, S. H., Nichols, P. L., Shaughnessy, P., & Kennedy, W. (1987). *Retardation in young children.* Hillsdale, NJ: Erlbaum.

Clark, E. A., & Hanisee, J. (1982). Intellectual and adaptive performance of Asian children in adoptive American settings. *Developmental Psychology, 18,* 595–599.

Darwin, C. (1859). *The origin of species.* London: Murray.

———. (1871). *The descent of man.* London: Murray.

Degler, C. N. (1991). *In search of human nature.* New York. Oxford University Press.

Devlin, B., Daniels, M., & Roeder, K. (1997). The heritability of IQ. *Nature, 388,* 468–471.

Diamond, J. (1997). *Guns, germs, and steel: The fates of human societies.* New York: Norton.

Draper, P. (1989). African marriage systems: Perspective from evolutionary ecology. *Ethology and Sociobiology, 10,* 145–169.

Ellis, L., & Nyborg, H. (1992). Racial/ethnic variations in male testosterone levels: A probable contributor to group differences in health. *Steroids, 57,* 72–75.

Eysenck, H. J. (1971). *Race, intelligence, and education.* London: Temple Smith.

———. (1975). The ethics of science and the duties of scientists. *British Association for the Advancement of Science,* New Issue, No. 1. (Reprinted in H. Gibson, *Hans Eysenck: The man and his work.* London: Owen.)

———. (1997a). Introduction: Science and racism. In R. Pearson, *Race, intelligence, and academe.* Washington, DC: Scott-Townsend.

———. (1997b). *Rebel with a cause* (revised and expanded ed.). New Brunswick, NJ: Transaction.

Frydman, M., & Lynn, R. (1989). The intelligence of Korean children adopted in Belgium. *Personality and Individual Differences, 10,* 1323–1326.

Galton, F. (1865). Hereditary talents and character. *Macmillan's Magazine, 12,* 157–166, 318–327.

———. (1869). *Hereditary genius.* London: Macmillan.

———. (1888). Head growth in students at the University of Cambridge. *Nature, 38,* 14–15.

Gordon, R. A. (1997). Everyday life as an intelligence test: Effects of intelligence and intelligence context. *Intelligence, 24,* 203–320.

Gould, S. J. (1981). *The mismeasure of man.* New York: Norton.

———. (1996). *The mismeasure of man* (revised and expanded ed.). New York: Norton.

Hamilton, W. D. (1964). The genetical evolution of social behaviour: I and II. *Journal of Theoretical Biology, 7,* 1–52.

Harvey, J., Persaud, R., Ren, M. A., Baker, G., & Murray, R. M. (1994). Volumetric MRI measurements in bipolars compared with schizophrenics and healthy controls. *Psychological Medicine, 24,* 689–699.

Herrnstein, R. J. (1973). *IQ in the meritocracy.* Boston, MA: Little, Brown.

Herrnstein, R. J., & Murray, C. (1994). *The bell curve: Intelligence and class structure in American life.* New York: Free Press.

Ho, K. C., Roessmann, U., Straumfjord, J. V., & Monroe, G. (1980). Analysis of brain weight. I and II. *Archives of Pathology and Laboratory Medicine, 104,* 635–645.

Hofman, M. A. (1993). Encephalization and the evolution of longevity in mammals. *Journal of Evolutionary Biology, 6,* 209–227.

Hunt, M. (1999). *The new know-nothings: The political foes of the scientific study of human nature.* New Brunswick, NJ: Transaction.

Itzkoff, S. W. (1994). *The decline of intelligence in America.* Westport, CT: Praeger.

Jensen, A. R. (1973). *Educability and group differences.* London: Methuen.

———. (1994). Psychometric g related to differences in head size. *Personality and Individual Differences, 17,* 597–606.

———. (1998). *The g factor.* Westport, CT: Praeger.

Jensen, A. R., & Whang, P. A. (1993). Reaction times and intelligence: A comparison of Chinese-American and Anglo-American children. *Journal of Biosocial Science, 25,* 397–410.

———. (1994). Speed of accessing arithmetic facts in long-term memory: A comparison of Chinese-American and Anglo-American children. *Contemporary Educational Psychology, 19,* 1–12.

Johanson, D. C., & Edey, M. A. (1981). *Lucy: The beginnings of human kind.* New York: Simon & Schuster.

Lynn, R. (1982). IQ in Japan and the United States shows a growing disparity. *Nature, 297,* 222–223.

———. (1991). Race differences in intelligence: A global perspective. *Mankind Quarterly, 31,* 255–296.

———. (1995, March 20). Wright and wrong. *National Review.*

———. (1997). Geographical variation in intelligence. In H. Nyborg (Ed.), *The scientific study of human nature.* Oxford: Elsevier.

Lynn, R., & Hattori, K. (1990). The heritability of intelligence in Japan. *Behavior Genetics, 20,* 545–546.

Maddox, J. (1992). How to publish the unpalatable? *Nature, 358,* 187.

Murray, C. (1996). Afterword. In R. J. Herrnstein & C. Murray, *The Bell Curve* (paperback ed.). New York: Free Press.

Osborne, R. T. (1980). *Twins: Black and white.* Athens, GA: Foundation for Human Understanding.

Pagel, M. D., & Harvey, P. H. (1988). How mammals produce large-brained offspring. *Evolution, 42,* 948–957.

Pearson, K. (1906). On the relationship of intelligence to size and shape of head, and to other physical and mental characters. *Biometrika, 5,* 105–146.

Pearson, R. (1997). *Race, intelligence and bias in academe* (2nd ed.). Washington, DC: Scott-Townsend.

Plomin, R., & Daniels, D. (1987). Why are children in the same family so different from one another? *Behavioral and Brain Sciences, 10,* 1–60.

Plomin, R., DeFries, J. C., McClearn, G. E., & Rutter, M. (1997). *Behavioral genetics* (3rd ed.). New York: W. H. Freeman.

Rauch, J. (1992). *Kindly inquisitors: New attacks on free thoughts*. Chicago: University of Chicago Press.

Rushton, J. P. (1989). Genetic similarity, human altruism, and group selection. *Behavioral and Brain Sciences, 12*, 503–559.

———. (1990). Race and crime: A reply to Roberts and Gabor. *Canadian Journal of Criminology, 32*, 315–334.

———. (1992a). Contributions to the history of psychology: XC. Evolutionary biology and heritable traits (with reference to Oriental-white-black differences): The 1989 AAAS paper. *Psychological Reports, 71*, 811–821.

———. (1992b). Cranial capacity related to sex, rank, and race in a stratified random sample of 6,325 U.S. militiary personnel. *Intelligence, 16*, 401–413.

———. (1994). Sex and race differences in cranial capacity from International Labour Office data. *Intelligence, 19*, 281–294.

———. (1995a). Race and crime: International data for 1989–1990. *Psychological Reports, 76*, 307–312.

———. (1995b). *Race, evolution, and behavior: A life-history perspective*. New Brunswick, NJ: Transaction.

———. (1997a). Cranial size and IQ in Asian Americans from birth to age seven. *Intelligence, 25*, 7–20.

———. (1997b). Race, intelligence, and the brain: The errors and omissions of the ''revised'' edition of S. J. Gould's *The Mismeasure of Man* (1996). *Personality and Individual Differences, 23*, 169–180.

———. (1998, March). The new enemies of evolutionary science. *Liberty, 11* (4) 31–35.

———. (1999). [Review of Jared Diamond's *Guns, germs, and steel: The fates of human societies*. New York: Norton]. *Population and Environment, 21*, 99–107.

Rushton, J. P., & Ankney, C. D. (1996). Brain size and cognitive ability: Correlations with age, sex, social class and race. *Psychonomic Bulletin and Review, 3*, 21–36.

Rushton, J. P., Fulker, D. W., Neale, M. C., Nias, D. K. B., & Eysenck, H. J. (1986). Altruism and aggression: The heritability of individual differences. *Journal of Personality and Social Psychology, 50*, 1192–1198.

Rushton, J. P., & Osborne, R. T. (1995). Genetic and environmental contributions to cranial capacity estimated in Black and White adolescents. *Intelligence, 20*, 1–13.

Sarich, V. M. (1982, October). *My adventures among the creationists*. Seminar presentation to Department of Anthropology, University of California, Berkeley, California.

———. (1995). In defense of *The Bell Curve*. *Skeptic, 3*(3), 84–93.

Seligman, D. (1992). *A question of intelligence: The IQ debate in America*. New York: Birch Lane.

Spencer, H. (1851). *Social statics*. London: Chapman.

Stringer, C., & McKie, R. (1996). *African exodus*. London: Cape.

Stringer, C. B., & Andrews, P. (1988). Genetic and fossil evidence for the origins of modern humans. *Science, 239*, 1263–1268.

Sulloway, F. J. (1996). *Born to rebel: Birth order, family dynamics, and creative lives*. New York: Pantheon Books.

Tooby, J., & Cosmides, L. (1990). On the universality of human nature and the uniqueness of the individual: the role of genetics and adaptation. *Journal of Personality, 58*, 17–67.

Van Valen, L. (1974). Brain size and intelligence in man. *American Journal of Physical Anthropology, 40*, 417–424.

Vernon, P. E. (1982). *The abilities and achievements of Orientals in North America.* New York: Academic.

Weinberg, R. A., Scarr, S., & Waldman, I. D. (1992). The Minnesota Transracial Adoption Study: A follow-up of IQ test performance at adolescence. *Intelligence, 16*, 117–135.

Weyl, N. (1989). *The geography of American achievement.* Washington, DC: Scott-Townsend.

Whitney, G. (1995). Ideology and censorship in behavior genetics. *Mankind Quarterly, 35*, 327–342.

Wickett, J. C., Vernon, P. A., & Lee, D. H. (1996). General intelligence and brain volume in a sample of healthy adult male siblings. *International Journal of Psychology, 31*, 238–239. (Abstract).

Wilson, D. S. (1994). Adaptive genetic variation and human evolutionary psychology. *Ethology and Sociobiology, 15*, 219–235.

Wilson, D. S., & Sober, E. (1994). Reintroducing group selection to the human behavioral sciences. *Behavioral and Brain Sciences, 17*, 585–654.

Wilson, E. O. (1975). *Sociobiology: The new synthesis.* Cambridge, MA: Harvard University Press.

Wilson, R. S. (1983). The Louisville twin study: Developmental synchronies in behavior. *Child Development, 54*, 298–316.

Winick, M., Meyer, K. K., & Harris, R. C. (1975). Malnutrition and environmental enrichment by early adoption. *Science, 190*, 1173–1175.

PART IV

SOCIOBIOLOGY AND THE CONCERNS OF SOCIOLOGISTS

Cultural and ethnic contrasts have already introduced us to such intimate concerns as intermarriage and sexual attraction. The next part of this volume deals with these more intimate details of the broader themes comprising earlier chapters. Ada Lampert takes an interesting perspective by turning to Darwin's own marriage to a first cousin as an illustration of the human tendency of attraction toward those who carry similar traits. This would apply, of course, to the inbreeding Nabulsi finds in tribal groups and to the broader themes covered in earlier chapters in this volume. Lampert's empirical study covers attraction between brothers who are either more alike or less alike in biological traits and finds that biological similarity increases attraction.

Harold A. Euler and Barbara Weitzel pursue the question of attraction between family members by examining grandparental solicitude under a variety of conditions. The findings of their study are quite consistent with an evolutionary analysis of genetic relationships and the potential arising from human culture and parental care. With elaboration they will also fit in nicely with recent studies explaining the cessation of reproduction at menopause in human females. This is a good example of the testing of a general theory of evolutionary biology applied specifically to humans and represents real progress toward E. O. Wilson's predictions of an evolutionary unification of the social sciences. Norma J. Schell and Carol C. Weisfeld provide another empirical investigation of patterns of spousal relationships within American culture as compared with a similar study of European findings. Ultimately we will achieve an overall picture that will bring about the integration of both biological and cultural factors within a Darwinian perspective. Finally M. L. Butovskaya and A. G. Kozintsev provide a further analysis of human males and females and the patterns found within different cultures, this time also showing how these patterns translate into the development of sex roles in childhood.

14

With Whom Was Darwin Supposed to Fall in Love?

Ada Lampert

Since Darwin understood that a living organism's traits were the product of natural selection, singling out those characteristics that might best increase the number of offspring in future generations, there came to his mind the very obvious insight that "mental," "psychological" traits are subject to this very same principle: they are selected according to their advantage in producing off-spring. Darwin himself wrote (1877) about emotions such as fear, jealousy, and aggression and about sex differences. We may consider falling in love to be an emotion even more directly and powerfully affecting the ability to produce off-spring. "Darwinistic falling in love" would therefore be of the kind that ensured children, grandchildren, great-grandchildren, and great-great-grandchildren.

Was Darwin behaving like a Darwinist when he fell in love? As far as I know, there was one woman in Darwin's life, to whom he was married for forty-five years and with whom he raised to maturity seven of ten children born. Their most famous son, Sir Francis Darwin, wrote (1949) of them: "His gentle and pleasant nature was revealed in its most beautiful form in his relations with my mother. In her he found his happiness, and thanks to her his life, which under different circumstances might have been overcast by a dark shadow, be-came a life of satisfaction and of quiet happiness." This, then, is successful falling in love. Who was the woman who succeeded so well? Darwin's biog-rapher (Hemleben, 1988) writes of her: "Fate did not make it difficult for him to find a wife. . . . Between the Darwins and the Wedgewoods there had always been close relations. . . . He chose Emma Wedgewood, the daughter of his be-loved uncle. . . . Naturally, Darwin and his cousin knew each other from child-hood" (p. 42). Two important features are mentioned in this description: (1) As a cousin, she had genetic similarity to Darwin. (2) As the daughter of a friendly family, she was familiar to him from childhood. In this chapter I wish to propose that among the mechanisms selected to guide attachment and falling in love between a man and a woman can be found the following two: (1) the search for genetic similarity between potential mates; (2) early childhood imprinting of familiar traits, which are searched for by adults in love relationships.

How would genetic similarity contribute to the successful production of off-spring? One may imagine a person falling in love with a cow. One could even find advantages in this kind of attraction: warmth, harmony, peace and quiet, and a refrigerator full of dairy products. But one thing would undoubtedly be missing: children. This nice couple is sentenced to childlessness because the genetic distance between father and mother prevents genetic transcription in common offspring.

In recent years the activity of the DNA "quality-control mechanisms" has begun to be deciphered (Radman & Wagner, 1988; Rennie, 1990). These are proteins that constantly monitor every new genetic sequence, discover transcrip-tion errors, and repair them. Also, this very same system probably prevents the successful integration of genetic sources from the father and from the mother and the growth of an offspring if and when the amount of differences between them surpasses the little allowed. Hence the opposite lesson: in order to ensure its continuation, the old and faithful DNA must uphold the existing sequence and bequeath itself as accurately as possible to next generations. When it needs partners in the reproduction process—and that is the meaning of sexuality—it will choose a homogeneous partner, similar to itself. If mother and father are similar, the chances of fertility rise, as does success in raising the offspring.

Falling in love with a gorilla is getting closer, but the genetic distance is still too great. A chimpanzee is a realistic possibility, and the two could probably have children. But children are not the goal, they are only the means: the goal is grandchildren, and here success is doubtful. The child of a man and a chim-panzee would be far from good-looking, nor could it be a great genius, and it is reasonable to assume that nobody would want it as a mate.

The evolutionary advantage is thus reserved for whoever limits himself to the human species, and even within these limits, an Englishman like Darwin would increase his fitness by choosing neither a Negro nor a Scot, but an English woman, and of all these, one most similar to himself. Indeed, a growing number of studies indicate that the greater the genetic similarity between partners, the greater the fertility, the harmony and stability in the marriage, and the mutual support, efficiency, and satisfaction with family life (Daly & Wilson, 1982; Russel, Wells, & Rushton, 1985; Weisfeld, Russel, Weisfeld, and Wells, 1992; Thiessen & Gregg (1980). Did the advantages of similarity push natural selection to shape a tendency of preferring those who are like us? Are we attracted to those who are genetically similar to us?

J. Philippe Rushton (1988) studied about one thousand cases of legal claims for parental recognition filed in courts in Canada and settled by means of blood antigen analysis. These were all white couples who had had intercourse resulting in the birth of a child. The genetic test was to determine whether the defendant was indeed the father. Rushton used this data to assess the degree of genetic similarity between the mother and father. Two important findings are relevant: (1) According to ten different genetic markers, the man and the woman who had sexual relations were more similar to each other than random matchings made by computer sampling from this same population. That is, their meeting, attraction, and relationship were somehow guided by the search for genetic similarity. When the man was declared the father, the genetic similarity between him and his partner was greater than that of a couple where the man was excluded from paternity. That is, greater genetic similarity ensures higher fertility. These findings support the idea that falling in love was selected as a behavior capable of identifying genetic similarity.

If we take this principle to its extreme, a man should be expected to fall in love with his mother, with whom he shares over 50 percent of his genes. If this were actually the case, and every man should be falling in love with and marrying his mother, we would face at least two new problems: (1) the increase in harmful recessive genes that damage the offspring's fitness; (2) the diminishment of variance in the genetic pool. Even if the problem of harmful recessive genes could be solved by purifying selection, variance is still an essential necessity whenever the environment presents new pressures. For example, in complex animals, in which a close "arms race" constantly takes place between viruses and bacteria, on the one hand, and an ever-inventive variety of antibodies, on the other hand, limitation of variance would give the enemy the upper hand. Genetic innovation in the immune system is a must (Hamilton, Axelrod, & Tanese, 1990).

This "arms race," which can be compared to the race between drivers' antiradar devices and cops' better radars, rises to higher and higher levels of sophistication to the point where it can be shown that the existence of single negative recessive genes in the population is more advantageous than a gene pool made up solely of uniform, healthy genes. An example in point is the story of malaria, one of the most devastating diseases of our times. More than 100 million people are infected with malaria every year, and about 1.5 million die (Rennie, 1992). One of the few human genes known to confer resistance to malaria is the gene for sickle-cell anemia. This gene causes the red blood cells to produce a defective form of the oxygen-carrying hemoglobin molecule. When one inherits one recessive copy of the sickle-cell-anemia gene, one enjoys better protection against malaria and good-enough health, which the second normal copy of the gene affords. If one inherits two copies of the sickle-cell-anemia gene, it spells death. People who inherit two healthy copies are defenseless against malaria. In malaria-infested areas of the world, the advantages of the "negative" recessive gene are greater than its drawbacks. In Nigeria, for instance, the gene is carried by 40 percent of the population. In malaria-free areas the drawbacks surpass the advantages, and of the black population of South

Africa, only 2 percent carry the gene. Natural selection, which guides people with whom to fall in love in order to ensure success for their children and grandchildren, the future inhabitants of Nigeria, will diligently and accurately combine the two principles: (1) falling in love with "one of us" in order to acquire the sickle-cell-anemia gene; (2) supposing the required gene has been found, additional mating among the immediate family must be avoided in order to prevent bequeathing two such genes to one child. In other words, falling in love will try to calibrate itself to "one of our own," but not to "our own flesh and blood."

The sickle-cell-anemia gene is but one example, albeit a sophisticated one, of the need to maintain genetic variance as a way to fight parasites. Further, we could say that the demand for genetic similarity that ensures former evolutionary advantages must stop before totality in order to enable new genetic inventions to be put to environmental trial and provide new, yet-known adaptations. Optimal falling in love will direct itself at one of the tribe but not at one of the family. This is why we have the phenomenon of "incest taboo." There is a stop sign on the continuum of genetic similarity ranging from cow to mother. We do not run to the end of the road, but stop somewhere before it. A No Entry sign is posted in front of mothers and sisters. Most people obey it, as do chimpanzees (Pusey, 1980), rhesus monkeys (Missakian, 1973), Japanese monkeys (Enomoto, 1974), Japanese quail (Bateson, 1982). Israeli babblers (Zahavi, 1990) and a wide variety of other creatures (Pusey, 1987; Pfennig & Sherman, 1995).

How can a creature know to direct its preference at the optimum, to seek genetic similarity, yet to avoid overdoing it? Natural selection of "correct" falling in love has in fact selected behavior mechanisms to implement it. One of these mechanisms is imprinting. Attraction to genetic similarity is fed by "positive imprinting," whereas avoidance of oversimilarity is ensured by "negative imprinting": Individuals with whom one spends one's childhood are not attractive to one. They are negatively imprinted on the mind. A good example of the relationship between the increase of genetic variance and the reinforcement of the immune system can be found in negative imprinting in mice. For young mice, the scent of the animals surrounding them is negatively imprinted. In adulthood, when they are seeking a mate, they avoid the mice whose scent is negatively imprinted. The gene for this scent is located in the chromosome near an important genetic complex of the immune system, the major histocompatability complex (MHC). When a mouse avoids a mate carrying a familiar scent, it ensures variation and innovation of the immune system in its offspring (Beauchamp, Yamazaki, & Boyse, 1985; Manning, Wakelaud, & Potts, 1992). Due to location proximity, both the identifying scent and the MHC are passed to the offspring.

In humans this mechanism is especially salient when it is extended to those who are not members of our family, but with whom we have spent our childhood. Famous examples are the avoidance of Israeli kibbutz-born youngsters from mating within the peer group (Shepher, 1971) and the failure of married life of couples who are matched from infancy and grow up together within the male's family in Taiwan (Wolf & Huang, 1980; Wolf, 1995). In both cases the special childrearing arrangements trigger the mechanism of negative imprinting in vain. Kibbutz children spent most of their days and nights together, not with their parents. The adopted girls in Taiwan grew up with their future grooms as if they were brother and sister. Blind negative imprinting was created to work in normal family life, where young children come to know their parents, brothers, and sisters, and to avoid them. The unusual environment of these children gave us the opportunity to look at the avoidance mechanism where it carried out extra duties.

When all is normal and children are raised with and by their parents, they are negatively imprinted and are not sexually attracted to them. What, then, happened to Oedipus? Oedipus fell in love with his mother and married her, and Freud turned him into a symbol of all children, doomed to fall in love with their mothers. But the story of Oedipus points to the very weakness in Freud's conclusion: Oedipus was taken away from his mother the day he was born and never knew her. He met her as an adult, and out of many pretty young girls he fell in love with her, probably due to his search for genetic similarity. The inhibition, the No Entry sign, was not evoked because of the lack of negative imprinting.

Imprinting, whether positive or negative, is made on a young and soft mind in its critical, sensitive period, when it exposes itself willingly to impressions from the environment, weaving them into a part of itself. Afterwards the images harden, and that which was assimilated becomes a lifelong legacy. Negative imprinting, which excludes sexual attraction to family members, takes place in the child's first years. The child's parents are long past their critical period; they no longer have soft minds. There are, therefore, many cases of a parent feeling sexually attracted to a child. This attraction arises due to genetic similarity and is not blocked by negative imprinting. Most parents nevertheless manage to deal with this problem. Some, like Freud, project the attraction they feel onto their children and call it an "Oedipus complex." Unfortunately, there are some parents, particularly fathers, who force themselves upon their children (Lampert, 1997).

Just as negative imprinting was selected for preventing too high a genetic similarity, so was positive imprinting selected as a mechanism to ensure attraction to similarity. One young man, a sworn lover of women, once told me that he liked women with thin, soft hair on their upper lip. Apparently, I could have assumed that the search for genetic similarity had gone too far, and that he was seeking a woman with a mustache, and thus one similar to himself, the man. Yet the reason he offered was different. He explained that his mother had had this soft hair. This, then, is positive imprinting: childhood experiences are wired into the primordial mind and become a desire we repeatedly attempt to satisfy all our lives. We devote ourselves to whatever we knew as children. The familiar is an addiction. Such are the imprintings left in us by the landscapes of our childhood, the tastes and smells of the past, the culture, language, and style in which we were raised, and the people, loving and beloved. Perhaps this is the explanation for Langlois and Roggman's (1990) data showing that members of a particular population judge a person as attractive when that person represents the average facial features of that same population: the length of the nose, the size of the eyes, the distance between them, and so on. The average in the population is the highest approximation to all the people surrounding every young child raised in it. If the child is attracted to the "beautiful," he or she is attracted to the weighted average of all the positive imprintings that he or she has assimilated from his or her human environment, and thus he or she also makes sure that the target of his or her attraction is genetically similar to himself or herself, since his or her human environment is composed of the tribe members with whom he or she shares a common genetic pool. At least, that is what the social environment used to be like before the age of big cities and social mobility that removed tribe members from each other. The natural selection of addiction to the familiar and the search for the beautiful occurred long ago in small familial groups (Lampert, 1997).

The positive imprinting mechanism does not begin or end with the matter of seeking out genetic similarity. Its purpose is to ensure matters of primary importance, such as the identification of mother, learning the mother tongue, learning the geographical map, and every other form of the child's adaptation to the realities of its life. This is a strongly potent and far-reaching mechanism, used to adapt the living organism to its environment by early exposure and assimilation and finally through repeated enacting. Alcohol, drugs, and tobacco are at first bitter and unsettling, but after the nervous system comes to know them, it seeks them until they turn into a condition needed for its well-being. The familiar is addictive, and this addiction is a mechanism that can help people choose whom to fall in love with. In the absence of a genetic test, the chooser relies on sight. Young women with a mustache are genetically similar to a mustached mother and therefore to her son as well.

But is attraction guided only by experienced imprinting, or can a person have an inborn ability to identify genetic similarity and be attracted to it? We could plan the following experiment in order to examine this question: Have some Eskimo children adopted by an African tribe, and see what they prefer in adulthood, the African average or the Eskimo average. This is, of course, an impossible experiment. In order to try to tackle the distinction between accumulated positive imprintings and possible inborn attraction to genetic similarity, I have conducted a research study with an all-male sample of brothers.

When brothers are raised together, they are imprinted upon each other. They are also offspring of the same parents, so they are genetically similar. The attraction to genetic similarity probably overflows into additional human contacts, other than those between men and women. Thus we find that pairs of friends are more similar to each other than random couples in the same population

(Rushton, 1989). If the attraction to genetic similarity works also on pairs that are not opposite-sex couples, then pairs of brothers enable a "natural" experiment that can differentiate between detection of and attraction to the genetically similar and addiction to the familiar. Genetic similarity between brothers can be very high, up to the level of identical twins, or minimal, like that between two individuals in the same community. On average, all pairs of brothers in the population share 50 percent of identical genes, but in the individual case of a certain pair, the identity may rise to 100 percent or fall to 0 percent because the allocation of one parental chromosome out of two to each offspring is random.

A sample of pairs of brothers who grew up together and still live together enables us to examine whether the level of friendship between them rises with genetic similarity. We neutralized imprinting and addiction to the familiar simply by sampling only brothers who grew up together, so that if there are differences between them in the levels of attraction and friendship, they are due to other factors and not to the presence or absence of imprinting. The sample included 268 pairs of adult brothers aged 20–55 who grew up together and still live together today. Every pair of brothers was born and raised in the same kibbutz in Israel. Every subject was measured on two scales: (1) level of friendship, involvement, and love between him and his brother, (2) genetic similarity assessed by physiological and genetic signs, such as blood group, height, weight, foot length, color and type of hair, and baldness. Comparison and/or correlation coefficients between the scales were used to examine whether the level of friendship increases with the increase in genetic similarity. Figure 14.1 describes the increase in identity between brothers in blood group along with the increase in their level of friendship. Figure 14.2 describes the increase in similarity between brothers in height, baldness, and foot length along with the increase in their level of friendship. These illustrations indicate that when brothers love each other very much, they probably have the same blood group, the same baldness, and the same foot length. The correlation between loving brothers in height comes close to the correlation between identical twins. The entire sample of 536 brothers consists of people who grew up and live together, yet among them there are some who love each other more and some who love less. We can, therefore, view attraction to the genetically similar as an independent factor affecting love between people apart from positive imprinting.

How is genetic similarity detected? People do not usually test blood group before they fall in love. This is a difficult and interesting question that has no answer yet. We must make do with the assumption that in some way we sense genetic similarity and are attracted to it, as are many animals and even plants (Pfenning [Pfennig] & Sherman, 1995.)

As opposed to negative imprinting, which disqualifies an entire person, positive imprinting, along with genetic-similarity recognition, emphasizes particular personal qualities with which we have become familiar in our childhood in mother, father, and the rest of the tribe, and that we seek later on. This is the reason for the similarity between characteristics one commonly finds in a man's

Figure 14.1
Percentage of Blood-Group Identity on Four Friendship Levels

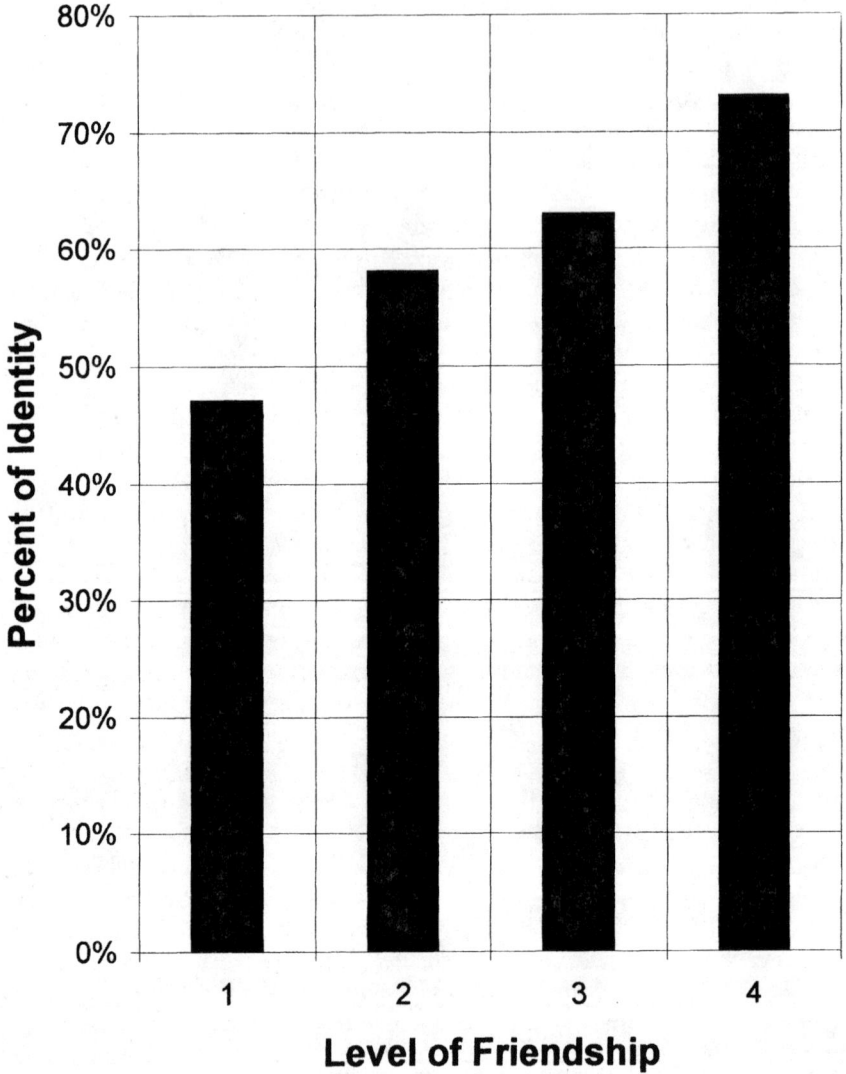

mother and his wife: the soft hair on the upper lip, the personality traits, the facial features, the name, the talents, and the warmth or coolness of their hearts. The inborn kin-recognition ability and the positive imprinting from childhood are both mechanisms aimed at assuring the attraction to our own "flesh-and-blood" pool of mates.

Emma Darwin indeed fulfilled the two requirements presented here for the

Figure 14.2
Correlations of Genetic Signs between Pairs on Four Friendship Levels

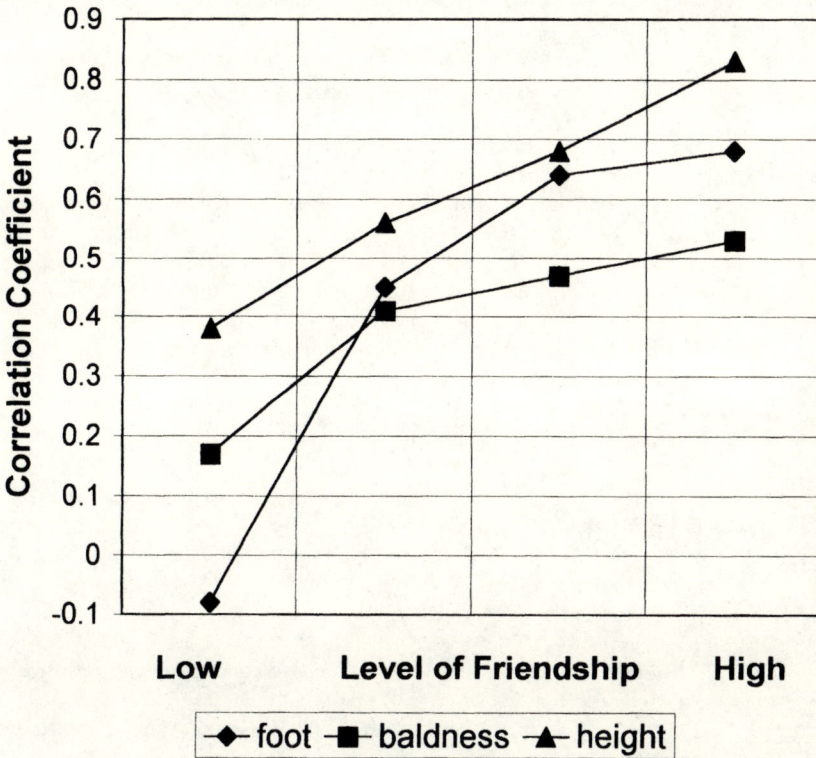

"correct" Darwinistic choice: she was familiar from childhood, and she was genetically close. Being a cousin, she was only one degree away from the stop sign preventing incest. Thus she could ensure for Darwin a stable, harmonious, fertile, satisfying, and happy married life. Incidentally, one who does not know English, like me, may read Emma as Eamma. In Hebrew "Eamma" is an important word: it means "mother" But Darwin was Darwinist enough to marry Emma and to avoid "Eamma."

REFERENCES

Bateson, P. P. G (1983). Optimal outbreeding. In P. Bateson (Ed.), *Mate choice* (pp. 257–278). Cambridge: Cambridge University Press.

Beauchamp, G. K., Yamazaki, K., & Boyse, E. A. (1985, July). The chemosensory recognition of genetic individuality. *Scientific American*, 66–72.

Daly, M., & Wilson, M. (1982). Whom are newborn babies said to resemble? *Ethology and Sociobiology, 3*, 69–78.

Darwin, C. (1877). A biological sketch of an infant. *Mind, 2*, 285–294.

Darwin, F. (1949). Memoirs of my father's day to day life. In C. Darwin, *Autobiography*. Tel-Aviv: Masada Publishers (Hebrew).

Enomoto, T. (1974). The sexual behavior of Japanese monkeys. *Journal of Human Evolution, 3*, 351–372.

Hamilton, W. D., Axelrod, R., & Tanese, R. (1990). Sexual reproduction as an adaptation to resist parasites. *Proceedings of the National Academy of Sciences, 87*, 3566–3573.

Hemleben, J. (1988). *Darwin*. Jerusalem: Keter (Hebrew).

Lampert, A. (1997). *The evolution of love*. Westport, CT: Praeger.

Langlois, J. H., & Roggman, L. A. (1990). Attractive faces are only average. *Psychological Science, 1*, 115–121.

Manning, C. J., Wakelaud, E. K., & Potts, W. K. (1992) Communal nesting patterns in mice implicate MHC genes in kin recognition. *Nature, 360* (640), 581–583.

Missakian, E. A. (1973). Genealogical mating activity in free-ranging groups of rhesus monkeys (*Macaca muylatta*) on Cayo Santiago. *Behavior, 45*, 225–241.

Pfenning [Pfennig], W. B., & Sherman. W. P. (1995, June). Kin recognition. *Scientific American*, 68–73.

Pusey, A. E. (1980). inbreeding avoidance in chimpanzees. *Animal Behavior, 28*, 543–552.

———. (1987). Sex biased dispersal and inbreeding avoidance in birds and mammals. *Trends in Ecology and Evolution, 2*, 295–299.

Radman, M., & Wagner, R. (1988, August). The high fidelity of DNA duplication. *Scientific American*, pp. 24–30.

Rennie, J. (1990, February). A DNA repair system stops species from interbreeding. *Scientific American*, 14–15.

———. (1992, January). Malaria: A case study of coevolution. *Scientific American*, January, p. 107.

Rushton, J. P. (1988). Genetic similarity, mate choice, and fecundity in humans. *Ethology and Sociobiology, 9*, 329–333.

———. (1989). Genetic similarity in male friendship. *Ethology and Sociobiology, 10*, 361–373.

Russel R. J. H., Wells, P. A., & Rushton, J. P. (1985). Evidence for genetic similarity detection in human marriage. *Ethology and Sociobiology, 6*, 183–187.

Shepher, J. (1971). mate selection among second generation kibbutz adolescents and adults: Incest avoidance and negative imprinting. *Archives of Sexual Behavior, 1*, 293–307.

Thiessen, D., & Gregg, B. (1980). Human assortative mating and genetic equilibrium: An evolutionary perspective. *Ethology and Sociobiology, 1*, 111–140.

Weisfeld, G. E., Russell, R. J. H., Weisfeld, C. C., & Wells, P. A. (1992). Correlates of satisfaction in British marriage. *Ethology and Sociology, 13*, 125–145.

Wolf, A. P. (1995). *Sexual attraction and childhood association: A Chinese brief for Edward Westermack*. Stanford: Stanford University Press.

Wolf, A. P., and Huang, C. S. (1980). *Marriage and adoption in China, 1845–1945*. Stanford: Stanford University Press.

Zahavi, A. (1990). Arabian babblers: The quest for social status in a cooperative breeder. In P. B. Stacey & W. D. Koening (Eds.), *Cooperative breeding in birds* (pp. 105–113). Cambridge: Cambridge University Press.

15

Grandparental Caregiving and Intergenerational Relations Reflect Reproductive Strategies

Harald A. Euler and Barbara Weitzel

If asked whether you had a favorite grandparent, we would assume that most of you would say yes, and further we would predict that the majority with a favorite, the maternal grandmother would be the most cherished grandparents. For example, the senior author here belongs to that majority. His maternal grandmother died over forty years ago, but she is still in his heart because she took care of him in a most loving way, although he was not her only grandchild. He grew up as a single child and has four children himself. His own mother, however, the paternal grandmother, of his offspring although once a loving mother, does not show much interest in her grandchildren, which he has often saddened him and for which he has reproached her at times. He tended to think that his mother was a special case, but found out that instead, this situation appears more common than might be assumed.

From the grandchild's viewpoint, discrimination between grandparents seems to be the rule. People often feel close to one grandparent, usually to the maternal grandmother. How can this discrimination be explained? Early childhood experience may be a possible answer. The influential attachment theory of the London psychiatrist John Bowlby (1969) specifies how persons become "mother figures," namely, through unconditional, responsive, and available care. But are grandparents themselves discriminatory in their love for grandchildren? Bowlby does not elaborate this point. In the ethological tradition, he considers the inclination for care of offspring as a general primate endowment. Discriminative caregiving, in his theory, is not part of this endowment, but instead is due to particular circumstances (Porter & Laney, 1980). So he probably would have argued that a grandparent might become the grandchild's favorite because he or she lived near by and became emotionally close to the grandchild due to cir-cumstances. Bowlby subscribes here to what Tooby and Cosmides (1992) call the Standard Social Science Model. Nature, in this view, provides the general endowment. Individual variations in preferences, aptitudes, and attitudes, how-ever, are due to cultural input.

Darwin also did not discuss discriminative solicitude toward offspring. Re-lationships between grandparents and grandchildren seem not to be a no topic in his major works. In regard to parents, he noted repeatedly that parental love is the strongest of all instincts (1871, 1872). Different "instinctive impulses have different degrees of strength," and the maternal instinctive impulse dis-criminates between own offspring and "mere fellow-creatures" (1872, p. 87). A more specific parental favoritism is not spelled out, although he was well aware of the existence of sex-specific infanticide.

The aim of our research is to apply sociological considerations to grandpar-ental solicitude. In this way we obtain specific predictions about grandparental solicitude with parsimoniously few principles. We test the deduced predictions with data obtained retrospectively from adult grandchildren.

Before we derive the predictions, allow us three remarks about the grandparent-grandchild relationship from an evolutionary standpoint. (1) The number of grandchildren is a better measure of reproductive success than the number of children. Care of grandchildren thus contributes to one's own repro-ductive success. (2) Despite the gradual disappearance of the three-generation family, increased mobility, and the alleged disintegration of family structures in modern societies, the relationship between grandparents and grandchildren re-tains its importance for both generations involved (e.g., Bengtson & Robertson, 1985; Brubaker, 1985; Kennedy, 1992; Rossi & Rossi, 1990; Swensen, 1994). (3) In view of the media-promoted youth ideal, one might assume that entrance into grandparenthood would mark a time of crisis. Instead, the opposite appears to be the case. The first grandchild is usually received with pride and joy, rather than a feeling of loss (Fischer, 1983). Middle-aged men may hide their growing bellies and enlarging baldness, but not their grandchildren. Women tend to hide their age, but not their grandmother status.

We derive our predictions about discriminative grandparental solicitude from a simple model with three factors: (1) sex-specific reproductive strategy, (2) paternity uncertainty of parents, and (3) paternity uncertainty of grandparents. These factors are well known (Dawkins, 1976), but we wish to clarify the first factor, reproductive strategy, when we talk about grandparents. Reproductive strategies are conditional strategies that enable persons to adopt particular be-havioral alternatives depending upon circumstances (Alexander, 1990). Al-though these strategies are genetically based, they are not immutable and statically fixed (Belsky, Steinberg, & Draper, 1991; Smith, 1987). Becoming a grandparent marks a change in reproductive strategy because the reproductive situation has changed. The new reproductive task is to aid the son or daughter in his or her reproductive strategy. Because the maternal strategy differs from

Table 15.1
Predictions about Grandparental Solicitude

Grandparent	Reproductive Strategy	Paternity Uncertainty
Maternal Grandmother	+	+ / +
Maternal Grandfather	+	− / +
Paternal Grandmother	−	+ / −
Paternal Grandfather	−	− / −

Note: Plus sign denotes relatively more solicitude, minus sign relatively less solicitude.

the paternal—the maternal being more restricted to child care (Daly & Wilson, 1983), the paternal having the option to gain additional descendants by mating with additional partners (Symons, 1979)—the grandparental reproductive effort should vary according to lineage. Maternal grandparents should therefore care more for the grandchild than paternal grandparents.

If we combine the three factors of (1) ontogenetically differentiated reproductive strategy, (2) paternity uncertainty of parents, and (3) paternity uncertainty of grandparents into a simple additive model with equal weight for each factor, we obtain an ordered prediction about discriminative grandparental solicitude, as shown in table 15.1. The plus sign denotes comparatively more solicitude, the minus sign comparatively less solitude. The most caring grandparent should be the maternal grandmother, followed by the maternal grandfather, the paternal grandmother, and last, the paternal grandfather.

We chose to test these predictions not by asking grandparents about their given discriminative caregiving, but by asking adult grandchildren about the current perception of their received grandparental solicitude. If one asks people rearing children whether they favor some children over others, self-descriptive statements about discriminative care are likely to be leveled due to equity norms that are themselves adaptive (Alexander, 1987). In fact, Fischer (1983) reports that the majority of grandmothers who had multiple grandchildren refused to name favorite grandchildren. Self-descriptive statements about received discriminative care, however, are presumably less influenced by equity norms. We therefore assume that ratings by adult grandchildren are a better indicator of discriminative grandparental solicitude received in childhood than ratings given by grandparents themselves.

Our total sample consisted of 1,875 persons aged 16 to 80 (720 male, 1,125

Table 15.2
Mean Solicitude Ratings (N = 603), Correlations between Residential Distance
and Solicitude (N = 208), and Correlations between Perceived Phenotypical
Similarity and Solicitude (N = 458)

	Solicitude	Correlation	Correlation
Grandparent	(Means)	distance/ solicitude	similarity/ solicitude
Maternal Grandmother	5.16	-.29	.37
Maternal Grandfather	4.52	-.34	.39
Paternal Grandmother	4.09	-.40	.42
Paternal Grandfather	3.70	-.41	.47

female, 12 unspecified). The participants younger than 40 years old were students in various undergraduate courses at the University of Kassel. The participants older than 40 years were recruited by the student participants. The participants were given a questionnaire that asked, among other questions, how much care each grandparent had provided for them up to the age of 7 years. The participants answered on a 7-point rating scale from 1 (not at all) to 7 (very much). For some analyses, we looked at the total sample of 1,857 cases. However, for the analysis of discriminative grandparental solicitude, we selected those 603 participants whose four (putative) biological grandparents were all living until the participant reached the age of 7 years. Step- and foster relations were not considered. The frequencies of living grandparents differ because of later parenthood and earlier death of men as compared to women. Without restricting ourselves to the complete cases, these frequency differences could cause unrecognized selection effects.

The results are shown in table 15.2. The first data column gives the means of received grandparental solicitude. The results confirm our hypotheses. The participants perceived as most caring maternal grandmother, followed by the maternal grandfather, the paternal grandmother, and the paternal grandfather. The analysis of variance with the variables sex of participant, sex of parent (i.e., maternal versus paternal grandparents), and sex of grandparent revealed highly significant main effects for the latter two variables. Maternal grandparents provided more care than paternal grandparents, with grandmothers providing more care than grandfathers in both lineages. The effects are considerable. Effect sizes, given as $\eta^{\circ}_{[alt]}$ (Tabachnik & Fidell, 1989, p. 55), are .11 for the lineage effect (sex of parent) and .17 for the effect of sex of grandparent. The term $\eta^{2}_{[alt]}$ denotes the variance attributable to the effect of interest divided by this variance plus error variance.

Of special theoretical interest here is the comparison of the maternal grandfather with the paternal grandmother. If grandparental caregiving were determined by a social role and child care were traditionally ascribed to the female role, grandmothers generally should have provided more care for grandchildren than grandfathers. However, maternal grandfathers cared significantly more than paternal grandmothers (t [602] = 3.79, p = .000, effect size d = .21, after Cohen, 1988). This difference is also significant for the older participants (40 years or more), with the magnitude of the difference being even more pronounced (4.47 versus 3.45). If grandparental solicitude were determined by traditional gender roles, we would expect the difference between maternal grandfather and paternal grandmother to be less pronounced for the older generation, assuming that traditional gender roles decreased in the second half of this century.

There are two possible confounding variables we need to look at, namely, residential proximity and age of grandparents. Occasions and requests for grandparental caregiving arise more frequently for those grandparents living close by, and the distances between the grandchild and the four grandparents or the two grandparent couples could differ. Maternal grandparents are on average the youngest grandparents, paternal grandfathers the oldest, and age could correlate with grandparental solicitude.

A subsample of 208 participants was asked for the residential distances in kilometers to their four grandparents, and another subsample of 297 participants was asked for their grandparents' years of birth. The means of the residential distances, however, did not differ significantly between the four grandparents. Neither did age correlate significantly with the solicitude ratings for any of the four grandparents. Therefore, neither residential proximity nor age can account for the discriminative solicitude.

The examination of the correlations between residential distance and solicitude produced an interesting result. The expected negative correlations differ, as it seems systematically, for the four grandparents (table 15.2). The effect, however, is not strong. The differences between the correlations just miss statistical significance, but it looks as if grandparental solicitude is an adaptation least facultative for the maternal grandmother and most facultative for the paternal grandfather. The numerical values of the correlation coefficients correspond to the theoretically derived gradations between the four grandparents.

The solicitude data considered so far were obtained from those participants whose four grandparents were all living and known during childhood. We therefore asked ourselves whether the amount of received grandparental solicitude differed for those grandchildren whose grandparents were not all known or living during childhood. Assuming that a grandchild tends to require a certain amount of grandparental care, it follows that a child with fewer living grandparents would require more from each grandparent than a child with more living grandparents. Likewise, from the grandparental viewpoint, the distribution of solicitude could depend on one's knowledge of being the sole grandparent or of being one among several, the diffusion of grandparental responsibilities being

Table 15.3
Solicitude Ratings for Grandparents Living Together with Spouse and for
Grandparents Living Separately from Spouse

Grandparent	Solicitude rating for grandparent		t-Test (p)
	living together with spouse	living separately from spouse	
Maternal Grandmother	5.09	5.06	ns
Maternal Grandfather	4.51	2.06	.000
Paternal Grandmother	4.20	3.25	.005
Paternal Grandfather	3.80	1.77	.000

presumably higher in the latter case. Whatever viewpoint is taken, if grandparental care depends on the number of still-living and known grandparents, the differences between grandparents should diminish with a decreasing number of grandparents. But a comparison of the data revealed that discriminative grandparental solicitude remained, irrespective of the number of other grandparents available. Also, a comparison of widowed grandparents (the spouse died before the grandchild's second birthday) with nonwidowed grandparents showed no significant differences. In short, the discriminative grandparental solicitude appeared to be a rather robust phenomenon.

Only separation from spouse was a variable we found effective in determining grandparental solicitude, again according to the well-known pattern, as shown in table 15.3. Separation or divorce allows men to drift away from kindred. Grandmothers are "kin keepers," as some family researchers call them (e.g., Troll & Bengston, 1992). Married grandfathers go along with grandchild care initiated by their wives. The high intracouple correlations in solicitude we found, namely, $r = .70$ maternally and $r = .74$ paternally, are to some extent due to the influence of grandmothers on their husbands.

Effective grandparental investment in grandchildren requires an ability to recognize kin. The more paternity uncertainty accrues to a grandparent, the more his or her solicitude should depend on the extent of grandchild resemblance. The paternal grandfather should rely most upon resemblance for his allocation of child care, the maternal grandmother the least. We asked our participants in several subsamples how similar in appearance and/or behavior they had been to each grandparent. The correlations between grandparental solicitude and perceived phenotypical similarity are shown in table 15.2. We again find the gradations from maternal grandmother to paternal grandfather. However, this effect is not strong. The differences between the correlation coefficients, tested after Steiger (1980), are marginally not significant ($p = .06$ for the difference between the highest and lowest correlation, $N = 191$). We realize that to ask adult grandchildren about similarity to grandparents is not the best measure, because for them the similarity does not matter as recipients of grandparental investment. It matters for grandparents, and it would have been better to have asked them.

Discriminative grandparental solicitude appears to be a rather solid and robust phenomenon. It proves to be statistically significant, covers a sizable share of variation in solicitude, and remains uninfluenced by several potentially care-relevant conditions like residential proximity, grandparent age, and availability of other grandparents.

Investment into descendants is an abstract concept based on a variety of concrete and operationally definable behaviors, of which care or solicitude is only one. Other possible behaviors are frequencies of contact, mourning upon a grandchild's death, expressing feelings of closeness, readiness to adopt, and willing property. Empirical findings about these behaviors support the hypothesis of discriminative grandparental solicitude (Berger & Schiefenhoevel, 1994; Daly & Wilson, 1980; Eisenberg, 1988; Fischer, 1983; Hartshorne & Manaster, 1982; Hoffman, 1978/79; Kahana & Kahana, 1970; Kennedy, 1990; Littlefield & Rushton, 1986; Matthews & Sprey, 1985; Rossi & Rossi, 1990; Russell & Wells, 1987; Smith, 1988).

The role of an ontogenetically differentiated reproductive strategy and of paternity confidence for grandparental caregiving is presumably underestimated by the data presented here for three reasons. First, we assume that children from complete families are overrepresented in our sample of students and their parents or acquaintances, and children of fathers who deserted the mother are underrepresented. The special life strategies that children of single-parent households acquire during childhood (Belsky, Steinberg, & Draper, 1991; Draper & Harpending, 1988) are less likely to provide access to German higher education. Second, the analysis of variance was based on the complete cases with all four grandparents known and living during childhood. Such a sample constitutes a further selection of children of fathers inclined toward family life and child care, because intact grandparenthood biases toward intact parenthood. Third, certainty of paternity is determined not only by one's sex and by the resemblance between progeny and self, but also by the subjectively evaluated risk of double mating (insemination by rival). Baker and Bellis (1989) showed sperm count to correlate negatively with the proportion of time spent together since previous copulation. If the wisdom of the body adjusts ejaculate content to the risk of double mating, why should the mind know nothing and be easily cuckolded? A look at the standard deviations of the solicitude ratings does not necessarily support this notion, but neither does it contradict it. The standard deviations are numerically higher for grandfathers than for grandmothers, but not significantly. It might well be possible that a subjectively assessed risk of double mating would explain a further proportion of variance in caregiving not allocated in our analysis.

The asymmetry of reproductive conditions results in a caregiving asymmetry between the four grandparents that is most conspicuous in the salient role of the maternal grandmother. The asymmetry in reproductive conditions causes further structural asymmetries not touched upon so far, but not irrelevant to grandparental solicitude, namely, asymmetries in dyadic relations between one's grandparents and one's parents. The relation between mother-in-law and daughter-in-law is peculiarly negative (Duvall, 1954), as we know from jokes, songs, and tales. Evolutionary theory offers a set of statements that as an ensemble predict an ordered gradation about who gets along how well with whom. The first is the obvious and trivial prediction that one's own child is closer than the child's spouse. Second, parents aid their children in the children's strategy. A daughter, more restricted than a son to child care as a reproductive strategy, is best aided by her parents in this strategy within the context of a good parent-daughter relationship. A poor parent-son relationship is comparatively less detrimental for a son's reproductive strategy of little investment after copulation. Third, uncertainty of paternity results in a generally better relationship between mother and children than between father and children. Thus grandmothers ought to have better relationships with their children than grandfathers.

Now to the in-laws. A daughter needs more stable partner support for her reproductive strategy than a son for his. The daughter is best aided if her parents welcome and relate well to her husband. A son, insofar as he is inclined toward polygyny, is comparatively less impeded by poor relations between his wife and his parents. Rejection of his partner by his parents may even be strategically supportive and in the grandparents' own reproductive interest.

Why then is the relationship between the mother-in-law and the daughter-in-law the poorest? When it was just postulated that mothers generally ought to have a better relationship with the next generation than fathers, would it not follow that the relationship between the father-in-law and the daughter-in-law would be the least positive? Here enters a last asymmetry, which elevates the relationship between father-in-law and daughter-in-law. This dyad is the only one of all eight that possesses a direct reproductive potential. The other seven are wrong matches because of incest barrier, same sex, or wrong age relation (mother-in-law to son-in-law). This direct reproductive potential hardly ever materializes, but it might nevertheless be psychologically relevant. The father-in-law might show off in public with his daughter-in-law and feel flattered if she is mistaken for his wife. Or he might, wholly without rivalrous feelings, imagine himself in his son's place and be thus entertained. A good relationship with his daughter-in-law is quite compatible with these inclinations, but a poor one would be detrimental.

Taken together, these considerations predict intergenerational dyadic relations in the following order, from best to worst: mother/daughter, father/daughter, mother/son, father/son, mother-in-law/son-in-law, father-in-law/son-in-law, father-in-law/daughter-in-law. The order implies different effect sizes, with own child versus in-law being the largest.

To sum up, the four grandparents take care of their grandchildren to a typically differentiated extent that cannot be sufficiently explained solely by individual circumstances, socialization, or cultural norms. Paternity certainty and reproductive strategy determine grandparental solicitude and other intergenerational relationships.

REFERENCES

Alexander, R. D. (1987). *The biology of moral systems*. New York: Aldine de Gruyter.

———. (1990). Epigenetic rules and Darwinian algorithms. *Ethology and Sociobiology, 11*, 241–303.

Baker, R. R., & Bellis, M. A. (1989). Number of sperm in human ejaculates varies in accordance with sperm competition theory. *Animal Behaviour, 37*, 867–869.

Belsky, J., Steinberg, L., & Draper, P. (1991). Childhood experience, interpersonal development, and reproductive strategy: An evolutionary theory of socialization. *Child Development, 62*, 647–670.

Bengtson, V. L., & Robertson, J. F. (Eds.). (1985). *Grandparenthood*. Beverly Hills, CA: Sage Publications.

Berger, C., & Schiefenhoevel, W. (1994, October). *Adoption in Tauwema, Trobriand-Inseln*. Poster session at the Congress "Anthropologie Heute" of the Gesellschaft für Anthropologie, Humboldt-University, Berlin, and University of Potsdam, Germany.

Bowlby, J. (1969). *Attachment*. London: Hogarth Press.

Brubaker, T. H. (1985). *Later life families*. Newbury Park, CA: Sage Publications.

Cohen, J. (1988). *Statistical power analysis for the behavioral sciences* (2nd ed.). Hillsdale, NJ: Lawrence Erlbaum.

Daly, M., & Wilson, M. (1980). Discriminative parental solicitude: A biological perspective. *Journal of Marriage and the Family, 42*, 277–288.

———. (1983). *Sex, evolution, and behavior* (2nd ed.). Belmont, CA: Wadsworth.

Darwin, C. (1871). *The descent of man, and selection in relation to sex*. London: Murray.

———. (1872). *The expression of the emotions in man and animals*. London: Murray.

Dawkins, R. (1976). *The selfish gene*. New York: Oxford University Press.

Draper, P., & Harpending, H. (1988). A sociobiological perspective on the development of human reproductive strategies. In K. B. MacDonald (Ed.), *Sociobiological perspectives on human development* (pp. 340–372). New York: Springer.

Duvall, E. M. (1954). *In-laws: Pro & con*. New York: Association Press.

Eisenberg, A. R. (1988). Grandchildren's perspectives on relationships with grandparents: The influence of gender across generations. *Sex Roles, 19*, 205–217.

Fischer, L. R. (1983). Transition to grandmotherhood. *International Journal of Aging and Human Development, 16*, 67–78.

Hartshorne, T. S., & Manaster, G. J. (1982). The relationship with grandparents: Contact, importance, role conceptions. *International Journal of Aging and Human Development, 15*, 233–245.

Hoffman, E. (1978/79). Young adults' relations with their grandparents: An exploratory study. *International Journal of Aging and Human Development, 10*, 299–310.

Kahana, B., & Kahana, E. (1970). Grandparenthood from the perspective of the developing grandchild. *Developmental Psychology, 3*, 98–105.

Kennedy, G. E. (1990). College students' expectations of grandparent and grandchild role behaviors. *Gerontologist, 30*, 43–48.

———. (1992). Quality in grandparent/grandchild relationships. *International Journal of Aging and Human Development, 35*, 83–98.

Littlefield, C. H., & Rushton, J. P. (1986). When a child dies: The sociobiology of be-
   reavement. *Journal of Personality and Social Psychology, 51*, 797–802.
Matthews, S. H., & Sprey, J. (1985). Adolescents' relationships with grandparents: An
   empirical contribution to conceptual clarification. *Journal of Gerontology, 40*,
   621–626.
Porter, R. H., & Laney, M. D. (1980). Attachment theory and the concept of inclusive
   fitness. *Merrill-Palmer Quarterly, 26*, 35–51.
Rossi, A. S., & Rossi, P. H. (1990). *Of human bonding: Parent-child relations across
   the life course.* Hawthorne, NY: Aldine de Gruyter.
Russell, R. J. H., & Wells, P. A. (1987). Estimating paternity confidence. *Ethology and
   Sociobiology, 8*, 215–220.
Smith, M. S. (1987). Evolution and developmental psychology: Toward a sociobiology
   of human development. In C. Crawford, M. Smith, & D. Krebs (Eds.), *Socio-
   biology and psychology: Ideas, issues, and applications* (pp. 225–252). Hillsdale,
   NJ: Lawrence Erlbaum.
———. (1988). Research in developmental sociobiology: Parenting and family behavior.
   In K. B. MacDonald (Ed.), *Sociobiological perspectives on human development*
   (pp. 271–292). New York: Springer.
Steiger, J. H. (1980). Tests for comparing elements of a correlation matrix. *Psychological
   Bulletin, 87*, 245–251.
Swensen, C. H. (1994). Older individuals in the family. In L. L'Abate (Ed.), *Handbook
   of developmental family psychology and psychopathology* (pp. 202–217). New
   York: John Wiley.
Symons, D. (1979). *The evolution of human sexuality.* New York: Oxford University
   Press.
Tabachnik, B. G., & Fidell, L. S. (1989). *Using multivariate statistics* (2nd ed.). New
   York: HarperCollins.
Tooby, J., & Cosmides, L. (1992). The psychological foundations of culture. In J. H.
   Barkow, L. Cosmides, & J. Tooby (Eds.), *The adapted mind* (pp. 19–136). New
   York: Oxford University Press.
Troll, L. E., & Bengtson, V. L. (1992). The oldest-old in families: An intergenerational
   perspective. *Families and Aging, 16*(3), 39–44.

16

Marital Power Dynamics: A Darwinian Perspective

Norma J. Schell and Carol C. Weisfeld

The research presented in this chapter attempts to address the question of how power dynamics in human marriage contribute to marital satisfaction. It began with two different theoretical views, one coming out of a Darwinian, evolutionary perspective and the other coming out of psychological exchange theory.

When Darwin described sexual selection in 1859, he suggested mechanisms that were likely to lead to a situation for humans in which the balance of power would favor males over females. Male-male competition is far keener than female-female competition. From the female point of view, she is interested in attracting a male who had proven his ability to garner and defend resources. One might argue that, for humans, their satisfaction with their marriage partners would depend to some extent on how well they were able to meet these selection criteria when finding a partner. Surely satisfaction with one's selection of a mate would serve biological goals; we know that couples who are able to stay together produce children who are more successful in every domain, from school performance to their own social relationships. It is likely, then, that a satisfying relationship with one's mate is likely to increase one's inclusive fitness. A modern researcher of marriage might wonder, though, whether satisfaction with one's marriage would continue to be linked to such factors as male dominance over the female spouse.

Some anthropologists (e.g., Friedl, 1984) have found a strong relationship across cultures between the degree to which a male controls key resources and the amount of power he has in his marital relationship. In our own research on British marriages (Weisfeld, Russell, Weisfeld, & Wells, 1992) we found a balance between the opposing forces of homogamy (like marrying like) and hypergamy (females marrying up) in many areas, including decision making. The most satisfied couples in our sample were those in which the husbands were mildly ascendant over their wives in decision making, as perceived by both of them. That is, the optimal situation was for the husbands to prevail a little bit more than the wives when it came to making important decisions. Overly dominant husbands made for slightly less happy marriages, and overly dominant wives were associated with the least happy marriages. The wives, interestingly enough, seemed to be most affected by this relationship. They seemed to want a slightly dominant husband more than the husband himself wanted to be dominant. This is consistent with the finding of David Buss (1994) that younger women continue to be attracted to slightly older men, and that the age difference is more important to the women than it is to the men. We have reasoned that a woman may feel that it is important to have a spouse on whose judgment she can rely. She doesn't submit to him because he is the husband; rather, she has chosen him as the husband precisely because she is comfortable relying on his judgment—but not all of the time.

From a psychological-exchange-theory perspective, the most satisfying marriages ought to be those in which benefits are exchanged. Scanzoni (1972) has suggested a money-for-sex bargain. Safilios-Rothschild (1976) has argued that love is an emotional resource that has been neglected by researchers. She has investigated the exchange of money for love. Safilios-Rothschild has also examined the balance of resources in marriage, arguing that as women become more gainfully employed, a more equitable balance is achieved, and that it is this equity that leads to happiness in modern marriage.

Our study focused on the dynamics of marital power in married American couples via respondents' self-reports, observed decision-making processes, and outcome. We then examined the correspondence of these measures with marital satisfaction. The variables we are looking at, then, are the following:

Partner's Resources

Education
Occupational status
Income

Love and affection

Marital Satisfaction

Is discrepancy or egalitarianism a better predictor?

Decision making

Perceived decision making
Observed decision-making process

Outcome: who won desired object
Nonverbal dominance

Marital Satisfaction

Is consistency or a particular (visual) pattern a predictor?

The instrument we used is the *Marriage Questionnaire* developed in 1986 by Russell and Wells (published in shorter form in 1993 as *The MARQ*). This is a multipurpose marriage questionnaire. The longer version, which we used, consists of some 170 items in matching booklets that are filled out independently by the husband and wife. Within each questionnaire is buried a set of items related to marital satisfaction:

Have you thought of divorcing your spouse?

Do you find sexual fulfillment in your marriage?

How often do you have a serious argument?

If you could choose, would you marry the same person again?

Is your spouse cruel to you?

In each couple, the husband's and wife's responses are summed and divided by ten to create a marital-satisfaction score for that couple. In similar fashion, we used questions to create subscales for resources (income, education, and the like), love, and decision making. These, then, were our self-report measures, which are really the husband's and wife's perceptions of what happens in the marriage.

In addition, we wanted to observe each couple in an actual decision-making task. We also wanted to be able to see which partner won the object being contested, and we wanted to be able to code for visual dominance in the interaction. So we constructed the experience for couples in this way: We administered the questionnaire in the home of the couple, and we included a small card at the back of the questionnaire. The card stated that we wanted to show our gratitude by giving them a gift of one feature film on videocassette. The card listed three films that we had available as gifts and asked each member of the couple to rank the films in order of preference. The three films were *Spartacus, When Harry Met Sally*, and *To Kill a Mockingbird*. The reason we chose these three films was that our survey of students and video-rental customers in Detroit had shown strong sex differences in popularity of the films: men liked the first, women liked the second, and the sexes liked the third equally well. Obviously, we were hoping to create a situation in which the husband and wife disagreed on the film choice and would have to come to a joint decision. When the couple had completed the questionnaire and the card, we asked them to participate in a short discussion on commitment, which we filmed using video recording equipment.

When the discussion was over, we told the couple about their film choices and asked them to decide on a film, as we could afford to give only one film per couple as the gift. We kept the camera running during the decision-making task and coded these films later, in slow motion. Thus we were able to obtain records of the decision-making process and visual dominance during the process,

Table 16.1
Data for 40 American Couples

Variable	Mean	Range
length of marriage	15.5 years	6 months to 54 years
husband's age	44 years	26 to 79 years
wife's age	41 years	26 to 68 years
number of children	1.9	0 to 9
husband's love (max 45)	38.77	31 to 45
wife's love (max 45)	39.6	34 to 45

below: 1 best, 5 worst

husband's marital sat.	1.79	1 to 3.4
wife's marital sat.	1.74	1 to 2.8

and we were also able to ascertain who prevailed and won the film of his or her choice.

A discussion of the results begins with a demographic profile of the couples. Our sample consisted of 40 middle-class Caucasian-American married couples, obtained as a convenience sample from among acquaintances and people responding to ads posted around campus. These tended to be stable marriages; in 30 of the couples (75), both spouses were in their first marriage. Other pertinent data are shown in table 16.1.

As table 16.1 indicates, the couples represent a broad range of younger and older spouses, and overall, husbands and wives are very close in measures of love and marital satisfaction. In fact, these seem to be fairly happy couples in the context of our larger American sample and our British sample. In terms of resources, again husbands and wives were fairly evenly matched. Approximately 80 percent of both husbands and wives had obtained college or postcollege degrees, and nearly 75 percent of both sexes reported full-time employment at the executive or managerial level. When it came to income, however, we did find a sex difference, as shown in table 16.2. (Income data were missing for one couple.) As table 16.2 indicates, husbands in these couples are contributing a higher percentage of the family income than their wives are, despite their similarity in professional preparation and type of occupation. One explanation might be that the wives in our sample were more likely to be employed in the social service areas, which are not well compensated in America. Husbands, in contrast, were more likely to be employed in business.

How does control over resources relate to marital satisfaction? We collapsed the three measures of education, occupation, and income into one category of resources, and correlational analyses found no pattern of relationship at all be-

Table 16.2
Contribution to Family Finances

Income for Family Finances	Husbands	Percent	Wives	Percent
100-76%	13	33.33%	0	0.00%
75-50%	24	61.54%	12	30.77%
49-25%	2	5.13%	16	41.03%
24-00%	0	0.00%	11	28.20%
TOTAL	39	100.00%	39	100.00%

tween resources and marital satisfaction. That is, it did not matter whether couples were egalitarian or discrepant in terms of control over resources. This had no bearing on their marital satisfaction. We recoded the data to see whether direction of discrepancy might matter: that is, whether it might be better for husbands to control more resources. It did not. Again, there was no relationship.

With regard to love, again, it did not matter whether couples were equally in love or discrepant with regard to reported love. We can, however, add a footnote. There was a trend for those couples equally in love at a very high level to show the greatest marital satisfaction. So both the amount of love and the balance of love may be important. Finally, there was some slight evidence in our sample of a resource-for-love exchange, as Safilios-Rothschild had suggested. In the 40 couples, the most common combination was for the husband to have more resources than the wife and for the wife to feel more love than the husband (14 couples).

The results for decision making can be broken down into the four categories presented earlier. First, with regard to perceptions, discrepancy in decision making in favor of the husband was related to higher marital satisfaction at a level of $p < .05$. That is, if a couple said that the husband had more influence in important decisions, the couple was likely to be more satisfied. This is consistent with our earlier findings with British couples, for whom a slight ascendance for males in decision making was associated with greater marital satisfaction.

When it came to the decision-making process, we were surprised that 20 of our couples spontaneously agreed on the same order of films. (There is a slight trend for these couples to be happier, and so we wonder whether we might have accidentally created a very simple measure of marital satisfaction.) Of the remaining 20 couples, we ended up with 15 films suitable for analysis. In general, the process was always polite. There was not a single instance of coercion; rather, couples exchanged information and stated preferences in a very civil and good-humored way. The end result was that the wife almost always took a leadership role in the actual decision-making process. The results are as follows: 1 couple compromised on the film choice (both agreed to take second choice);

Table 16.3
Results of Decisions about Videotapes

Number of Couples	Decision-maker	Objective Winner
1	both	neither
6	wife	wife
5	wife	husband
3	husband	wife

6 wives chose their own films, with no contest from the husband; 5 wives chose their husbands' films; and 3 husbands chose their wives' films. These results are recast in table 16.3 in terms of the observed decision maker and the objective winner (the one who ended up with the desired film).

As table 16.3 indicates, there are a couple of very interesting findings. One is that it often happened that one spouse would take a leadership role but choose the film that the other person wanted. For example, the wife would say, "We'll take the one he wants." This represents an act of magnanimity that may be fairly common in an ongoing relationship. Obviously, the magnanimous spouse is then owed something, in the reciprocal altruism sense, and one can count on having an opportunity to get paid back. Table 16.3 also shows 6 cases in which wives simply chose their own films, and the husbands did not object or defend their own interests. But there was no case in which a husband did that. No husband acquired the video of his choice by his own influence or persuasion. The only way a husband acquired the video of his choice was by means of his wife's deciding to take that video. What is even more interesting is that none of this has anything to do with marital satisfaction. Marital satisfaction did not depend on who made the decision or whose film they ended up with, or whether those two things were correlated with each other.

Marital satisfaction was related, however, to visual dominance. We analyzed the films, using Ellyson's (1974) ratio of looking while speaking to looking while listening. The general results were as follows: 8 husbands were visually dominant, 3 wives were visually dominant, and 4 couples showed no difference in visual dominance. There was a modest ($p < .10$) trend for visually dominant husbands to have higher marital satisfaction and for husbands of visually dominant wives to be less satisfied. We found no evidence that visual dominance on the part of the husband was related to dominance in any other domain. It was possible for a man to be visually dominant in the interaction with his wife and not be dominant in other aspects of their relationship.

In summary, then, we found that resources had very little to do with satisfaction for these middle-class couples. For them, marital satisfaction was correlated with perception of decision making in the marriage and with visual dominance in an actual decision-making task. If couples said that the husband made more decisions, or if dominance was reflected in his nonverbal signalling, the couple tended to be happier. What they actually did, at least in this particular task, seemed to be unrelated to perception and to marital satisfaction. This is consistent with what Olson and Rabunsky (1972) found, a lack of consistency between what couples do and what they say they do. We admit that our decision was a fairly trivial one. But somehow our couples are happier if they feel that the husband is dominant, or if he is dominant in his nonverbal behavior. These two aspects of his dominance may be more public, and that fact may make them more important to the wife than what actually happens in a private, fairly trivial act of decision making. That is, it may be important to the wife that her husband feels that he is dominant and that her husband maintains a physical demeanor that gives a dominant impression to the outside world where he competes for resources. We interpret this as evidence for the continuing importance of traits that have been selected in the course of human evolution.

REFERENCES

Buss, D. M. (1994). *The evolution of desire*. New York: Basic Books.

Darwin, C. (1859). *On the origin of the species by means of natural selection, or preservation of favoured races in the struggle for life*. London: Murray.

Ellyson, S. L. (1974). Visual behavior exhibited by males differing as to interpersonal control orientation in one- and two-way communication systems. Unpublished dissertation, University of Delaware.

Friedl, E. (1984). *Women and men*. Prospect Heights, IL: Waveland Press.

Olson, D. H., & Rabunsky, C. (1972). Validity of four measures of family power. *Journal of Marriage and the Family, 34*, 224–234.

Russell, R. J. H., & Wells, P. A. (1993). *Marriage and relationship questionnaire*. Sevenoaks, Kent, England: Hodder & Stoughton.

Safilios-Rothschild, C. (1976). A macro- and micro-examination of family power and love: An exchange model. *Journal of Marriage and the Family, 38*, 353–362.

Scanzoni, J. (1972). *Sexual bargaining: Power politics in the American marriage*. Englewood Cliffs, NJ: Prentice-Hall.

Weisfeld, G. E., Russell, R. J. H., Weisfeld, C. C., & Wells, P. A. (1992). Correlates of satisfaction in British marriages. *Ethology and Sociobiology, 13*, 125–145.

17

Sexual Dimorphism and the Evolution of Gender Stereotypes in Man: A Sociobiological Perspective

M. L. Butovskaya and A. G. Kozintsev

Sociobiology has made considerable progress in explaining basic principles of animal social organization and social behavior. Whether it can also promote the understanding of social relationships in humans is a matter of debate (Rose, 1980; Washburn, 1980). While social anthropologists tend to underestimate the role of biological factors (such as natural selection, inclusive fitness, or cost/benefit ratio) in human societies, there is a growing body of evidence suggesting that some cultural practices and norms have derived from adaptive strategies that increased the probability of survival in specific environments (Irons, 1980; Reynolds, 1984). Darwin (1859) was among the first to observe that sexual dimorphism decreased in the course of human evolution, an idea that was later upheld by paleoanthropological data (Wolpoff, 1976, Trinkaus, 1980). In the present chapter we discuss certain parallels between the reduction of sexual dimorphism and the evolution of human social behavior (division of labor, dominance versus subordination relationships between the sexes, gender differences in mental qualities, partner choice, parental investment, and the like).

First, we should not forget that humans are primates. We are still biological beings, regardless of our unique cultural environment. From the standpoint of inclusive fitness, men and women radically differ in their reproductive potential, life strategies, and orientations (Wilson, 1975; Dawkins, 1976). Thanks to the works of Williams (1966), Trivers (1972), and Symons (1979), the fact that the parental investment of females is much larger than that of males has become not just common knowledge, but part of the scientific paradigm.

This, indeed, is where the interests of both sexes frequently come into conflict. Because the two sexes differ in their reproductive potential, females in most primate species tend to be coy and apparently show less interest in sex, while males, as Darwin noted, are "very eager" and sexually oriented. In fact, social relationships in primate groups have largely been shaped by this distinction.

It is an almost universal rule that social relationships among female conspecifics center around food resources, which are so vital for pregnant females and the offspring (Wrangham, 1986). As suggested by socioecological models developed by van Schaik (1989), the type of social organization (female bonded or non-female bonded) depends on the quality and predictability of food resources and on their territorial and seasonal distribution. While females compete for resources, they themselves are the main resource and the main object of competition for males. Males always "go where females are" (Boekhorst, 1991) and try to monopolize fertile females whenever possible. There is little doubt that while the evolution of the males actually focused on competition for scarce female eggs, females have evolved to compete for scarce male investment (Wright, 1994).

Female competition is often indirect, and in many cases a balance is struck between cooperation and exploitation (Hrdy, 1981). In some primate species, females openly compete for a better reproductive partner. We have witnessed such examples in pigtailed macaques (*Macaca nemestrina*) and rhesus monkeys (*Macaca mulatta*). In these species, consisting of relatively stable multimale, multifemale groups, high-ranking estrous females actively try to monopolize male leaders. In stump-tailed macaques (*Macaca arctoides*), a species lacking seasonality of reproduction, high-ranking females try to prevent the male leaders from copulating with other females. Social power is the crucial factor determining whether or not the transmission of the female's genes to the next generation will be successful. In female-bonded species, competition between the females for power results in the system of rank inheritance and in the development of strategies aimed at establishing friendly bonds with the alpha male (Fedigan, 1982; Rhine, 1992).

Our own species is no exception in this battle for successful reproduction. Success here is most often closely correlated with the woman's ability to attract men, a skill especially adaptive in species with a high male parental investment. The human species is just such an example. Since the earliest stages of human evolution, females have been selected for their ability to attract successful and skillful partners ready to invest in childrearing. It is unlikely that physical strength and young age were crucial for this choice. As data on modern hunter-gatherers suggest, it was, rather, social status and the related ability to control resources that really mattered. Natural selection favored females who were able to correctly predict the paternal investment potential of their would-be mates. Traits like generosity, reliability, and readiness to establish a long-lasting alliance with the female were the most attractive and thus the more adaptive ones (Wright, 1994). Today as well, the principal factors responsible for the male's inclusive fitness are his social success and parental potential. Due to the availability of contraceptives and the decrease of the birth rate, paternal investment strategies have become the most reliable means for the males to maximize their inclusive fitness.

Sexual dimorphism in humans is more pronounced than in gibbons, but less so than in common chimpanzees or bonobos, and much less than in gorillas. Along with behavioral data, these facts have provided some ground for speculations concerning the models of early hominid pair bonding. It is likely that our ancestors practiced moderate polygamy, and a long-lasting attachment to a single mating partner was unusual. The analysis of divorces in fifty-eight countries representing different political and economic systems, based on United Nations demographic data for 1947–81, has demonstrated that obligatory lifelong monogamy is not a predominant mating strategy anywhere (Fisher, 1989). Interestingly, the peak of divorces coincides with the fourth year of marriage, when males are usually twenty-five to twenty-nine and females twenty to twenty-nine years of age. The study provides an evolutionary explanation of this fact. In modern hunter-gatherers, four years is the usual interval between conception and the end of breastfeeding. This could be the duration of the male's active investment in the female and their child. After that, the marital union would be less crucial as a means of survival for the child (and thus for the male's genes).

It is quite probable that culture-specific gender stereotypes formed under the influence of ecological conditions in a way that has enabled socially successful individuals (first and foremost, males) to spread the maximal number of their genes. But having become adaptive in a certain environment, gender differences in social behavior eventually became fixed by tradition and were culturally transmitted from one generation to another regardless of environmental changes (Silk, 1990; Irons, 1991). In most societies, the biological predisposition to men's higher reproductive potential was mirrored by marital and parental norms. As mentioned by Wright (1994), polygyny was practiced in 85 percent of societies for which respective information was available (980 out of 1,154). However, 43 percent of them were described as "occasionally polygynous." As demonstrated by Fisher (1989), even when polygyny is allowed, it is actually practiced by about 10 percent of men, since it correlates with high social status and wealth. A cross-cultural study by Daly and Wilson (1983) suggests that both men and women in any society usually prefer to have a stable sociosexual and economic alliance with just one partner. So serial monogamy is (and apparently was) the most typical marriage strategy in humans (Kinzey, 1987; Fisher, 1989).

The transition from lifelong monogamy to serial monogamy in modern Western society is sometimes regarded as potentially dangerous (Tucker, 1993). We disagree, since humans are evolutionarily adapted to this type of bonding. After all, it is widely known that extramarital sexual unions were common even in societies in which they were strongly condemned by religious and moral traditions (but, in the case of men, tacitly protected by the double standard).

Neither all women nor all men follow the same strategy to achieve maximal adaptive success. Trivers (1972) claims that two options are available for girls who have attained sexual maturity. The first one is to marry up, having found a man of higher socioeconomic status. Attractiveness is the prerequisite in this case, since pretty women are more likely to encourage high-ranking men to make a long-term parental investment. Buss (1989), using data from thirty-seven societies, has demonstrated that women universally prefer older men, while men prefer younger women. This, he believes, is explained by sex differences in reproductive possibilities. Younger females are not just more attractive but more fertile, while no strong correlation between age and fertility exists in males; also, for evolutionary reasons, the potential partners' appearance is less important for females. In modern society, this strategy strongly encourages women to take care of themselves (using makeup, going in for sports, and the like).

The second strategy, according to Trivers, offers reproductive chances to less beautiful women. Its features are high sexual activity and low selectiveness. Being promiscuous, a woman can obtain small portions of resources from several men. The number of extramarital children is known to have considerably increased in the lower strata of American society. Because a girl from a low-class family has few chances to marry up and is quite likely to get connected with an unemployed or degraded man, the role of an unmarried mother appears more attractive to her, the more so because governmental support and progress in public health care in Western countries offer her a fairly good chance of raising a healthy offspring.

Both strategies are apparently practiced worldwide, and there is little doubt that our ancestors practiced them as well. Females were always less available than men, so each woman had a chance to marry (theoretically, at least).

The high-fertility paradigm typical of preindustrial societies is becoming part of the past. The high reproductive potential of the males is no longer the crucial factor of gene transmission. Today, when both sexes have approached parity in their parental investment, the anticipated trend of further social evolution is the decrease of male supremacy, paralleling the principal trend of human evolution: reduction of sexual dimorphism (Cliquet, 1984). Muscular strength, dexterity, aggressiveness, and other male attributes that could have been highly adaptive in the past are no longer necessary or sufficient for surviving in industrial society.

To maximize their inclusive fitness, present-day males must follow the K-strategy of fatherhood; they prefer to have fewer children, but invest more in each of them. Marital relations in developed countries largely depend on the father's participation in childrearing (Snarey, 1993). Women appreciate high paternal investment, and it is highly essential for them that their spouses spend as much time at home as possible, since the men's departures often result in the decrease of family income and thus reduce the resources available for the children. Also, contrary to popular predictions, success in a man's career is correlated with the amount of time he spends with his children (Snarey, 1993). The more a man is satisfied with his social position, the better climate he generates at home, the more satisfaction he obtains from rearing his children, and, eventually, the more of his genes will survive.

Although in most societies gender stereotypes are closely related to the principal sexual and reproductive paradigm, it may be asked if industrialization has in any way affected the process of the reduction of sexual dimorphism. Has this process accelerated? Have social orientations of males and females become different? The standards of female beauty are known to have changed dramatically over the last centuries and even since the early twentieth century. A woman commonly regarded as attractive in an industrial society is something of an outgrown adolescent, tall, lean, and having certain markedly unfeminine features like a narrow pelvis, small breasts, and angular shoulders. Most likely, this complex would be strongly nonadaptive in the ancestral environment. Behaviorally, the modern ideal of female attractiveness includes high general and social activity, brightness, and high intellectual level. Most of these features, too, hardly maximized women's fitness in preindustrial populations.

The important tendency of the last decades has been the active intrusion of women into political, economic, and educational domains, where many of them attain high positions, get high pay, and become economically independent. Under such conditions, husbands more and more often have to play the role of baby-sitters and become more engaged in household activities,. But this is not surprising, given the increased role of the K-strategy of paternal investment. The shift in reproductive paradigms, then, is paralleled by changes in gender stereotypes.

Finally, we would like to discuss certain ontogenetic trends that shape gender stereotypes in modern society. Our own data on different social strategies and individual practices in six- and seven-year-old primary schoolchildren of both sexes were studied using ethological methods. Fifty-six characteristics of social behavior, including frequencies of performed and received aggression, help, avoidance, demonstrations, friendly contacts, smiling, and laughing, as well as the number of active and passive partners for each behavior, were assessed by means of principal-components analysis (table 17.1). The first principal component (hereafter PC1), accounting for 26 percent of the total variation, is positively correlated with all characteristics. Traits with the highest loadings are received attention (number of times the child was looked at) and the number of smiles others addressed him or her, as well as aggression (both performed and received), avoidance by others, received help, and the like. So PC1 measures a child's general involvement in social life. To some extent it can also be viewed as an objective measure of leadership. The difference between PC1 scores in boys and girls is insignificant, suggesting that leadership in this group was not a prerogative of boys and that its criteria were unrelated to gender (table 17.2).

PC2 and PC3, on the other hand, show strong dependence on gender (figure 17.1). Each of them describes an independent strategy used by girls (boys can evidently afford not to use them). The first strategy of girls, revealed by PC2, is to look at more children and laugh together with others; such behavior is

Table 17.1
General Status in a Junior-School Class (Principal-component Analysis of 56 Behavioral and 5 Anthropometric Traits of 13 Girls and 9 Boys Aged 6–7)

	Factor Loadings
I. Behavioral traits	Highest loadings
Contact aggression, performed (N recipients)	0.74
Contact aggression, received (N performers)	0.71
Noncontact aggression, performed (frequency)	0.67
Noncontact aggression, performed (N recipients)	0.67
Avoiding, received (frequency)	0.77
Avoiding, received (N performers)	0.75
Giving orders, performed (N recipients)	0.67
Giving orders, received (N performers)	0.70
Help, received (N performers)	0.72
Initiating friendly contacts, received (frequency)	0.73
Smiling, received (frequency)	0.82
Smiling, received (N performers)	0.62
Directed laughing, performed (N recipients)	0.60
Directed laughing, received (frequency)	0.73
Directed laughing, received (N performers)	0.78
Attention, received (frequency)	0.87
Attention, received (N performers)	0.73
Subjective popularity, received (N performers)	0.62
II. Anthropometric traits	
Body length	0.05
Body weight	−0.10
Number of permanent teeth erupted	−0.05
Muscle component	0.39
Fat component	−0.01

rewarded by the decrease of received aggression. The second strategy (PC3) is to perform more demonstrations and avoidance; as a reward, more help is received.

Aggressive interactions were more frequent compared to help in boy-boy, boy-girl, and girl-boy dyads (figure 17.2). The ratio of aggression was higher for situations where boys were active aggressors. Only in the case of girl-girl dyads was help more frequent than contact aggression, suggesting that friendly motivation was higher in interactions between girls.

Correlation between friendly preferences and similarity in the general profile of social behavior revealed some principal differences between the sexes (table 17.3). Girls are evidently inclined to select friends among children with similar social characteristics, while no such tendency was found in boys.

Table 17.2
Factor Scores of First Principal Component,
Reflecting General Social Status in Junior
School Children (from Highest to Lowest)

Names	Factor Scores
NIKITA G.	7.1
Olesya	5.5
SEREZHA	4.9
Nastya	4.3
NIKITA B.	3.3
Lyuda	3.0
ANTON	2.6
SASHA G.	2.6
Valya	2.3
Ira Sh.	2.1
Rita	0.2
ILIA G.	−0.8
Marina	−1.8
Sasha J.	−1.9
ELDAR	−2.6
Nadya	−2.6
ILIA B.	−3.1
Ira D.	−3.3
Lera	−4.2
Xenia	−4.5
Masha	−6.3
LESHA	−6.8
Total	
Boys	0.74 ± 1.49
Girls	−0.51 ± 1.01
(insignificant)	

Note: Names of boys are given in capital letters. When
more than one child has the same name, an extra
letter is given for differentiation of the two.

Our findings demonstrate that the traditionally dichotomic interpretation of
gender is just a crude attempt at optimizing the utilization of innate biological
qualities. Girls are expected to be more prepared for childrearing and cooking,
while boys are said to be able to do work demanding physical strength in the
household. Both boys and girls expressed a desire to master some profession
and get a permanent job after having grown up.

Some of our results support the idea that the males' behavior is becoming
more girlish. Specifically, six- and seven-year-old boys spend much time playing

Figure 17.1
The Structure of Social Behavior in a Moscow Junior-School Class (17 Main Traits, 56 Characteristics, Results of Principal-Components Analysis)

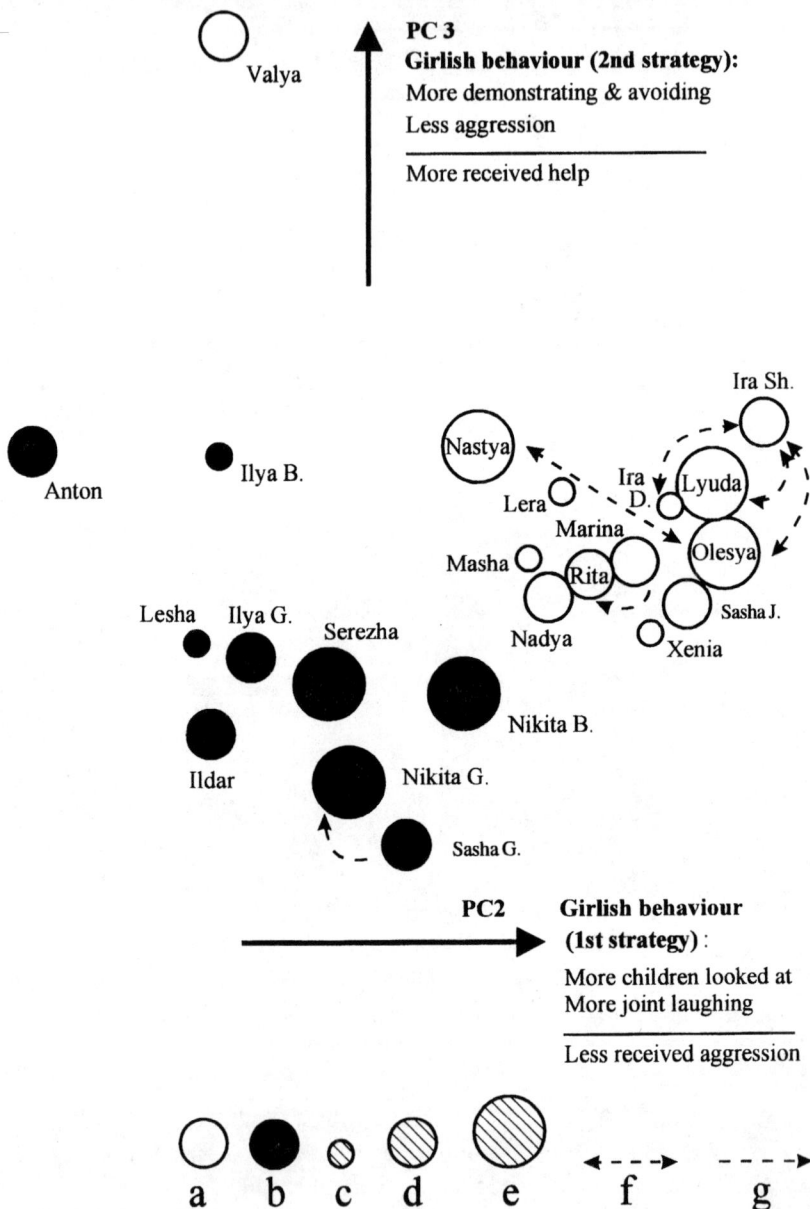

Valya

PC 3
Girlish behaviour (2nd strategy):
More demonstrating & avoiding
Less aggression

More received help

Ira Sh.

Nastya

Anton

Ilya B.

Lera

Ira D.

Lyuda

Marina

Masha

Rita

Olesya

Lesha   Ilya G.

Serezha

Nadya

Sasha J.

Xenia

Ildar

Nikita B.

Nikita G.

Sasha G.

PC2

Girlish behaviour (1st strategy) :
More children looked at
More joint laughing

Less received aggression

a   b   c   d   e   f   g

Notes: PC1 (27.9%) is general social status; PC1 (15.6%) is girlish behavior (first strategy): more children looked at, more joint laughing, less received aggression; PC3 (9.2%) is girlish behavior (second strategy): more demonstrating and avoiding, less aggression, more received help; a, white circles, indicates girls; b, black circles, indicates boys; c, small circles, indicates low-ranking individuals; d, medium circles, indicates middle-ranking individuals; e, large circles, indicates high-ranking individuals; shadings show that circle sizes are not connected with the sex of individuals—f, symmetrical friendly ties, and g, asymmetrical friendly ties.

Figure 17.2
Distribution of Aggression Counterhelp in the Group of Moscow Junior Schoolchildren, Aged 6–7

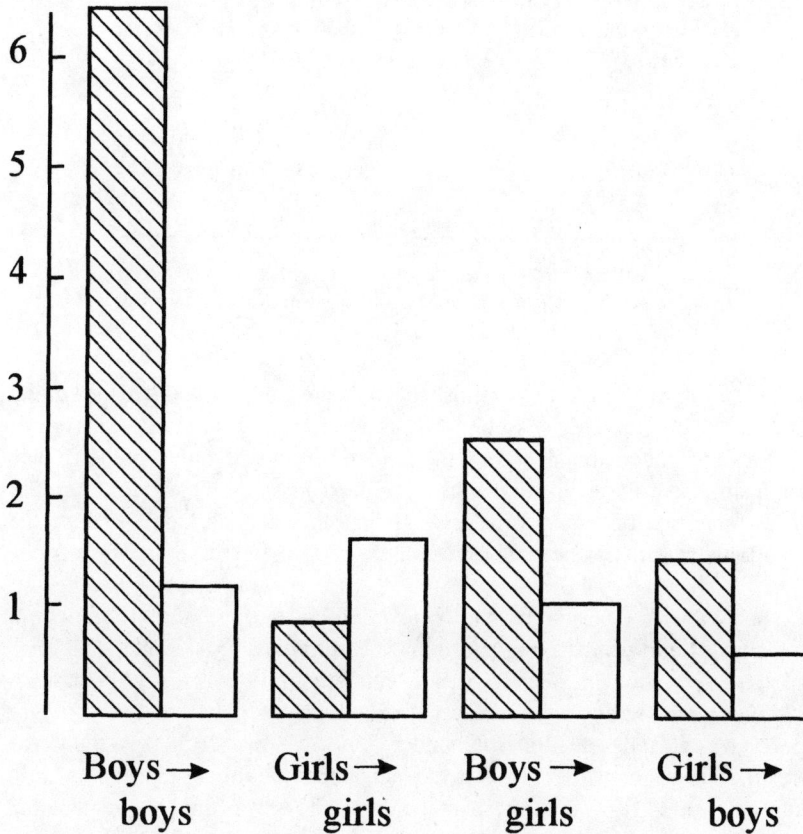

with toy animals and not only with toy cars or guns; a large proportion of social play occurred in bisexual teams (games like "family," "shop," "ringlet," and "robber Cossacks"), while boys participated in typically girlish activities like playing "elastic" and "check hopping"; no significant sex differences were observed in drawing most attractive objects or choice of future occupation.

Already in their early years girls spontaneously begin to master the strategies that enable them to get incorporated into the social system and regulate social relationships without the use of physical force. Social strategies of boys, on the other hand, seem to be formed largely by adults at a somewhat later stage of development.

Complex and painful initiation rituals existed in traditional societies, marking the transition from the status of a boy to that of an adult man. Although males and females differ in their innate predisposition to certain behavioral models

Table 17.3
Friendly Ties and Overall Similarity in Social Behavior
(Correlation between Two Interindividual Matrices: [a]
Based on frequency of friendly contacts and [b] based on 56
characteristics of social behavior [First six principal
components, 72.5% of variation])

1. Entire group (n=22):	Rs= 0.24, p=0.005*
2. Boys (n=9):	Rs= 0.11, p=0.183
3. Girls (n=13):	Rs= 0.33, p=0.017*

Note: Probabilities were assessed using matrix-of-permutation tests (1,000
permutations made in each case). Significant correlations are marked
by asterisks.

(Eibl-Eibesfeldt, 1989), culture in most societies apparently hypertrophies initial differences by accentuating masculine features of boys' behavior. Because initiation rites have been abandoned by the modernized world and only their much milder forms have survived, like those practiced in some closed communities of males, such as the army, prisons, and boarding schools, and because the coeducational system has become universal, innate genetic dissimilarity between the sexes is gradually diminishing. As a result, new stereotypes of male and female behavior are emerging, aimed at leveling out sharp sexual differences in social behavior and professional and domestic activities.

Unlike the situation in the past, gender-related variation in many behavioral traits shows a continuous pattern rather than a clear-cut dichotomy, the primary reason being relaxed selection for gender identity. The second reason is the general rise in personal identity and independence. A woman can rear a child without being helped by a partner. Interestingly, however, the role of maternal relatives is still considerable, unlike that of paternal relatives, who are becoming less and less important. Life in an urban society demands a new type of socialization oriented on a more feminine style of behavior in some aspects (less xenophobia, more friendliness, and ability to cope with social tension) and a more masculine style in others (orientations toward success in future careers, self-confidence, economic independence, and technical skills). This trend parallels the main tendency in morphological evolution of modern man.

ACKNOWLEDGMENTS

This study was partly supported by a grant from the George Soros Foundation. We gratefully acknowledge the financial aid that was offered to us by the Royal Society of Great Britain, which enabled us to attend the Cambridge meeting of the European Sociobiological Society in 1995. We are especially grateful to Robin Allot for his generous help.

REFERENCES

Boekhorst, I. J. A. te (1991). *Social structure of three great ape species: An approach based on field data and individual oriented models.* Utrecht: Riksuniversiteit Utrecht.

Buss, D. M. (1989). Sex differences in human mate preferences: Evolutionary hypotheses tested in 37 cultures. *Behavioral and Brain Sciences, 12,* 1–4.

Cliquet, R. L. (1984). The relevance of sociobiological theory for emancipatory feminism. In J. Wind (Ed.), *Essays in human sociobiology* (Vol. 1, pp. 117–128). London: Academic Press.

Daly, M., & Wilson, M. (1983). *Sex, evolution, and behavior* (2nd ed.). Boston: Willard Grant.

Darwin, C. (1859). *On the origin of species by means of natural selection.* London: Murray.

Dawkins, R. (1976). *The selfish gene.* Oxford: Oxford University Press.

Eibl-Eibesfeldt, I. (1989). *Human ethology.* New York.: Aldine de Gruyter.

Fedigan, L. (1982). *Primate paradigms: Sex roles and social bonds.* Montreal: Eden Press.

Fisher, H. E. (1989). Evolution of human serial pair bonding. *American Journal of Physical Anthropology, 78,* 331–354.

Hrdy, S. B. (1981). *The woman that never evolved.* Cambridge, MA: Harvard University Press.

Irons, W. (1980). Is Yomut social behavior adaptive? In G. W. Barlow & J. Silverberg (Eds.), *Sociobiology: Beyond nature/nuture?* (pp. 417–473). Boulder, CO: Westview Press.

———. 1991. Anthropology. In M. Maxwell (Ed.), *The sociobiological imagination* (pp. 71–90). Albany: State University of New York.

Kinzey, W. G. (1987). *The evolution of human behavior: Primate models.* Albany: State University of New York Press.

Reynolds, V. (1984). The relationship between biological and cultural evolution. In J. Wind (Ed.), *Essays in human sociobiology* (Vol. 1, pp. 71–80). London: Academic Press.

Rhine, R. J. (1992). Dominance life history of the alpha matriarch of a colony of stump-tailed macaques (*Macaca arctoides*): Establishment of long-term dominance. *Primate Report, 32,* 107–118.

Rose, S. (1980). "It's only human nature": The sociobiologist's fairyland. In M. F. A. Montagu (Ed.), *Sociobiology examined* (pp. 158–170). Oxford: Oxford University Press.

Silk, J. B. (1990). Human adoption in evolutionary perspective. *Human Nature, 1,* 25–52.

Snarey, J. (1993). *How fathers care for the next generation: A four-decade study.* Cambridge, MA: Harvard University Press.

Symons, D. (1979). *The evolution of human sexuality.* New York: Oxford University Press.

Trinkaus, E. (1980). Sexual differences in Neanderthal limb bones. *Journal of Human Evolution, 9,* 377–398.

Trivers, R. L. (1972). Parental investment and sexual selection. In B. Campbell (Ed.), *Sexual selection and the descent of man* (pp. 136–179). Chicago: Ardine Press.

Tucker, W. (1993, October 4). Monogamy and its discontents. *National Review.*

van Schaik, C. P. (1989). The ecology of social relationships amongst female primates. In V. Handen & R. Foley (Eds.), *Comparative socioecology: The behavioral ecology of humans and other mammals* (pp. 95–218). Oxford: Basil Blackwell.

Washburn, S. L. (1980). Human behavior and the behavior of other animals. In M. F. A. Montagu (Ed.), *Sociobiology examined* (pp. 254–282). Oxford: Oxford University Press.

Williams, G. C. (1966). *Adaptation and natural selection: A critique of some current evolutionary thought.* Princeton, NJ: Princeton University Press.

Wilson, E. O. (1975). *Sociobiology: The new synthesis.* Cambridge, MA: Belknap Press of Harvard University Press.

Wolpoff, M. H. (1976). Some aspects of the evolution of early hominid sexual dimorphism. *Current Anthropology, 17,* 579–606.

Wrangham, R. W. (1986). Ecology and social relationships in two species of chimpanzee. In D. I. Rubenstein & R. W. Wrangham (Eds.), *Ecological aspects of social evolution: Birds and mammals* (pp. 352–378). Princeton, NJ: Princeton University Press.

Wright, R. (1994). *The moral animal.* New York: Pantheon Books.

PART V

EVOLUTIONARY PSYCHOLOGY AND PSYCHIATRY

The final part of this volume faces the issues of sociobiology and psychology in five chapters dealing with the domain of the emotions, a topic that has been difficult to pin down without a biological orientation. Michael Bradie takes us back to Darwin's own approach to the understanding of human morality and shows how Darwin's understanding is rooted in a tradition of thought in British philosophy in which sympathy with the other is a significant factor in the construction of moral systems. Charles Elworthy helps us see how the social sciences and specifically evolutionary psychology can be fitted within the broader frame spelled out in previous chapters.

John S. Price relates the evolutionary phenomenon of sexual selection to an underlying competition for reproductive success understood as an aspect of sexual selection in males and females. This results in winners and losers, and we can thus see some possible origins of depression, here understood within the context of the European humanistic tradition. A scenario of evolutionary biology thus can be related to the domain of literary traditions, somewhat in the spirit of the chapter by Constable but tied now to such phenomena as psychiatric symptoms. These wide-ranging potentialities of sociobiology are given a sharper focus by Daniel Wilson, Sean Stanton and Sandra Wilson. They direct their chapter to psychiatric disorders and contemporary aspects of neurotransmitters and provide the excitement of seeing how medical and psychiatric sciences may be viewed from a sociobiological perspective and may lead through such an account to the integration of modern medicine with the broader themes of evolutionary thought. The closing chapter by Glenn Weisfeld considers an overview of human emotions and looks specifically at the emotional complex of pride and shame, relating these emotions to interpersonal conflict in the evolutionary past. While existing universally across cultures, they also are expressed in different circumstances and thus mediate the particulars of social existence.

18

Darwin and the Eighteenth-Century British Moral Tradition

Michael Bradie

Charles Darwin's treatment of morality is a reflection of and response to concerns that emerge in Francis Hutcheson and other eighteenth-century British moralists concerning the reconciliation of self-interest and benevolence. In essence Darwin can be construed as providing an evolutionary answer to a problem posed by Hutcheson: ''[I]t is true indeed, that the actions we approve in others, are generally imagin'd to tend to the natural good of Mankind, or that of some Parts of it. But whence this secret chain between each person and Mankind? How is my interest connected with the most distant parts of it?'' (Hutcheson, 1969, vol. 2, p. 111).

THE EIGHTEENTH-CENTURY LEGACY IN BRITISH MORAL PHILOSOPHY

Two problems form the core of the eighteenth-century discussions on moral motivation. The first, ''Why be moral?'' is really two questions, ''Why are we moral?'' and ''Should we be moral, and if so, why?'' The first, descriptive question is one of moral motivation. That we should be moral and that being moral involved acting to promote the common good were presupposed by these writers. As to why we should be moral, although the eighteenth-century British moralists rejected a radical Hobbesian position, more often than not they wound up defending a version of the view that we should be moral because it is in our self-interest (properly understood) to do so.

The second problem, then, was how to reconcile self-interest and benevolence (which in the nineteenth century became the problem of reconciling self-interest with altruism). This problem was bequeathed to the eighteenth century by a Hobbesian view of man that forced moralists either to deny that real benevolence existed or to show either (1) how benevolence could overcome self-interest or (2) why the two often, if not usually, coincided.[1]

James Mackintosh provides a direct link between the eighteenth-century British moral tradition and Charles Darwin. Mackintosh, Darwin's uncle by marriage, wrote a well-known treatise on moral philosophy (Mackintosh, 1834) that Darwin read carefully. In it, Mackintosh argued against the reductionist thesis that all passions or interests could be reduced to self-love. Human morality, for Mackintosh, was characterized by the striving for universality, necessity, and immutability in moral judgments. With Joseph Butler (1983), Mackintosh postulated conscience as a supreme moral principle. Conscience is a universal, independent principle that guides our voluntary actions (Mackintosh, 1834, pp. 380–381). At the same time, Mackintosh argued for the contingent malleability of conscience. Our conscience morally approves as virtuous those actions that promote the general good, although the general good is rarely if ever our deliberate aim. Our moral sentiments "coincide" with the dictates of conscience. But, Mackintosh asked "whence arises the coincidence between [conscience] and the moral sentiments? It may seem at first sight, that such a theory rests the foundation of morals upon a coincidence altogether mysterious, and apparently capricious and fantastic" (1834, p. 382).

Mackintosh was as hard pressed as was Butler to account for the "social dimension" of morality. Darwin, who expressed admiration for Mackintosh's mental powers, was unhappy with Mackintosh's attempt to explain the coincidence between the universal judgments of conscience and socially useful behaviors by appeal to "mental contiguity." Mackintosh's solution, to which Darwin objected, is that "it may truly be said, that if observation and experience did not clearly ascertain that beneficial tendency is the constant attendant and mark of all virtuous dispositions and actions, the same great truth would be revealed to us by the voice of conscience. The coincidence, instead of being arbitrary, arises necessarily from the laws of human nature, and the circumstances in which mankind are placed" (Mackintosh, 1834, p. 383).

In his notebooks, Darwin posed the fundamental problem of morality as the attempt to reconcile and explain the "coincidence of morality with individual interest." Darwin mused that conscience, à la Mackintosh, is supreme only insofar as it is part of our inherited habitual natures (Gruber, 1974, pp. 398–399).[2] Darwin's objection hinges on his evolutionary perspective, which denies the universality or immutability of human nature and sees the circumstances in which humans find themselves both changing and modifying human beings and their natures. For Darwin, Mackintosh's "solution" merely reposes the problem of coincidence.

DARWIN'S MORAL THEORY

"A moral being is one who is capable of comparing his past and future actions and motives,—of approving of some and disapproving of others; and the fact that man is the one being who with certainty can be thus designated makes the greatest of all distinctions between him and the lower animals" (Darwin, 1981, vol. 2, p. 392). Darwin is quick to point out, however, that in his view this great difference is still only a difference in degree and not in kind.

Darwin summarizes his view of the evolution of the moral sense in the last chapter of part 2 of *The Descent of Man* (1981, vol. 2, pp. 389ff.). He reminds us of the main conclusion of the book. Human beings are descended from some "less organized from." Mental faculties as well as physical characteristics are variable. These variations are inherited. He conjectures that the increase in relative brain size that is a feature of human beings is a result in part of the development of language and the subsequent interaction of mental faculties with the physical brain. The "culture of the mind" contributes, in both the race and the individual, to the development of the higher intellectual powers in man.

The foundations of morality lie in the social instincts. For human beings, Darwin notes, these include family ties, love, and sympathy. These foundations are shared by other social animals, although, given their intellectual limitations, they do not have morality in any true sense of the word. These instincts "have in all probability been acquired through natural selection" (Darwin, 1981, vol. 2, p. 391).[3]

The moral sense in human beings evolved through the interaction of the social instincts and the higher mental powers. The crucial intellectual capacities for the development of morality, in Darwin's view, are memory, anticipation, and the power of reflection. We remember deeds done or undone in the past, and upon reflection these memories give rise to feelings of satisfaction or dissatisfaction. The feelings of dissatisfaction, in turn, give rise to feelings of regret, and we resolve to act differently in the future. Such is the origin of conscience. Our sense of obligation arises from the presence of permanent, stronger, and more enduring instincts. These turn out, for Darwin, to be the social instincts.

We share with other social animals certain social instincts that manifest themselves in a general desire to aid members of our community. In the lower animals, these desires manifest themselves in "hard-wired" actions or responses. Human beings, on the other hand, are "soft wired." They have a (variable) general desire to aid members of their community but no built-in responses or specific patterns of behavior. The implementation and satisfaction of these general desires is enhanced by the human capacity for language. Through language, human beings can make their desires and needs clearer to others who might be in a position to satisfy them. Praise and blame become possible and, once instituted, become selective shapers of behavioral repertoires. In addition, the formulation of moral rules and codes becomes possible, and this, in turn, contributes to the spread of moral practices and the development of "higher" moralities, that is, moralities that embrace wider and wider human communities.[4]

Four propositions lie at the heart of Darwin's ethical theory:

1. The foundation of morality lies in the social instincts.

2. The evolution of mentality/intelligence leads to the evolution of the social instincts.

3. The development of language leads to the development of common opinions and the development of public approval and disapproval based upon considerations of mutual sympathy.

4. The evolution of morality is aided and abetted by the formation of habits (cf. Kropotkin, 1947, pp. 33–34).

The evolution of intelligence leads to the evolution of conscience or a moral sense. Feelings of regret, shame, remorse, repentance, and guilt are part of our moral conscience and evolve as part of a mechanism to ensure automatic acceptance of certain rules of conduct that are necessary for group survival. These sentiments arise from a conflict between instincts. All organisms with a sufficiently developed social nature are subject to these conflicts. Instincts are often in conflict and have different strengths. The migratory instinct in birds sometimes overcomes the maternal instinct when, for example, a mother bird will abandon her late-hatched and underdeveloped chicks to fly south (Darwin, 1981, vol. 1, p. 82). Of two instincts, in general, natural selection will make the one that promotes the good of the species more "potent" since individuals with instincts that promote the good of the species would tend to survive in greater numbers (but, as Darwin notes, the migratory/maternal conflict in birds is an exception to this rule). Human beings, unlike other animals, have the ability to reflect upon and judge their intentions and actions. This ability for second-order reflection results in *conscience*. But how are we to account for it? Darwin distinguishes between social instincts and self-directed instincts. When we act against the inclinations of our social instincts we experience regret. But why? If we attempt to explain action in the presence of conflicting instincts by appealing to the strength of instincts and if we assume that the social instincts are "higher" because of their greater strength, we are unable to explain why humans often act counter to their social inclinations. On this simple model, our other-directed motives ought to outweigh our self-directed motives every time.

Butler, in the eighteenth century, faced a similar problem in trying to account for the nature and effectiveness of conscience. Butler had distinguished between the *strength* and the *authority* of an instinct and asserted that, in effect, our social instincts have a higher authority. He argued that the principles and propensities that motivate us can be hierarchically organized. The hierarchical ordering is based upon two criteria: (1) strength and (2) authority. The idea then seems to be this: we have a postulated hierarchical model of motivation that consists of two factors, benevolence and self-love in some mixture, with an appropriate strength and authority. Every individual action is presumed to be the product of a "mixed" motivation of the same factors. We can, in principle, compare the "motivational profile" of a given real-world action against the standard model—"the correspondence of actions to the nature of the agent renders them natural: their disproportion to it, unnatural" (Butler, 1983, p. 31; also cf. pp. 32f–39–40). But Darwin could not accept such an answer, as his notebooks reveal.

James Rachels argues that Darwin's solution is to appeal to a modified model based on strength considerations alone. The strength of our instincts or motives varies from time to time and circumstance to circumstance. When we act against our social instincts, we do so because at that moment our self-directed instincts are stronger. We act and the moment passes. As it does, the strength of our self-directed instincts wanes with respect to the strength of our other-directed instincts. We reflect again and experience remorse, guilt and the like. The "authority" of the social instincts rests, in Darwin's view, solely on their longevity and permanence (Rachels, 1990, pp. 7–8). But why have the social instincts evolved to be more permanent and lasting than the self-directed motives or instincts? Darwin's answer is that these instincts emerge as more "potent." Natural selection singles them out because they are conducive to group survival.

Darwin accepted a version of the inheritance of acquired characteristics in the form of the "use and disuse" principle.[5] Virtue practiced becomes virtue habituated. Habits turn into instincts and are inherited. The moral fiber of humankind becomes stronger and more advanced. "The moral nature of man has reached the highest standard as yet attained, partly through the advancement of the reasoning powers and consequently of a just public opinion, but especially through the sympathies being rendered more tender and widely diffused through the effects of habit, example, instruction, and reflection. It is not improbable that virtuous tendencies may through long practice be inherited" (Darwin, 1981, vol. 2, p. 394).[6]

What does this higher morality look like? Darwin is careful not to suggest that any specific moral codes can be derived from evolutionary considerations. Indeed, although the instincts upon which morality rests may have evolved through natural selection, the development and practice of specific virtues owes much to education, enculturation, and social contingencies.[7] Nonetheless, "[A]s all men desire their own happiness, praise or blame is bestowed on actions and motives, according as they lead to this end; and as happiness is an essential part of the general good, the greatest-happiness principle indirectly serves as a nearly safe standard of right and wrong" (Darwin, 1981, vol. 2, p. 393).

Was Darwin a utilitarian? More important, did he hold that the theory of natural selection serves as an explanation or justification for the greatest-happiness principle? Many nineteenth-century commentators interpreted him this way. Schurman (1893) was one such. Schurman considered the struggle for life and survival of the fittest as the two fundamental components of Darwin's theory (1893, p. 116). The first he took to derive from Malthus and hence from political economy; the second he took to derive from utilitarianism. "No one uninfluenced by the ethics of the school of Hume and Bentham would have ventured to interpret the evolution of life as a continuous realization of utilities" (1893, p. 117). Schurman argues that "[n]atural selection rests upon a biological utilitarianism, which may be egoistic [corresponding to individual selection] or communistic [corresponding to group selection], but never universalistic" (1893, p. 118). This is then turned by Schurman into a critique of both natural selection as an explanation of morality and of utilitarianism. The evolution of man by natural selection has produced bodily changes, along with the evolution of mental powers that render human beings "independent of nature." As such, the ethical systems that men have developed are not completely explicable as the result of the working of natural selection.

"We began by remarking that the biological theory borrowed the notion of utility from empirical morals; but we must now confess the loan has been so successfully invested that there is some ground for believing the proceeds suffice, not only to wipe out the obligation, but even to make the ethics debtor to biology" (Schurman, 1893, p. 122). Schurman argues that utilitarianism, in some form, is a necessary implication of Darwinism. But there is more to morality than can be accounted for by utilitarianism; hence there is more to morality than can be accounted for by an appeal to natural selection. Biology, he concludes, does not afford a "complete scientific explanation of the phenomena of morals" (p. 127).

There is, however, other evidence in *Descent* that Darwin construed his evolutionary theory of morality as quite distinct from utilitarianism and as neither entailing it nor being entailed by it. At best, he saw utilitarianism as a useful guide for determining right and wrong. To see where Darwin considered his view as different from utilitarianism, we have to examine the argument for the development of the social instincts in somewhat more detail.

The heart of Darwin's argument is contained in chapters 3 and 5 of volume 1, part 1, of *Descent*. For Darwin, the foundation of morality lies in sympathy. But what is the origin of sympathy? Sympathy is distinct from love. How is it acquired? In the view of Adam Smith and Alexander Bain, sympathy is induced by the recollection of past suffering plus association with the misfortune of others. But, says Darwin, this does not explain why we feel a stronger sympathy for those more closely related to us (1981, vol. 1, p. 81). Darwin's answer is that human survival depends on cooperation. Rudimentary forms of cooperation exist in lower animals and are selectively advantageous. These form the basis of the social instincts.

Darwin's argument in *The Descent of Man* is that we have three options in explaining the evolution of social instincts that (when conjoined with the power to reason and reinforced by habit) form the basis of the moral sense or conscience. The first is by trying to understand their evolution as a result of selfishness. Darwin rejects this as the basis of an appropriate analysis on the grounds that animals exhibit phenotypic other-regarding behavior that cannot be understood as self-regarding or selfish.

The second option is the greatest-happiness principle. This Darwin rejects on the ground that natural selection is not primarily concerned with the pleasure and pain of organisms but rather with producing structures and traits that are conducive to the promotion of reproductive success. The social instincts, as he sees it here, have evolved because of their contribution to the common good, but that good is not the sum of pleasures or happiness in the community but rather the development of a social lifestyle that enhances the reproductive success of its members. Pleasure and happiness are, at best, derivative ends. For Darwin, X promotes pleasure because it is good (i.e., it has been selected because it promotes reproductive success). For the utilitarians, from whom Darwin is here trying to distance himself, X is good because it promotes pleasure (or the greatest happiness or whatever). For Darwin, the utilitarians have the argument backwards.

The third option for Darwin is that the evolution of the social instincts is driven by selection for the common good or the common welfare. This is to be understood as sketched earlier. Natural selection does not "see" the happiness of organisms or the total happiness of the communities that they form. Selection "sees" the potentialities for differential reproductive success and rewards them. Those instincts emerge and survive that tend to promote reproductive success. The engendered behaviors may result in pleasure or happiness for the organisms that pursue them, but the evolutionary ("ultimate") rationale for performing them is not that they are pleasure inducing but that they promote reproductive success. The organism's immediate ("proximate") rationale may indeed be the anticipated pleasure that will be enjoyed, but the evolutionist is unhappy with leaving the explanation there. Why are these activities pleasure inducing and those not? The answer must be that the former are fitness enhancing and the latter not. I take it that the evolutionary argument is committed to denying that there can be systematic pressure for the evolution of characteristics that, however pleasure inducing, are not contributing to enhanced reproductive success or correlated with behaviors that have that result or are in some way implicated in the enhancement of reproductive success. Of course, on some mosaic interpretation of behavioral repertoires it may be difficult in practice to isolate just which behaviors are enhancing what and to what extent. The point here is both empirical and conceptual. The conceptual point is that it is enhanced reproductive success that drives the evolution of the social instincts, not the correlated enhanced pleasure or happiness that may or may not accompany their exercise.

Darwin's model is an improvement, from a naturalistic point of view, over Butler's and Hutcheson's. It provides a naturalistic account of the emergence of conscience and attempts to account for the "authority" of considerations of conscience (and the interests of others) over the considerations of self. But it falls somewhat short of being completely successful. Explanations of abilities of any sort in terms of natural selection are quite limited. First, they do not explain the emergence of particular traits. Thus, even if we were to accept the idea that having the capacity to be moral was selectively advantageous, that in itself does not explain why morality as such evolved and not some equally enhancing alternative mechanism. Second, when Darwin distinguishes between "higher" and "lower" moral values, he classifies the social (other-regarding) values as higher and the individual (self-regarding) values as lower (1981, vol. 1, p. 100). But why identify "moral" with "social"? And why give the social values higher priority than the individual values? (cf. Murphy, 1982.) Virtue-based moralities include self-regarding virtues as well as social virtues in the realm of the moral. Darwin, as the inheritor of the eighteenth-century moral debate, is in accord with one of the central conclusions of the post-Hobbesian tradition in British moral philosophy in locating social values as "higher" than self-regarding values. But does it follow from any evolutionary argument that social values are "higher" (i.e., more moral) than self-regarding values? Even if it could be shown that evolution leads to a strengthening of the social values vis-à-vis self-regarding values, the question of the moral authority of such values remains.

The gap between evolved strength and moral authority is the chasm between the natural and the moral. For those contemporary evolutionary ethicists for whom this gap is unbridgeable, such as Michael Ruse and E. O. Wilson (1986), the conclusion to be drawn from the evolutionary argument is the rejection of the search for ultimate justifications in questions of morality. For those naturalists for whom the gap is bridgeable, such as Robert Richards (1987), the task remains to show how, if at all, the evolutionary argument can be molded to provide a justification of moral principles.

NOTES

1. The details of the eighteenth-century tradition that Darwin inherited are more fully explored in Bradie (1994).

2. See Manier (1978) for a discussion of the influence of Mackintosh on Darwin.

3. It is well to note that the reason Darwin gives for holding that the social instincts have "probably been acquired by natural selection" is that "they are highly beneficial to the species" (Darwin, 1981, vol. 2, part 2, p. 391).

4. "As the reasoning powers advance and experience is gained, the more remote effects of certain lines of conduct on the character of the individual, and on the general good, are perceived; and then the self-regarding virtues, from coming within the scope of public opinion, receive praise, and their opposites receive blame. But with the less civilized nations reason often errs, and many bad customs and base superstitions come within the same scope, and consequently are esteemed as high virtues, and their breach as heavy crimes" (Darwin, 1981, vol. 2, p. 393).

5. Darwin writes: "It is well known that use strengthens the muscles in the individual, and complete disuse, or the destruction of the proper nerve, weakens them. When the eye is destroyed the optic nerve often becomes atrophied. When an artery is tied, the lateral channels increase not only in diameter, but in the thickness and strength of their

coats. . . . Whether the several foregoing modifications would become hereditary, if the same habits of life were followed during many generations, is not known, but is probable" (1981, vol. 1, pp. 116–117).

6. Darwin, of course, had no means of distinguishing between gene-based inheritance and cultural inheritance of habits, customs, and lifestyles. See the editor's introduction to Darwin (1981).

7. Murphy (1982, pp. 82f–83) argues on the basis of this that Darwin recognized, as Freud did not, that there are elements to morality that may lie beyond the scope of his theory. That is, the theory of natural selection is not the be-all and end-all of moral phenomena.

REFERENCES

Bradie, M. (1992). Darwin's legacy. *Biology and Philosophy, 7,* 111–126.
————. (1994). *The secret chain.* Albany: State University of New York Press.
Butler, Joseph. (1983). *Five sermons.* Indianapolis: Hackett.
Darwin, Charles. (1936). *On the origin of species, by means of natural selection.* New York: Modern Library.
————. (1981). *The descent of man, and selection in relation to sex.* Princeton, NJ: Princeton University Press.
Gruber, Howard. (1974). *Darwin on man: A psychological study of scientific creativites.* London: Wildwood Houses.
Hucheson, Francis. (1969). *An inquiry concerning the original of our ideas of virtue and moral good.* New York: Gregg International Publishers.
Kropotkin, P. (1947). *Ethics: Origin and development* (Louis S. Friedland & Joseph R. Piroshnikoff, Trans.). New York: Tudor Publishing Company.
Mackintosh, Sir James. (1834). *A general view of the progress of ethical philosophy.* Philadelphia: Carey, Lea, & Blanchard.
Manier, Edward. (1978). *The young Darwin and his cultural circle.* Boston: D. Reidel.
Midgley, Mary. (1979). *Beast and man: The roots of human nature.* New York: Harvester Press.
Murphy, Jeffrie. (1982). *Evolution, morality, and the meaning of life.* Totowa, NJ: Rowman & Littlefield.
Rachels, James. (1990). *Created from animals: The moral implications of Darwinism.* Oxford: Oxford University Press.
Richards, Robert J. (1987). *Darwin and the emergence of evolutionary theories of mind and behavior.* Chicago: University of Chicago Press.
Ruse, Michael, & Wilson, E. O. (1986). Moral philosophy as applied science. *Philosophy, 61,* 173–192.
Schurman, J. G. (1893). *The ethical import of Darwinism.* New York: Scribner's.

19

Evolutionary Psychology: The Appropriate Disciplinary Link between Evolutionary Theory and the Social Sciences

Charles Elworthy

Despite the enormous power of evolutionary theory and associated empirical research, it has as yet had relatively little impact on the analysis of human behavior in the orthodox social sciences. It is illuminating to assess the reasons for this weakness of the link between biological theory and social behavior in terms of supply and demand.

SUPPLY

1. Many of the sociobiological explanations that have been offered to social scientists have been adaptivist rather than adaptationist in character (Symons, 1990, 1992). Given that much modern behavior may not be reproductively adaptive, this has tended to lead to evolutionary approaches being regarded as erroneous and simplistic.

2. The majority of empirical studies have restricted themselves to the testing of quite specific hypotheses, often related to clearly biological themes such as mating and reproduction. This has tended to reduce their relevance for social scientists attempting to explain a wide range of behavior with one generally applicable model.

3. Different subdisciplines associated with biology such as primatology, paleontology, and evolutionary theory have often attempted to deduce their own implications for human behavior, leading to a corresponding confusion about a unified "sociobiology."

DEMAND

1. There is no unified demand for biological theory and empirical research from social scientists, because there is no recognized paradigm and thus no generally accepted

appreciation of how such a disciplinary interface should function and what it should contain.

2. The nearest to a general methodology for the analysis of social behavior is neoclassical economic theory, which has advocates throughout the social sciences (e.g., Smelser & Swedberg, 1994; Radnitzky & Bernholz, 1987; Hirshleifer, 1985). The following characteristics of this economic methodology offer an attractive interface to evolutionary explanations of human behavior: (a) It requires a model of human psychology that describes the essential features of human objectives and motivations. Whether these are conscious or unconscious is immaterial. (b) It is most effective if the psychological processes described can be assumed to apply to a large class of individuals, such as all humans or all adults. (c) An assumption of maximizing behavior allows the application of the classic technical solution for analyzing and predicting behavior: maximization subject to constraints.

In order to reconcile this difference between the supply of new insights deriving from the biological sciences with the largely latent demand in the social sciences, an alternative model of human characteristics should be provided that is not only superior to *Homo oeconomicus, Homo sociologicus*, and their kin, but can be used within the existing theoretical structures of economics, political science, sociology, or international relations. I argue that the appropriate solution to this problem is a *Homo biologicus* (1993), a model of human behavior based on evolutionary psychology. This is an intermediate discipline between evolutionary theory and the social sciences, which investigates and specifies the characteristics of the human "cognitive level" (Cosmides & Tooby, 1987, pp. 283–284). Evolutionary psychology concentrates on the cognitive adaptations that were shaped by evolutionary processes and that in turn shape behavior. Cosmides and Tooby, for example, have investigated social exchange from a Darwinian perspective, showing that specific psychological processes, such as the "look for cheaters algorithm," evolved to ensure that social exchange was successful (Cosmides, 1989; Cosmides & Tooby, 1989, 1992). The application of Darwinian principles to the study of the function, contributing mechanisms, phylogeny, and ontogeny of these "Darwinian algorithms" means that the powerful tools of the biological sciences can be applied to the study of the human mind.

The knowledge thus obtained may appear less relevant to the understanding of human social behavior than predictions directly from evolutionary theory, for example, that individuals assist kin in proportion to their degree of relationship. Yet such arguments present two major difficulties that have inhibited the acceptance of sociobiology in the social sciences. The first is that they may not in fact be valid: as Vining has shown with respect to status and reproductive success, in a modern world full of artificial cues humans may not maximize the transmission of their genes to the next generation (1986, p. 167). The second reason is that explicitly biological themes are not directly relevant to many social scientists, who desire an understanding of such fields as attitudes to risk or inheritance behavior. Evolutionary psychology, in contrast, allows social scientists to replace their current assumptions about human behavior with models based on evolutionary theory while retaining established analytical frameworks.

RISK AND UNCERTAINTY FROM AN EVOLUTIONARY PERSPECTIVE

In order to examine the practical application of this approach, it is instructive to consider its application to particular aspects of behavior. One appropriate domain is behavior under conditions of risk and uncertainty, which has applications throughout the social sciences and has been the subject of investigation by a wide variety of disciplines over several centuries.

Panic as Extreme Risk Aversion

Evolutionary theory indicates that in order to preserve their own lives and resources, individual organisms should, other things being equal, avoid situations that have high expected costs. Whereas situations offering exceptionally high benefits can be expected to have been relatively rare, immediate and large costs would have been encountered frequently, such as those following the failure to detect a predator. The most obvious high-cost situations are those in which an individual's life is threatened and humans are thus expected to take immediate and extreme measures in order to escape from such mortal dangers.

Nesse has postulated that such a historically adaptive response underlies the phenomenon of panic, in which the goal of escape overrides all other behavioral motivations (1987). He uses the functionality of real panic in investigating the psychiatric condition known as "panic disorder," which generally takes the following form: "The problem typically begins with a sudden attack of overwhelming anxiety that occurs 'from out of the blue' while the person is away from home. A feeling that death is imminent and an overwhelming need to 'get out of here' displace all other thoughts. Shortness of breath and pounding of the heart precede actual movement" (p. 75S). In order to "escape," individuals seek to return to the security of family relationships and known physical environments, the similarity of the physical and mental conditions with those associated with flight from danger, together with numerous other aspects of these conditions, supports Nesse's hypothesis that "the abnormality in people with panic disorder is not the occurrence of panic per se, but rather that the attacks occur when there is no real danger" (p. 74S).

These hypotheses are much more consistent with the associated physiological and behavioral phenomena than those derived from alternative approaches based on Freudian analysis, learning theory, neurochemical mechanisms, or genetic imperfections, although elements of such approaches will in all probability be included in explanations of the underlying mechanisms (Nesse, 1987, 76S–78S). The cognitive programs involved in escaping from danger are probably related, for example, to other cognitive programs such as those preventing separation from the mother; they are ultimately instantiated in neurochemical mechanisms; they are likely to involve learning from experience and the environment; and they will vary with genetic differences in the population. The compatibility of these specific hypotheses with Nesse's explanation results from the generality of phylogenetic and functional explanations, which provide an interpretative framework in which the various proximate and ontogenetic factors can be understood.

Young Male Risk Seeking

Panic involves the protection of life and limb through flight from a threatening situation, and individuals exhibiting panic disorder and agoraphobia can be interpreted as the most risk averse of the population in these domains. Just as such extremely risk-averse reactions are unintelligible to the majority of people, for whom open spaces as such provide "no reason" to panic, so other forms of behavior are interpreted as senseless or irrational because of their risk-seeking qualities. Consider the following account:

One time I was in the men's room of the bar and there was a guy at the urinal. He was kind of drunk, and said to me in a mean-sounding voice, "I don't like your face. I think I'll push it in."

I was scared green, I replied in an equally mean voice, "Get out of my way, or I'll pee right through ya!"

He said something else, and I figured it was getting pretty close to a fight now. I had never been in a fight. I didn't know what to do exactly, and I was afraid of getting hurt. I did think of one thing: I moved away from the wall, because I figured if I got hit, I'd get hit from the back, too.

Then I felt a sort of funny crunching in my eye—it didn't hurt much—and the next thing I know, I'm slamming the son of a gun right back, automatically. It was remarkable for me to discover that I didn't have to think; the "machinery" knew what to do.

"OK. That's one for one," I said. "Ya wanna keep on goin'?"

The other guy backed off and left. We would have killed each other if the other guy was as dumb as I was. (Feynman, 1986, p. 177)

Feynman's perception is correct: such "trivial altercations," which typically take place between men and often involve apparently senseless disputes in bars and streets, "constitute the most prevalent variety of homicide in the United States" (Daly & Wilson, 1998, p. 125). Wilson and Daly have termed this form of behavior "the young male syndrome" because it is typically exhibited by young males engaged in status competition with each other (Wilson & Daly, 1985). The functionality of this behavior in the evolutionary past arises from the fundamentally different characteristics of the reproduction of human males and females. As with many other species, including the majority of animals, females have evolved to invest their somatic resources in their offspring, which are limited in number by physiological constraints, such as the gestation period. Human males, however, are not limited in their reproductive success by such physical characteristics, but rather by the number of females to whom they gain sexual access. It is possible for one male to utilize the reproductive resources of numerous females, whereas females normally have little if any gain from multiple mates.

The consequence is that in humans, as in many other species, the variance of male reproductive success was higher than that of females in the evolutionary past: "A male hunter/gatherer with one wife (who will probably produce four or five children during her lifetime) may increase his reproductive success an enormous 20 to 25 percent if he sires a single child by another woman during his lifetime. If he can obtain (and support) a second wife, he may double his reproductive success. The reproductive realities facing his wife are very different. She will bear four or five children during her lifetime whether she copulates with one, ten, or a thousand men" (Symons, 1979, p. 207). This history of generally polygynous sexual relationships is supported by anthropological research on marriage patterns among human societies.[1] The high variance in male reproductive success that typically characterizes polygynous societies implies intense competition among men in the evolutionary past for resources that provided reproductive opportunities, the essence of sexual selection (Trivers, 1972, p. 141).

If human societies were effectively polygynous in the environment of evolutionary adaptation, so that males displayed high variance in the number of genetic replicators they contributed to succeeding generations, then sexual selection would be expected to have provided mechanisms that led to success in intermale competition. The existence of sexual size dimorphism between human males and females that correlates with the degree of societal polygyny and that parallels similar systematic dimorphism among other orders lends support to this hypothesis (Alexander et al., 1979, pp. 415–417). Equivalents in psychological processes to this morphological dimorphism are also expected, and one set of such cognitive programs is the young male syndrome, which was presumably involved in Feynman's fight (Wilson & Daly, 1985, p. 66). Support for the hypothesis that such psychological processes are the result of inter-male competition is provided by the social and demographic characteristics of those who most often express this form of behavior—young low-status males (Wilson & Daly, 1985, pp. 62–66; Daly & Wilson, 1988, pp. 168–173). Further evidence is offered by the very high rate of car accidents experienced by young men, "affairs of honour" such as duels, and cross-species comparisons (Wilson & Daly, 1985, p. 68; Axelrod, 1986, p. 1095; Wrangham & Peterson, 1996).

CHOICE UNDER CONDITIONS OF UNCERTAINTY

In the seventeenth century mathematicians such as Pascal assumed that the attractiveness of an uncertain choice is given by the product of the probability of the events occurring and the value of the outcomes (Daston, 1988; Machina, 1987, p. 122). This is the basis for Pascal's famous wager that "no matter how small we make the odds of God's existence, the pay-off is infinite: infinite bliss for the saved and infinite misery for the damned" (Gigerenzer et al., 1989, p. 1). The general model of classical probability led to the specification of what has become known as the St. Petersburg paradox by Nicholas Bernoulli in 1713:

Pierre and Paul play a coin toss game with a fair coin. If the coin comes up heads on the first toss, Pierre agrees to pay Paul $1; if heads does not turn up until the second toss, Paul receives $2; if not until the third toss, $4, and so on. Reckoned according to the standard method, Paul's expectation (and therefore the fair price of playing the game) would be

$$E = (\tfrac{1}{2} \times \$1) + (\tfrac{1}{4} \times \$2) + (\tfrac{1}{8} \times \$4) + \ldots + [((\tfrac{1}{2})^n \times \$2^{n-1})] + \ldots$$

Since there is a small but finite chance that even a fair coin will produce an unbroken run of tails, and since the pay-offs increase in proportion to the decreasing probabilities of such an event, the expectation is infinite. (Gigerenzer et al., 1989, p. 14)

The paradoxical nature of this problem does not derive from the mode of calculation or the infinite answer, both of which are in accord with conventional mathematics. It results rather from the empirical observation that reasonable people would pay only a relatively small sum to participate in such a gamble. The contrast between the logical answer and observed behavior provided an impetus to further research, and the "solution" to this paradox provided by Nicholas's cousin David Bernoulli established the pattern that has dominated the literature in the succeeding centuries. Rather than restricting himself to the objectively observable probabilities of the classical perspective, David Bernoulli hypothesized the existence of unobservable subjective influences, the predecessors of utility, which caused diminishing marginal satisfaction with decreases in wealth (1738).

The modern treatment of decision making under uncertainty is closely associated with the classic treatment by von Neumann and Morgenstern, whose intention was to create a mathematically tractable model of idealized decision makers (1953, p. 28). This "expected utility" model formed the basis for postwar research on decision making, but the clarity of the formulation rapidly brought about results similar to that experienced by Pascal's classical probability: observed behavior could not be reconciled with the mathematical specification. A bewildering succession of alternations and amendments has taken place in the last decades, with subjective treatments of probability joining the subjective estimation of utility. Although these formulations have varied widely, the general pattern has been often repeated of formal models having been specified, and then empirical research having shown that, in terms of those models, significant numbers of subjects have behaved "irrationally."[2]

The state of research has now developed to the stage where each of the four central principles of cancellation, transitivity, dominance, and invariance has been systematically falsified in experimental tests. The result of this enormous disparity between the theoretical specification of choice under uncertainty and the observed behavior of human decision makers led Tversky and Kahneman to renounce hope of reconciling theory and observation: "Because framing effects and the associated failures of invariance are ubiquitous, no adequate descriptive theory can ignore these phenomena. On the other hand, because invariance (or extensionality) is normatively indispensable, no adequate prescriptive theory should permit its violation. Consequently the dream of constructing a theory that is acceptable both descriptively and normatively appears unrealizable" (1986, p. S272).

Evolutionary psychology suggests that the search for a general "choice under uncertainty" psychological capacity is profoundly misguided. Such an ability is unlikely to exist; instead, many domain-specific cognitive programs can be expected to provide "satisficing" solutions to forms of choice problems that were relevant in the evolutionary past.[3] The seemingly irrational behavior described in the "biases and heuristics" literature can in many cases be understood when the importance of context and content are recognized, and experimental evidence supports the conclusion that individuals can provide plausible answers to realistically framed problems under conditions of uncertainty.[4]

CONCLUSIONS

The examples of risk assessment and choice under conditions of uncertainty have been examined in some detail in order to demonstrate how hypotheses generated by evolutionary psychology differ from more traditional analyses in psychology and the social sciences. A more extended review of applications has been edited by Barkow, Cosmides, and Tooby, including family and kin relations, the appreciation of the environment, the development of language, and reproductive strategies (1992).

By describing the "cognitive level," evolutionary psychology can thus filter evolutionary theory and empirical research so as to provide an appropriate interface for social scientists. The following list summarizes the principal characteristics of evolutionary psychology that support such an approach:

Adaptationist: Evolutionary psychology addresses and describes the adaptations that underlie behavior, rather than attempting to model and predict behavior itself.

Cognitive: Noncognitive adaptations may constrain behavior in certain respects—humans cannot fly without assistance, for example. Psychological adaptations determine behavior in a more fundamental sense, however, and are thus the appropriate channel to its understanding and modeling.

Functional: Evolutionary psychology describes the (often unconscious) objectives of those cognitive processes that underlie behavior, such as reproduction or resource acquisition.

Eclectic: Diverse perspectives that shed light on the phylogeny, ontogeny, function, and mechanisms of psychological adaptations are used by evolutionary psychologists. Theoretical and empirical developments in the natural and social sciences may be used to illuminate and describe cognitive processes.

Hierarchical: Researchers may initially specify general functional capabilities, such as language usage or parental care, without understanding the psychological mechanisms that underlie these processes. These "black boxes" may then become better understood, through neurophysiological advances, for example, enabling in turn the function of the process to be more accurately described.

Maximizing: Evolutionary selection is a maximizing, but not generally an optimizing, principle. It leads to the climbing of peaks in the adaptive landscape but not necessarily to the choice of the highest peak. Even though psychological adaptations may not currently lead to adaptive behavior, they can still be interpreted as maximizing in other domains and thus offer an elegant foundation for methodologies based on maximization subject to constraints.

Evolutionary psychology thus overcomes the principal impediments that have prevented the (largely latent) demand from the social sciences for a foundation in biology from being satisfied by the existing sociobiological approaches. It is appropriate while honoring Darwin's work to remember that in his later life he was extremely interested in applying his theories to psychological processes such as emotions (1965), and he can be described as the first evolutionary psychologist (Ghiselin, 1973). A further theme has been the failure of a generalized acceptance of sociobiology by the social sciences, more than twenty years after the publication of Wilson's path-breaking book (1975). I hope that I have convincingly indicated how a redirection of sociobiological research efforts, from behavior itself to the psychological processes underlying behavior, may encourage social scientists to base their analyses on a Darwinian foundation.

NOTES

1. Murdock in a sample of 849 human societies found occasional or usual polygyny in 708, monogamy in 137, and polyandry in 4 (1967). See also Betzig (1986, pp .79–86).

2. "It is hard to think of a more notorious, long-standing, and often outright confused controversy in modern decision theory than the continuing debate on the meaning of 'rationality' in choice under uncertainty" (Machina, 1981, p. 163). Nine different models are reviewed by Schoemaker (1982), summarized in table 1, p. 538.

3. Simon has long been one of the foremost exponents of the importance of "bounded rationality." He coined the term *satisficing* to emphasize that humans do not possess psychological processes of unlimited power (1955, 1959).

4. Evolutionarily informed researchers have been remarkably successful in reinterpreting the judgment under uncertainty literature (e.g., Gigerenzer, 1991; Cosmides & Tooby, 1996, but cf. Kahneman & Tversky, 1996). Researchers have also demonstrated that "fast and frugal" mechanisms may lead to satisficing results with very simple algorithms (Gigerenzer & Goldstein, 1996).

REFERENCES

Alexander, R. D., Hoogland, J. L., Howard, R. D., Noonan, K. M., & Sherman, P. W. (1979). Sexual dimorphisms and breeding systems in pinnipeds, ungulates, primates, and humans. In N. A. Chagnon and W. G. Irons (Eds.), *Evolutionary biology and human social behavior: An anthropological perspective*. North Scituate, MA: Duxbury Press.

Axelrod, R. (1986). An evolutionary approach to norms. *American Political Science Review, 80*(4), 1095–1111.

Barkow, J. H., Cosmides, L., & Tooby, J. (Eds.). (1992). *The adapted mind: Evolutionary psychology and the generation of culture*. Oxford: Oxford University Press.

Bernoulli, D. (1738). Specimen theoriae novae de mensura sortis. *Commentarii Academiae Scientiarum Imperialis Petropolitanae, 5*, 175–192.

Betzig, L. L. (1986). *Despotism and differential reproduction: A Darwinian view of history*. Hawthrone: Aldine de Gruyter.

Cosmides, L. (1989). The logic of social exchange: Has natural selection shaped how humans reason? Studies with the Wason Selection Task. *Cognition, 31* (3), 187–276.

Cosmides, L., & Tooby, J. (1987). From evolution to behavior: Evolutionary biology as the missing link. In J. Dupré (Ed.), *The latest on the best: Essays on evolution and optimality*. Cambridge, MA: MIT Press.

———. (1989). Evolutionary psychology and the generation of culture, Part II. Case study: A computational theory of social exchange: *Ethology and Sociobiology, 10*, 51–97.

———. (1992). Cognitive adaptations for social exchange. In H. Barkow, L. Cosmides, & J. Toobey (Eds.), *The adapted mind: Evolutionary psychology and the generation of culture*. Oxford: Oxford University Press.

———. (1996). Are humans good intuitive statisticians after all? Rethinking some conclusions from the literature on judgment under uncertainty. *Cognition, 58*, 1–73.

Daly, M., & Wilson, M. (1988). Homicide. In S. B. Hrdy & R. W. Wraungham (Eds.), *Foundations of human behavior*. New York: Aldine de Gruyter.

Darwin, C. (1965). *The expression of the emotions in man and animals*. Chicago: University of Chicago Press.

Daston, L. (1988). *Classical probability in the Enlightenment*. Princeton, NJ: Princeton University Press.

Elworthy, C. (1993). *Homo biologicus: An evolutionary model for the human sciences*. Sozialwissenschaftliche Schriften Heft 25. Berlin: Duncker & Humblot.

Feynman, R. P. (1986). *"Surely You're Joking, Mr Feynman!" Adventures of a curious character*. London: Unwin Paperbacks.

Ghiselin, M. T. (1973). Darwin and evolutionary psychology. *Science, 179*, 964–968.

Gigerenzer, G. (1991). How to make cognitive illusions disappear: Beyond "heuristics and biases." *European Review of Social Psychology, 2*, 83–115.

Gigerenzer, G., & Goldstein, D. G. (1996). Reasoning the fast and frugal way: Models of bounded rationality. *Psychological Review, 103*, 650–669.

Gigerenzer, G., Swijtink, Z., Porter, T., Daston, L., Beatty, J., & Krüger, L. (1989). The empire of chance: How probability changed science and everyday life. In W.

Lepenies, R. Rorty, J. B. Schneewind, & Q. Skinner (Eds.), *Ideas in context*. Cambridge: Cambridge University Press.

Hirshleifer, J. (1985). The expanding domain of economics. *American Economic Review, 75* (6), 53–68.

Kahneman, D., & Tversky, A. (1996). On the reality of cognitive illusions. *Psychological Review, 103*, 582–591.

Machina, M. J. (1981). Rational decision making versus rational decision modelling. *Journal of Mathematical Psychology, 24* (2), 163–175.

———. (1987). Choice under uncertainly: Problems solved and unsolved. *Journal of Economic perspectives, 1* (1), 121–154

Murdock, G. P. (1967). *Ethnographic atlas*. Pittsburgh: University of Pittsburgh Press.

Nesse, R. M. (1987). An evolutionary perspective on panic disorder and agoraphobia. *Ethology and Sociobiology, 8* (3S), 73S–83S.

Radnitzky, G., & Bernholz, P. (Eds.). (1987). *Economic imperialism: The economic approach applied outside the field of economics*. New York: Paragon House.

Schoemaker, P. J. H. (1982). The expected utility model: Its variants, purposes, evidence, and limitations. *Journal of Economic Literature, 20* (2), 529–563.

Simon, H. A. (1995). A behavioral model of rational choice. *Quarterly Journal of Economics, 69* (1), 99–118.

———. (1959). Theories of decision-making in economics and behavioral science. *American Economic Review, 49* (3), 253–83.

Smelser, N. J., & R. Swedberg, (Eds.) (1994). *The handbook of economic sociology*. Princeton, NJ: Princeton University Press.

Symons, D. (1979). *The evolution of human sexuality*. New York: Oxford University Press.

———. (1990). Adaptiveness and adaptation. *Ethology and Sociobiology, 11* (4–5), 427–444.

———. (1992). On the use and misuse of Darwinism in the study of human behavior. In J. H. Barkow, L. Cosmides, & J. Tooby, (Eds.), *The adapted mind: Evolutionary psychology and the generation of culture*. Oxford: Oxford University Press.

Trivers, R. L. (1972). Parental investment and sexual selection. In B. G. Campbell (Ed.), *Sexual selection and the descent of man, 1871–1971*. Chicago: Aldine.

Tversky, A., Kahneman, D. (1986). Rational choice and the framing of decisions. *Journal of Business, 59*, S251–278.

Vining, D. R. Jr. (1986). Social versus reproductive success: The central theoretical problem of human sociobiology. *Behavioral and Brain Science, 9* (1), 67–216.

von Neumann, J., & Morgenstern, O. (1953). *Theory of games and economic behavior* (3rd ed.). Princeton, NJ: Princeton University Press.

Wilson, E. O. (1975). *Sociobiology: The new synthesis*. Cambridge, MA: Belknap Press of Harvard University Press.

Wilson, M., & Daly, M. (1985). Competitiveness, risk taking, and violence: The young male syndrome. *Ethology and Sociobiology, 6*, 59–73.

Wrangham, R. W., & Peterson, D. (1996). *Demonic males: Apes and the origins of violence*. Boston: Houghton Mifflin.

20

Implications of Sexual Selection for Variation in Human Personality and Behavior

John S. Price

Darwin made it clear that natural selection is based on differential ability to deal with the physical environment, including predator and prey relations with other species, but at the same time he recognized that selection occurs as a result of interactions with members of the same species. In *On the Origin of Species* he wrote, "This form is selection depends not on a struggle for existence in relation to other organic beings or the external conditions, but on the struggle between individuals of one sex, generally the males, for the possession of the other sex" (Darwin 1859, p. 69).

In 1871 Darwin published *The Descent of Man, and Selection in Relation to Sex*, which was devoted to a meticulous analysis of sexual selection. In this book he introduced the term "sexual selection" for the first time, and he pointed out that it has two components:

Sexual selection depends on the success of certain individuals over others of the same sex, in relation to the propagation of the species; whilst natural selection depends on the success of both sexes, at all ages, in relation to the general conditions of life. The sexual struggle is of two kinds; in the one it is between individuals of one sex, generally the male, in order to drive away or kill their rivals, the female remaining passive; whilst in the other, the struggle is likewise between the individuals of the same sex, in order to excite or charm those of the opposite sex, generally the females, which no longer remain passive but select more agreeable partners. This latter kind of selection is closely analogous to that which man unintentionally, yet effectually, brings to bear on his domesticated productions, when he preserves during a long period the most pleasing or useful individuals, without any wish to modify the breed. (1871, p. 916)

Darwin used the theory of sexual selection to explain the evolution of bizarre male adornments such as the tail of the peacock and the antlers of the stag. He pointed out that sexual selection and natural selection tend to operate in different directions, so that the characters that make for success in sexual selection may make it more difficult for a bird, say, to fly in search of food, and they may also make it more conspicuous to predators. On the other hand, the two types of selection may work in the same direction, as when an individual who is good at getting food and therefore is well nourished may be more able to compete with members of the same sex and be more attractive to members of the opposite sex. Those animals who are successful in social competition may obtain large territories and therefore be less susceptible to starvation, predation, and disease; in such cases a positive feedback cycle operates between natural selection and sexual selection.

These observations of Darwin raise some important questions about human social life, for instance:

1. How much of human differential reproduction can be accounted for by sexual selection?

2. What are the criteria for selection? In other words, what is the human equivalent of the stag's antlers and the peacock's tail?

3. How can we classify the criteria for sexual selection into physical and psychological?

4. What is the nature of the social process by which the selection is made?

5. What is the fate (or social role) of those who are not selected?

It is the last question in which I as a psychiatrist am most interested, but since the answer depends to some extent on the previous question about the nature of the selection process, I should devote some time to that issue.

PROCESSES OF SEXUAL SELECTION

Darwin recognized two types of sexual selection, which came to be known as *intrasexual selection*, in which the members of the same sex compete together for access to the other sex, and *intersexual selection*, in which members of one sex choose members of the opposite sex as partners in mating (Ryan, 1985; Bradbury & Davies, 1987; Andersson, 1994). In some species these processes are not entirely separate, as when a female sheep encourages two rams to fight and then mates with the winner (Geist, 1971). Those structures that are attractive to members of the opposite sex are sometimes intimidating to members of the same sex. The term "intersexual selection" has been criticized with some justification by Cronin (1991), who prefers "epigamic selection" or "mate choice."

INTRASEXUAL SELECTION

The evolution of human social life can be seen as the evolution of ever more sophisticated and effective methods of sexual selection, and it might be helpful to enumerate some of the possible stages, concentrating mainly on the intrasexual component of sexual selection.

Unritualized Social Competition

Many insects kill members of the same sex, some worms plug each other's sexual orifices, some beetles spray each other with antiaphrodisiac gas, and various forms of sperm competition are practiced by different species; this category includes any action to reduce the other's viability or fertility over which the victim has no role in "consenting." Possibly the suppression of sexual development by pheromones in some rodents and New World monkeys comes into this category. Otherwise, it does not occur in vertebrates.

Ritual Agonistic Behavior

In ritual agonistic behavior the loser, being unharmed, must consent to lose. He has the option (at an unconscious level) of not consenting and can be said to choose between a consenting strategy (in which he submits or flees) and a nonconsenting strategy (in which he retaliates).

In evolutionary terms, ritual agonistic behavior seems to be performing two rather separate functions. To the extent that it takes the form of intergenerational conflict, it serves to delay reproduction until later in the life span. To the extent that it is intergenerational conflict, it serves to create lifelong variation in social status and hence in fertility within a socially interacting cohort of conspecifics. It is this second function that subserves intrasexual selection and that concerns us here.

According to the simplest view, each individual has to choose between two strategies, a dominant strategy in which he reproduces more and a subordinate strategy in which he reproduces less. The dominant strategy is designed to maximize his own reproduction; the subordinate strategy is designed to "make the best of a bad job" in terms of his own reproduction, to avoid the mortality that dominant strategists inflict on each other, and to maximize the reproduction of his close kin (and raise the kin-selection component of inclusive fitness). For individuals in this system, the overall strategy is similar to Maynard Smith's (1982) "assessor" strategy; on at least one occasion in their ontogeny they have to assess their chances and choose between the dominant or "hawk" substrategy, on the one hand, and the subordinate or "dove" substrategy, on the other.

How do they make this choice? There are a number of possibilities that have not in fact evolved. They could leave it entirely to the opposite sex and adopt the strategy "If chosen as a mate, adopt dominant strategy; if not chosen, adopt subordinate strategy,'' thus relying entirely on the intersexual component of sexual selection and eliminating the intrasexual component. Or they could do it by counting heads, such as, "If the home/nest contains more than x individuals when you reach age y, adopt subordinate strategy; otherwise, adopt dominant strategy." What has in fact evolved is a form of social comparison that bears a certain resemblance to coconsultation. We can imagine a primitive vertebrate scratching its head and wondering whether to adopt a dominant or subordinate strategy, so it chooses a consultant and says, please help me make up my mind. The consultant says, use me as a yardstick. If you find yourself superior to me, your chances are good and you should adopt the dominant strategy; otherwise, you should play safe and adopt the subordinate strategy. The consultant takes the form of a fight in which our indecisive individual uses the strength of the consultant as a yardstick to estimate his own strength, and after the consultation he either says to himself, "I am a strong person," and adopts a dominant strategy, or "I am a weak person," and adopts a subordinate strategy. Of course, the interaction is symmetrical and the "consultant" is making a similar decision (it is a coconsultation). This could be called *dyadic comparison* because, in each comparative episode, each individual compares himself with one other. It is the main form of vertebrate social comparison and is called ritual agonistic behavior.

Group living provides the opportunity for more sophisticated social comparison, and probably the selection mediated by ritual agonistic behavior in groups is more effective than that in territorial species. The opportunities for effective comparison in a single fight are limited, but if animals live in groups, they have extended time in which to evaluate each other's strengths. Fights can follow a long period of mutual assessment, can be protracted, and can be divided into bouts. As a result, the rank order in a group should reflect small differences in strength, skill, intelligence, and courage. Ritual agonistic behavior amplifies these small differences into gross social disparity.

External Mediation of Intrasexual Selection

In human evolution there has been a major change in ranking behavior. Instead of two rivals A and B fighting it out between themselves, the choice between A and B is made by C, D, E, and others. This is the change that Gilbert, Price, and Allan (1995) drew attention to as important in evolution. In order to achieve greater social success than B, A has to make himself attractive to C, D, and E rather than make himself intimidating to B. Selection is now by external judges rather than by interaction between the rivals themselves. The scope for greater efficiency of selection and for cultural variation in the criteria of selection opens up an entirely new "ball game" of the sexual selection process. In fact, this development must have been about as important as the development of sexual selection itself. To distinguish it from the dyadic comparison that occurs in ritual agonistic behavior, the evaluation of A and B by C, D, E, and others could be called *polyadic comparison*. Of course, it is seldom as simple as that, and in most cases everyone is evaluating everyone else. In the same way that a person "cannot not communicate," he or she cannot not evaluate or be evaluated.

Does this kind of sexual selection occur in animals as well as man? In macaques, baboons, and chimpanzees the outcome of ritual agonistic behavior is affected by alliances with same-sex conspecifics, so that the capacity for alliance formation is being selected as well as fighting ability. The choice between two potential allies offers a primordium of polyadic comparison, in that the criterion of choice is not so much "Does he intimidate me?" as "Is he likely to intimidate the other fellow (and, if so, is he likely to favor me)?"; this is still some way from "Which of the two is more attractive?" but it is a major advance from the evaluation of others entirely in terms of dyadic comparison. In chimpanzees, in addition, the influence of female group members affects the rank order in males, and this is a further step toward intrasexual polyadic comparison; in fact, it is similar to the situation reported to occur in at least one tribe of American Indians in which only males are allowed to run for office and only females are allowed to vote.

In human society polyadic comparison has been enormously increased in importance, particularly due to language and the opportunity this gives for the comparers to discuss those being compared and to make a careful allocation of prestige; it also gives the group members the opportunity to discuss the criteria for the allocation of prestige. But it has not replaced the other forms of social competition, and so we see them operating side by side.

THE FATE OF SEXUAL SELECTION'S UNSELECTED AND DESELECTED: MILTON FILLS IN FOR DARWIN

Darwin wrote little about the fate of those unselected by sexual selection or about those who became deselected after once being selected. He spoke of them being killed or driven away, but he does not appear to have speculated on the behavior of those who were driven away. He did not recognize the need for a behavioral strategy to deal with this group, or any psychological or emotional state that might pertain to them. In particular, he did not consider them as a special case in his book on the expression of the emotions (1872).

It is ironic to note, therefore, that his favorite poet, John Milton, who had been at the same Cambridge college 200 years previously, and a copy of whose *Paradise Lost* Darwin took with him on his voyage in the *Beagle*, was at least interested in and one might almost say obsessed by the fate of those who were "driven away." In two of his major poems, *Paradise Lost* and *Samson Agonistes*, he examines the situation of someone who is defeated by overwhelming force.

In *Paradise Lost* the rebel angel Satan, together with Beelzebub and his other followers, has been cast out of heaven because he challenged God, who

> Hurled [them] headlong flaming from the'ethereal sky,
> With hideous ruin and combustion, down
> To bottomless perdition, there to dwell
> In adamantine chains and penal fire.
>
> (Gilfillan, 1853, vol. 1, p. 5, ll. 6–9)

The action of the poem opens as they regroup themselves in Hell and consider their options. There is no hint of remorse or submission in the mind of Satan, who mixes his "deep despair" with "obdurate pride and steadfast hate." Reconciliation with his victor is rejected:

> What though the field be lost?
> All is not lost; th'unconquerable will,
> And study of revenge, immortal hate
> And courage never to submit or yield
> And, what is else, not to be overcome;
> That glory never shall his wrath or might
> Extort from me: to bow and sue for grace
> With suppliant knee, and deify his power.
>
> (Gilfillan, 1853, vol. 1, p. 6, l. 32, p. 7, l. 6)

Tauntingly, he asks his followers if they have "sworn to adore the conqueror." Even though his first lieutenant, Beelzebub, points out tactfully on two occasions that God must be omnipotent to have defeated the rebel army, Satan determines to fight on with "force and guile," determined that it is "better to reign in hell, than serve in heav'n" (Gilfillan, 1853, vol. 1, p. 11, l. 25).

In *Samson Agonistes*, Samson, betrayed by Dalilah, blinded and imprisoned by the followers of the god Dagon, is visited by his father, who is planning to arrange a ransom and who tells him to keep on fighting. But Samson rejects this advice and expresses his depressive position:

> So much I feel my genial spirits droop,
> My hopes all flat, nature within me seems
> In all her functions weary of herself,
> My race of glory run, and race of shame,
> And I shall shortly be with them that rest.
>
> (Gilfillan, 1853, vol. 2, p. 91, ll. 27–31)

Samson's father then tells him to be calm and to accept healing words from his friends, but Samson does not follow this advice; he expresses the idea that his mental torment is even worse than his physical torment, and he contemplates the idea of suicide:

> Sleep hath forsook and given me o'er
> To death's benumbing opium as my only cure:

Thence faintings, swoonings of despair,
And sense of heav'n's desertion.

(Gilfillan, 1853, vol. 2, p. 92, ll. 28–31)

Then Samson's father leaves, and he is visited by Harapha, a champion of the Philistines who was not involved in the previous battles with Samson. Here Samson is roused out of his depression and challenges Harapha, finally dismissing him with the words

Go, baffled coward, lest I run upon thee,
Though in these chains, bulk without spirit vast,
And with one buffet lay thy structure low,
Or swing thee in the air, then dash thee down
To the hazard of thy brains and shatter'd sides.

(Gilfillan, 1853, vol. 2, p. 111, ll. 2–6)

The chorus then counsels Samson to patience. But patience, acceptance, and reconciliation are not a part of Samson's reaction to defeat, and the poem concludes with his splendid act of vengeance in which he destroys both himself and his conquerors.

In both these poems Milton is exercised about the reaction of the man who is defeated and cast down. Does he fight back in spite of his depression and his chains? Or does he accept his lot, in the one case to accept the advice of Beelzebub that his opponent is omnipotent, and in the other to accept the advice of the chorus to be patient? In both poems Milton portrays a fallen hero who is chained and in deep despair, but in both cases the despair does not inhibit pride or the determination to retaliate. He is portraying a society that does not admit voluntary submission and reconciliation.

The message seems to be that the ancient way of man, illustrated in the tales of gods and heroes, is one of unmitigated fighting and retaliation—the only way to keep a defeated enemy down is to bind him in adamantine chains, and if, as in the case of Satan, this is not enough, to "transfix him with linked thunderbolts to the bottom of the gulf." The new way, characterized by Christianity, is one of forgiveness, repentance, voluntary submission, and reconciliation. Only briefly, in the case of Samson, does Milton consider that the aggression of a defeated champion may be inhibited by a depressed mood.

SEXUAL SELECTION AND PERSONALITY VARIATION

Intrasexual selection is mediated by agonistic behavior in most vertebrates, and the result of agonistic behavior is social asymmetry, which takes the form of owner/nonowner in territorial species and high/low social rank in group-living species. These asymmetries are associated with conditional behavioral strategies that we have called the high– and low–self-esteem strategies, because at the intrapsychic level the role of owner or high-ranking person is associated with high self-esteem, whereas the role of nonowner or low-ranking person is associated with low self-esteem (Price, Sloman, Gardner, Gilbert, & Rohde, 1994). This variation probably accounts for much of the variation in what has been called upperness/lowerness, or the vertical dimension of two-dimensional personality space (the other dimension being closeness/distance) (Birtchnell, 1993).

SEXUAL SELECTION AND MENTAL ILLNESS

When someone has been selected and has enjoyed the privileges of ownership and high rank, the change to the role of being deselected is likely to be a drastic one and to result in a major change in behavior. This, we think, is what we observe in depressive illness (Price et al., 1994).

Biologists, on the whole, do not share Milton's preoccupation with the underdog. One might almost suspect that researchers are obeying an innate human instruction to "attend to the winner." An exception is a comment by Welch (1967) that may be the earliest suggestion that human depression might be related to social subordination. It occurs in the discussion following a paper on dolphins by Bartholomew. Welch says: "Some animals are pushed aside, prohibited from participation. . . . Because social hierarchies exist, this happens to some extent in all societies. Some animals, usually the dominants, do very well both behaviorally and physiologically, while some, usually the subordinates, do very poorly. Whether the differences are maintained by physical contact and actual expressions of physical aggression, or whether they are maintained and reinforced simply by ritualised behavioral signals, the subordinate animals do, in fact, appear to recognise when they have been reaffirmed in their status. . . . This emphasises in a very real way the dictation, by social pressure, of different states of being: the state of being in a position of subordination, or in a position of dominance, and the various gradations in between." He then discusses the physiological abnormalities that occur in subordinates, including the fact that they are "psychologically sterilised, although not physically abused," and adds, "Might not the depressive psychotics in our human population—particularly abundant at very high population densities and characterised by high activation of sympathetico-pituitary-adrenocortical activities—produced by extreme social environments and reinforced in social subordination, be analogous to them?" (pp. 239–240).

In fact, the Norwegian zoologist who first discovered animal social hierarchies commented on the "depressed" behavior of low-ranking birds, particularly those who had lost former high rank (Schjelderup-Ebbe, 1935); the fact that Schjelderup-Ebbe made this discovery while still a schoolboy may reflect the fact that hierarchies are much more visible to low-ranking than to high-ranking people. Abraham Maslow, one of the earliest scientists to work with monkey hierarchies, reported that a subordinate rhesus monkey was more like another subordinate monkey than like himself when dominant, and it is probably no coincidence that Maslow went on to study the self-esteem of American women and discovered the enormous variation in self-esteem that occurs from one woman to another (Maslow, 1937). Depressive behavior has been reported not only in low-ranking birds and monkeys, but also in low-ranking reptiles and in birds of territorial species who have failed to obtain territories (references are given in Price et al., 1994). In the laboratory, social defeat produces depressive behavior and lasting physiological changes such as increased secretion of corticosteroids, which is also seen in depressed patients (Price, 1995).

SOME CONSEQUENCES OF POLYADIC COMPARISON

I think that it would be difficult to overestimate the importance of the switch from dyadic to polyadic sexual selection. For the first time in evolution, selection can be determined by cultural factors, which can themselves be the subject of selection. The criteria of selection, being determined by culture, can allow for the selection of characters that were previously unselectable. The switch to polyadic selection alters the mathematics that have been applied to the evolution of altruism, the detection of free riders, and the possibility of selection at the between-group level. We should note that the switch to polyadic selection applies also to epigamic selection in that in most human societies mates are chosen at least partly by the parental generation rather than by the mating dyad, and many parents know to their cost that the criteria used by the two generations are not the same. The evolution of polyadic comparison and its coexistence in human societies with dyadic comparison have other consequences, some of which I will now discuss.

Proscription of Agonistic Behavior by Society

Groups practicing polyadic comparison would have an enormous advantage over groups still limited to dyadic comparison (agonistic behavior). Culturally they would be at an advantage because their leaders would have those characteristics that are the criteria for the allocation of prestige, and in most human groups these appear to be a combination of competence and dedication to the interests of the group. Groups with such leaders should outperform groups whose leaders were selected for power to intimidate. Genetically, the polyadic groups would tend to have more members with qualities of competence and unselfishness because there is a correlation in most human groups between prestige and reproduction (Perusse, 1993); therefore we have probably experienced a gene/culture coevolution for competence and group loyalty.

Among those groups practicing polyadic comparison, there would be an advantage to those groups in whom selection was entirely by polyadic comparison, and therefore there would be an advantage in preventing agonistic behavior as much as possible. Thus we can expect ritual agonistic behavior to be proscribed by groups, both in their childrearing practices and in their code of behavior for

adults. In childhood there is an enormous parental influence toward nonintimi-
dating behavior; see, for instance, the life histories described by Vaillant in his
Adaptation to Life (1977), in which a cohort of American college men report
severe sanctions on aggressive behavior during their childhoods. The proximate
reason for parents stopping their children from quarrelling may well be that they
find the noise irksome, or that they consider it bad manners, or that they think
that the children should spend the time improving themselves in some way; but
the ultimate, evolutionary reason may be that they want to decide the children's
rank order themselves by the giving and withholding of praise and criticism,
and so they do not want the rank order decided by the children themselves in
the course of quarrelling (ritual agonistic behavior). Also, they want to develop
in their children the mentality that looks for self-esteem in the form of praise
rather than in the form of the submission of others, so that when they leave
home they will still be oriented toward polyadic comparison. The widespread
existence of bullying in school playgrounds (Rivers & Smith, 1994) might seem
to gainsay this thesis, but it is probably due to the fact that there were no schools
in our "environment of evolutionary adaptedness." Glantz and Pearce (1989)
have pointed out that in hunter-gatherer societies children are always with adults
who prevent bullying.

In adult life, fighting between same-sex adults is also proscribed. Duelling
was forbidden by monarchs, not because of the fear of loss of life (which was
slight), but because the king wanted prestige to go to people he approved of
rather than to those who were skilled with the sword or pistol. What dyadic
competition is allowed between adults is governed by society's rules rather than
by nature's. Fine differences in ability can be assessed by pitting individuals
against each other in sport and in intellectual tests, but these are polyadically
controlled dyadic comparisons. Prestige is allocated not only for performance
but also for sportsmanship, and bad marks are allocated to those who are seen
to cheat or who do not accept the decision of the referee.

Because of this proscription, ritual agonistic behavior is only seen in situations
over which society has little control: in prisons, on street corners, in the school
playground, in the family, and in situations in which master and servant are
alone together. Also, society does not proscribe ritual agonistic behavior in marriage; in fact, sayings abound to the effect of "Never interfere between husband
and wife." This may well be because the rank order within marriage does not
affect the rank order in the group as a whole, and therefore it affects neither the
choice of leaders nor the correlation between prestige and reproduction. This
leaves us with the paradoxical situation that agonistic behavior, which evolved
to subserve competition between males, is now mainly seen to occur between
male and female in their roles as husband and wife.

Societal prohibition of agonistic behavior complicates the study of human
aggression and anger. He (or she) who fights has two adversaries, one being the
opponent, the other being society as a whole. It is often difficult to tell whether
the expression of anger has been inhibited because of intimidation by the op-

ponent or because of obedience to an internalized social norm forbidding angry expression.

Development of a Latency Period

Students of baboon social life have pointed out that the brief period of immaturity before the adolescents join the adult dominance hierarchy is a time in which they evaluate each other, and each group of peers has worked out its rank order by the time the canine teeth have developed. The human latency period allows a much-extended time of mutual evaluation by the peer group. It also allows the previous generation to play an important part in the evaluation, and of course in human life we see a whole professional class of evaluators ranking our adolescents according to adult standards. Therefore, whereas the accepted function of the latency period is to allow more learning, we can add the additional function of allowing ranking according to ability to learn and according to other prosocial attributes manifested during development.

Religion and War as Projective Tests

Society wants individuals who are assertive and capable and yet have the capacity for submission of their individual goals to those of the group. The induction of children into religious practices allows an evaluation of this capacity for submission and also provides a test of memorizing capacity by requiring the child to learn scripture and ritual.

The wars of primitive man are ritualized, and the death rate is low. There is much observation of individual fighting attributes. In this way society can allocate prestige to those who will risk their lives for the sake of the group. This is a possible explanation for the universality of religion and war in human groups: those groups that lacked these aids to polyadic comparison did not survive.

Why People Are Nice

On the whole, society allocates prestige to people who are nice. "Nice" means that they are decent, honest, reasonable, cooperative people who put the good of the group before their own selfish interests; they are also likeable and interested in their fellow human beings. Thanks to polyadic comparison, human groups have been selecting for niceness for millions of years, and we have become very good at it. Therefore, we have to some extent overcome the legacy of dyadic comparison, which is to select for intimidating, selfish bullies. The genes may be selfish, but the people are unselfish, and it is the people we have to interact with, not the genes. I think that in this sense the message of evolutionary biology is an encouraging one. We are nice because, for a very long time, we have selected each other to be nice.

ALTERNATIVE SUBORDINATE STRATEGIES

The subordinate strategy has two components. The first is the decision to be subordinate, and this decision is taken in negotiation with one or more fellow group members, as described earlier. The second component is the decision as to how to conduct oneself as a subordinate. There are a number of choices, depending on the ecology of the species. A major issue is whether to remain close to, or avoid, the dominants, and there is variation here both between and within species (Price, 1995). Some animals make a virtue of the fact that they do not have the responsibility of defending a territory through what may be a hard and even lethal winter, so the low–self-esteem individuals who have lost out in the battle for territories do something entirely different, such as migrating to a warmer climate or going into a state of hibernation, and often when they come back, they find the territory owners dead. The variation between strategies in these partially migrating and partially hibernating species is maintained by negative frequency-dependent selection.

The human problem of how to conduct oneself as a subordinate is the subject matter of much of philosophy and religion. These disciplines usually counsel patience and self-abnegation, as did the chorus to Samson. But there is another way, which was taken by both Satan and Samson and appears also to have been taken by Milton and Darwin.

MILTON AND DARWIN AS NONYIELDING REBELS

In order to understand human subordination, it is necessary to appreciate that a decision between the dominant and subordinate strategies is taken relatively independently at two levels of the human brain (Stevens & Price, 1996). There is a lower, reptilian level (MacLean, 1990) at which there is a decision to provide or withdraw the basic materials needed for fighting; here, the dominant strategy takes the form of an elevation of mood, giving energy, optimism, and sense of ownership, while the subordinate strategy of depression takes away these armaments, leaving the individual tired, pessimistic, and with no sense of entitlement. At a higher level, in the neomammalian brain, another type of decision is made, and this is conscious, rational, voluntary, and deliberative and takes the form of deciding whether to give in or fight on. Even the individual who suffers the incapacity and torment of depression (metaphorically expressed by Milton as "adamantine chains and penal fire") can fight on by an act of will, even though willpower itself is sapped by the depression.

Milton rebelled against the state (he was the principal Roundhead pamphleteer, attacking the monarchists) and lost; Darwin rebelled against the church (the doctrine of the Creation), and although he did not actually lose, his diaries and letters reveal his constant anticipation of losing, as a result of which he withdrew from the London arena and delayed publication of his theory for twenty years, suffering almost constant nervous symptoms. In spite of their real and imagined defeats, they both fought on, Milton writing pamphlets and poetry, Darwin elaborating his theory of natural selection. Their heads were "bloody but unbowed." Acts of submission at the higher, neomammalian level would have preempted or relieved their suffering, but their resources of ambition, pride, and courage enabled them both to bend their adamantine chains and to make their unique contributions to the human record. It is this triumph of the will over the flesh that Milton celebrates in the first books of *Paradise Lost*.

REFERENCES

Andersson, M. (1994). *Sexual selection*. Princeton, NJ: Princeton University Press.

Birtchnell, J. (1993). *How humans relate: A new interpersonal theory*. Westport, CT: Praeger.

Bradbury, J. W., & Davies, N. B. (1987). Relative roles of intra- and intersexual selection. In J. W. Bradbury & M. B. Andersson (Eds.), *Sexual selection: Testing the alternatives* (pp. 143–163). Chichester: John Wiley.

Cronin, H. (1991). *The ant and the peacock*. Cambridge: Cambridge University Press.

Darwin, C. (1859). On the origin of species. London: John Murray.

———. (1871). *The descent of man, and selection in relation to sex*. London: John Murray.

———. (1872). *The expression of the emotions in man and animals*. London: John Murray.

Geist, V. (1971). *Mountain sheep*. Chicago: University of Chicago Press.

Gilbert, P., Price, J., & Allan, S. (1995). Social comparison, social attractiveness, and evolution: How might they be related? *New Ideas in Psychology, 13*, 149–165.

Gilfillan, G. (1853). Milton's poetical works, with life, critical dissertation and explanatory notes. 2 vols. Edinburgh: James Nichol.

Glantz, K., & Pearce, J. K. (1989). *Exiles from Eden: Psychotherapy from an evolutionary perspective*. London: W. W. Norton.

MacLean, P. D. (1990). *The triune brain in evolution*. New York: Plenum Press.

Maslow, A. H. (1937). Dominance-feeling, behavior, and status. *Psychological Review, 44*, 404–429.

Maynard Smith, J. (1982). *Evolution and the theory of games*. Cambridge: Cambridge University Press.

Perusse, D. (1993). Cultural and reproductive success in industrial societies: Testing the relationship at the proximate and ultimate levels. *Behavioral and Brain Sciences, 16*, 267–322.

Price, J. S. (1995). The rat resident/intruder paradigm: A model for the involuntary subordinate strategy (ISS). *ASCAP (Across Species Comparison and Psychopathology) Newsletter (Newsletter of the Society for Psychophysiological Integration), 8*, 12–16.

Price, J. S., Sloman, L., Gardner, R., Gilbert, P., & Rohde, P. (1994). The social competition hypothesis of depression. *British Journal of Psychiatry, 164*, 309–315.

Rivers, I., & Smith, P. K. (1994). Types of bullying behavior and their correlates. *Aggressive Behavior, 20*, 359–368.

Ryan, M. J. (1985). *The Tungara frog: A study in selection and communication*. Chicago: University of Chicago Press.

Schjelderup-Ebbe, T. (1935). Social behavior of birds. In C. Murchison (Ed.), *A Handbook of social psychology* (pp. 947–972). Worcester, MA: Clark University Press.

Stevens, A., & Price, J. (1996). *Evolutionary psychiatry: A new beginning.* London: Routledge.

Vaillant, G. (1977). *Adaptation to life: How the best and brightest come of age.* Boston: Little, Brown.

Welch, B. L. (1967). Discussion following "Aggressive behavior in Cetacea." In C. D. Clemente & D. B. Lindsley (Eds.), *Aggression and defense* (pp. 239–240). Berkeley: University of California Press.

21

Serotonin, Dopamine, and the Evolution of Sociophysiological Neurotransmission

Daniel R. Wilson, Sean Stanton, and Sandra Wilson

Darwin, in the good intellectual company of Hume, Mill, and Bentham, had a sound general view of moral theory as it essentially applied to the evolution of moral capacities (Ruse, 1986). As Ruse further notes, each of these men struggled to derive ethics from the muck of nature and nature's law and, therefore, made no appeal to divine inspiration as the source of or guiding hand for ethical interests. Yet beyond these strong affinities there were important, if subtle, distinctions in the character of their thought. Unlike the others, Darwin was greatly influenced by natural theology, especially that of James Mackintosh. Mackintosh emphasized that the human tendency to a moral purpose is God given. He further emphasized this tendency caused humans' sentiments to warm as feelings were first engendered, then possibly reciprocated and, in due course, gratified by the engagement of the moral ethic of doing good unto others (Richards, 1987). Darwin, replacing Mackintosh's God with a more reducible Nature, saw a mechanism of community selection as a link between moral standards and the motive of biological imperative. He thus anticipated by more than a century elements of the definite work of Hamilton (1963, 1964a, 1964b) and Trivers (1971). The others, meanwhile, whatever else their philosophic advances, failed to note this convergence between motives and standards in natural systems.

Despite this robust and most promising start, both Darwinian ethics and related concepts of kinship selection long lay dormant. Neither initiative could prosper in the absence of that cogent grasp of the details of kinship genetics (Hamilton, 1963, 1964a, 1964b) and neuroscience (Nestler & Hyman, 1993) that has been attained only in the past generation. Yet with these details manifest, it is now possible to reformulate certain of Darwin's early ideas in the more satisfactory light of contemporary neuroscience. Finally, a far more integrated perspective is emerging (Richards, 1987).

THE SOCIAL-COMPETITION HYPOTHESIS

In the new field of evolutionary psychology there have been developments that parallel the triumph of kinship genetics and neuroscientific progress. Among these parallel developments is the specification of the "social-competition hypothesis" of mood disorders, in which has been marshalled a great deal of evidence concerning the evolved basis of moral and social behavior in humans (Price, Sloman, Gardner, Gilbert, & Rhode, 1994). The lines of evidence buttressing this social-competition hypothesis began with the social psychology of mood disorders but is poised to progress briskly with an infusion of research concerning the evolution of neurotransmitter systems. It is these systems that most proximally subserve not only behavior generally, but natural ethics and morality in particular.

At this juncture it is useful to summarize the essential elements of the social-competition hypothesis before relating it to kinship genetics or neuroscience. There is general agreement that mood disorders constitute some contemporary phenotypic expression of a phylogenetic healthy behavioral repertory (Lewis, 1934; Hill, 1968; Beck, 1967; Glantz & Pearce, 1989; Gilbert, 1992; Wilson, 1993). Clearly, some variant phenotypes of current mood disorders performed adaptive functions among ancestral hominids. Thus, beyond the constraints of clinical theory and practice, the phylogenetic precursors of what are now diagnosed as "mood disorders" can be better appreciated to have performed some function in connection to social competition (Price, 1967; Sloman, 1976; Gardner, 1982; Gilbert, 1992; Wilson, 1993).

The essential result of social competition in any taxon is that winners behave differently than losers (Price et al., 1994). Yet phylogeny appears to have modulated competitive drives as a key step in the emergence of sociality. Indeed, recent trends in studies of behavioral ecology posit that most social mammals cooperate or compete in one of two basic social modes (Chance, 1988). In clinical terms these two modes, the agonic and hedonic, respectively, relate closely to syndromes of depression and mania. Clinical depression is readily identified in evolutionary psychological terms as a strategy of deescalation or loss, while mania is a strategy of winning via engagement and assertion (Price et al., 1994).

A further component of the hypothesis is that humans retain, among other possibilities, a phylogenetically structured capacity to yield (become depressed) or assert (become expansive) in the face of socially competitive situations. Such social competition can be specified at several levels, and the hypothesis linking mood disorders to social competition may be expressed in hierarchical fashion. The hierarchy subsumes sexual selection, social stratification, ritual antagonistic behavior, resource-holding potential, and social-attention-holding potential, among other domains. These are skillfully detailed elsewhere (Gilbert, 1992; Price et al., 1994).

In the most fundamental sense, mood disorders appear to have emerged (1) in taxonomic terms at the reptilian-mammalian interface, (2) in neurobiological terms at the subcortical-limbic interface, and (3) in psychological terms at the egoic-superegoic interface. Yet such interfaces are not, in themselves, the defining aspects of mammaloprimatoid psychology. Instead, the attainment of social dominance via social appeal and prestige rather than hierarchical threat has come to characterize behavior in the hominoid line (Barkow, 1989; Gilbert, 1992). Put differently, influence rather than raw intimidation is more commonly successful as a means to social dominance among primates. These observations are confirmed in both ethological and socioanthropological field studies (Paglia, 1994). Moreover, this ethological consensus accrues however much the culturogenic-socioconstructivist view has been favored by academes (Paglia, 1994).

Nonetheless, sociocultural factors most certainly do shape phenotypic expression of human mood states and disorders. In humans, social competition may induce either yielding or assertion. Such yielding or assertion may cause emotional dysregulation, especially when there is a crescendo of interpersonal affects. Such escalations or deescalations may have been quite favorable in the environment of evolutionary adaptation. However, any further dysregulations may, in the context of contemporary culture, constitute syndromes of clinical depression or mania. Consequently, the neurotransmitter genetics and physiology of mood disorders are relevant in that these molecules subserve crucial variations in the phenotypic expression of behavior.

EVOLUTIONARY EPIDEMIOLOGY OF THE NEUROTRANSMISSION OF MORALITY

Depression and mania, as is now more widely agreed, are diseases having a fundamentally neurobiological and genetic foundation. Therefore, it is useful to attempt some connection between the social-competition hypothesis and recent trends in neuroscientific research.

Much of what is known about neurotransmitter biology has accrued in the past generation. While even twenty years ago it was obvious that dopamine and serotonin were of special interest to the explanation of mood disorders, only with the advent of molecular analytic techniques in the past few years has a deeper appreciation of the physiological and genetic complexity of these neurotransmitter systems emerged. Owens & Risch (1995) have summarized recent developments quite cogently. Relevant aspects of their summary are detailed here in the specific context of phylogeny. Moreover, it is important to realize serotonin and dopamine have second-level correlations and interactions in the functional anatomy of the brain which yet lie beyond the comprehension of neuroscience. Generally, they have reciprocal physiological effects. Dopamine is an essentially stimulating neurotransmitter, and serotonin is inhibitory, as is perhaps most clearly reflected by pharmacological manipulations: serotonergic compounds show antidepressant effects, whereas dopamine antagonists are potent antipsychotic (antimanic) agents. Finally and for reasons that are not yet fully clear, serotonin and dopamine subsystems appear to have coevolved in the course of mammaloprimatoid brain and social evolution.

Dopamine, long thought to exist in only two types, is now known to be quite heterogeneous (with at least five variant loci, some with considerable point polymorphism; see Owens & Risch, 1995). So too is serotonin more complex than was previously thought. Still, it is important to note that each of these neurotransmitter families can be still further grouped into two clusters of similar function and distribution. The five dopamine receptors aggregate in line with either the original D1 or D2 typology. This aggregation reflects similarities in both anatomy and physiology. The D2 cluster is of special interest to the present discussion because its subtypes (D2, D3, D4) appear to be phylogenetically later and designed to subserve the evolution of higher-level cognitive-affective expressions selected in response to mammalian social evolutionary pressures. The D1 cluster, meanwhile, serves extrapyramidal-tract and midbrain functions. These capacities arose as solutions in the face of more primitive selective pressures at the time of major reptilian evolutionary advancement.

The distribution of the mRNA of each protein type can serve as an index of neuronal receptor networks in brain tissue (Chio, Hess, Graham & Huff, 1990). Differences in distribution of the subtypes are borne out in mRNA assays. The mRNA densities of the D2 cluster localize more robustly to the hypothalamic, limbic, and frontocortical areas than do the densities for the D1 cluster. This difference appears to account for the fact that it is within the D2 family that the most effective and least toxic antipsychotic drugs are to be found (Owens & Risch, 1995). Moreover, the distinctive anatomical and functional subtypology of dopamine receptors supports the notion that the subtypes derive from a phylogenetically common stock. Each seems to have branched off to resolve specific selective pressures extant at different times in the environment of evolutionary adaptation (MacLean, 1990). This branching appears to have occurred in conjunction with changes in the mammaloprimatoid brain necessary to accommodate social behavior.

The overall picture for serotonin resembles that of dopamine (Owens & Risch, 1995). Serotonin is, however, more ubiquitous in the brain and so has perhaps less varied distribution than dopamine. Nevertheless, serotonin has now some four family clusters with a total of more than a dozen subtypes. Indeed, the rapid pace of research makes a fuller account of the current state of knowledge concerning serotonin receptor subtypes specially difficult to summarize and truly beyond the scope of the present discussion. Suffice to say assays of serotonin receptor mRNA densities vary considerably across brain regions. These assays together establish a clear basis for the mediation by serotonin of widely diverse cognitive and behavioral activities. Of special interest to the present discussion are serotonin projections of the dorsal raphe-striatum. These are phylogenetically earlier features of reptilian brain evolution (a mode similar to D1) than are serotonin projections that mediate the medial raphe-hippocampal system. Such robust anatomical links to limbic functions as these clearly arose, in good measure, to subserve cognitive-affective demands of mammalian social evolution (MacLean, 1990).

COORDINATED ACTIONS OF DOPAMINE AND SEROTONIN

Taken as a whole, research concerning neurotransmitter subtypology required a revised dopamine hypothesis in which it is posited that normal thought and mood are contingent upon orchestrated actions of dopamine and serotonin (and probably other agents such as neurohormones and endogenous opioids). There is a particular focus on how the conjoint actions of dopamine and serotonin appear to mediate normal patterns of mood and thought as well as psychosis, abulia, and extrapyramidal effects.

Psychotic and mood symptoms are thought to arise when both the prefrontal-cortex and mesolimbic systems become hyperdopaminergic (Janssen et al., 1988; Davis, Kahn, Ko, & Davidson, 1991; Meltzer, 1992). Serotonin depletion can aggravate symptoms as well (P. Keck, personal communication, 1997). This induces initial escalation of mood and activity, often followed by increasing irritability and/or thought disorder that can be further accentuated by any reductions in serotoninergic activity. Moreover, compounds with a high ratio of serotonin distributed through the raphe-hippocampal-limbic circuit, S_{irhl}, to dopamine of the limbic type, D_2 (S_{rhl}/D_4), may be especially efficacious in the treatment of psychotic and mood disorders, including severe manic depression (Meltzer, 1992).

Serotonin inhibits the dopamine projections to the striatum, and serotonin reuptake inhibitors raise mood but also can induce extrapyramidal side effects. Moreover, lesions of the serotonergic dorsal raphe diminish neuroleptic-induced catalepsy. Serotonin also inhibits dopaminergic limbic structures, notably the accumbens. These serotonin effects on dopamine are modulated by receptors of the 1_A, 2, and 3 serotonin subtypes. The increased affinity of clozapine for limbic D_4 receptors and its antagonism of $5\text{-}H_2T$ likely account for its unique profile, both in vitro and in vivo. The $5\text{-}HT_2$ moieties can elicit dopamine release "downstream" in the striatum and thus limit extrapyramidal symptoms induced by dopamine antagonists operating elsewhere in the brain. Abnormalities of the largely mammalian capacities of thought and mood (positive and negative symptoms) can thereby be ameliorated without significant alteration of lower reptilian circuits (involuntary motor symptoms). Thus the novel antipsychotic risperidone improves psychosis and mood with few motor effects ($5\text{-}HT_2$, antagonism in tandem with D_4; Janssen et al., 1988).

The conjoint activity of the dopamine and serotonin systems is also of central interest to the elucidation of substance abuse (Gawin & Ellenwood, 1988; Wyatt, Karoum, Suddath, & Hitri, 1988; Kosten, 1993). This, in turn, is relevant to the present discussion in that cocaine use is mimetic of both manic and depressive symptoms. Cocaine administration acutely increases synaptic dopamine via the inhibition of presynaptic dopamine reuptake. This produces escalation of mood and activity. If overstimulation is chronically sustained, D_2 receptors eventually downregulate, leading to physical dependence. Thereafter, abrupt cessation of cocaine depletes dopamine. This induces acute agitated depression with a chronic syndrome of anhedonia, dysphoria, lethargy, somnolence, and apathy persisting for up to a year. For its part, serotonergic agents moderate both the predependent appetite for cocaine and the postdependent withdrawal syndrome. Cocaine physiology in humans, then, is a fair model of manic-depressive, hawk-dove, escalation-deescalation neuromental states. Such significant inter-play between social-rank hierarchy and particulars of serotonin-dopamine neuro-transmission is further borne out in animal studies of escalation and deescalation.

MANIC ESCALATION AND DEPRESSIVE DE-ESCALATION AS HAWK DOVE NEUROMENTAL STATES

In most mammals, reproductive success depends on success in various social roles. One key role is in besting others who are pursuing the same resources (Barash, 1977; Gilbert, 1989; Krebs & Davies, 1981; Trivers, 1971). As noted earlier, primate social hierarchy is a key factor in reproductive and social success. Those higher up have more breeding opportunities and often make the more attractive allies than those lower down. Thus navigating hierarchical re-lationships wherein some individuals are more powerful than others has been a selective pressure for millions of years. Social neuromentalities evolved in the context of conspecific behaviors of either competition or cooperation. Further, these evolved neuromentalities propitiate the adoption of predictable, context-dependent roles and behaviors that modulate both competition and cooperation (Nesse, 1990). In some respects, social neuromentalities are synonymous with the everyday term ''state of mind'' to the extent each embodies algorithms of social roles. Gradations in states of mind are to be expected between individuals across the social hierarchy. That is, the state of mind of a dominant differs from those of hierarchical subordinates. Being neuromentalities, these differences in state of mind have proved to be measurable psychobiological variables. The consequences can be profound.

Most remarkably, in some fishes inhibition of subordinates can cause change of sex (significantly, the subordinate's genes can thus be maintained by mating with dominants; Keenleyside, 1979). Likewise, dominant naked mole rats and New World monkeys have been observed to directly suppress sexual reproduc-tive physiology in subordinates (Abbott, Barrett, Faulkes, & George, 1989). Studies have also documented major biological distinctions between dominant and subordinate baboons, notably stress hormones that have been linked to the pathophysiology of affective and anxiety disorders (Sapolsky, 1989, 1990a, 1990b; Ray & Sapolsky, 1992). Biological profiles of subordinates differ from those of more dominant animals in other species (Henry, 1982; Henry & Stephens, 1977).

Pharmacological probes are also revealing. Drug effects can vary between dominant and subordinate animals, sometimes quite markedly. For example, the response of rhesus monkeys given amphetamine depends on social rank. Dominants showed increased threat, chase, and attack behaviors, whereas subordinates showed increased submissive behavior, for example, fear grimaces and turning away (Harbour, Barchas, & Barchas, 1981). Harbour, Barchas, and Barchas (1984) reported a case in which amphetamine was given to a female rhesus monkey that moved between two groups. In the first group, she was highly subordinate, isolate, and fearful, with quite limited affiliation. Amphetamine accentuated all of these submissive behavior patterns quite dramatically. However, when moved to a new group, she was conspicuously favored by the alpha male. Her behavior changed markedly with this increased rank. When she was administered amphetamine as the dominant female in this novel social milieu, her threats increased dramatically. She did not evidence submissive behavior as previously, but instead increased her social approach.

Several empirical studies have confirmed that social competition psychobiologically modulates mood states in primates. A recent review of neuroendocrine correlates of social rank confirms that endocrine states are highly sensitive to social feedback (Sapolsky, 1993). Hormone levels in both dominant and subordinate animals vary as a function of group stability. However, instability of the social-rank hierarchy affects individuals differently with respect to directional trends in rank (going up versus down) as well as basic temperament (Sapolsky, 1994). In this sense dominance and subordination are largely controlled by sociophysiological feedback in the environment.

Perhaps the most significant evidence to date is the correlation of the primate social-rank spectrum with serotonin parameters that mirror human mood swings. Physiological changes can be both the cause and the consequence of rank changes (Raleigh, McGuire, Brammer, & Yuwiler, 1984; Hartmann, 1992). Blood levels of serotonin are significantly higher in dominant male vervet monkeys as compared to their subordinates (Raleigh et al., 1984). Sham deposition of the alpha male was followed by a sharp fall in his serotonin blood level but a sharp rise in that of the subordinate male who was raised to dominant status. Restoration some weeks later of the formerly dominant male was marked by a reversal of these serotonin findings as well as his renewed tenure as the group alpha. Significantly, this restoration led to a fall in both social rank and serotonin level of the "new" dominant below that of his original beta baseline. Social ascendance with subsequent fall in rank produced sociophysiological changes that reduced the subordinate to a lower biosocial status than he would have had if no change had occurred.

Sex hormones similarly reflect such social contexts. Androgen levels in humans, both male and female, vary at non-stressed homeostatic baseline depending on the degree of social competition or cooperation and also vary directly with success in competitive games (Kemper, 1984). These rises can occur even in nonphysical contests such as chess (Mazure, Booth, & Dabbs, 1992). Thus neurotransmitter and endocrine parameters not only reflect general features of social status, affect, and mood, but directly link to reproductive biology itself.

A CALL TO FIGURATIVE SHOVELS: THE ARCHAEOLOGY OF BRAIN AND MIND

Thus we come full circle. Major subtypes of neurotransmitter receptors, particularly dopamine and serotonin, have unique anatomical distributions and functional consequences. These consequences are quite germane to, among other concerns, the resolution of stress engendered by social competition. This resolution is most often accomplished via innovative capacities for cognitive-affective assessment and behavior. These capacities increasingly typified the brains and psychology first of reptiles, then of social mammals, and most distinctively of the primates (Dunbar, 1988; Mithen, 1996).

The functional anatomy of neurotransmitters as sketched here is intended as a link, in a plausible but only preliminary manner, to the pathological sociophysiology of manic-depressive neuromentality. This pathophysiology arises as the phylogenetic hierarchy set in the triune brain encounters contemporary ontogenic demands. Thus, in turn, manic depression seems to be an expression of an enduring population polymorphism of the human genome that is increasingly mismatched in the contemporary environment.

In any case we begin to unearth biopsychosocial "fossils" that enliven the evolutionary ethical discourse that so fascinated Darwin. Even if the present "dig" does not in itself discover the missing links between neurobiology and evolutionary psychology, the time is at hand for a thoroughgoing "archaeology" of brain and mind. To this end, further research in neuropharmacology would do well to heed evolutionary science and vice versa.

REFERENCES

Abbott, D. H., Barrett, J., Faulkes, C. G., & George, L. M. (1989). Social contraception in naked mole-rats and marmoset monkeys. *Journal of Zoology, 219*, 703–710.
Barash, D. P. (1977). *Sociobiology and behavior.* London: Heinemann.
Barkow, J. (1989). *Darwin, sex, and status: Biological approaches to mind and culture.* Toronto: University of Toronto Press.
Beck, A. T. (1967). *Depression: Clinical, experimental, and theoretical aspects.* New York: Hoeber.
Chance, M. (Ed.) (1988). *Social fabrics of the mind.* Hove: Erlbaum.

Chio, C. L., Hess, G. F., Graham, R. S., Huff, R. (1990). A second molecular form of D2 dopamine receptor in rat and bovine caudate nucleus. *Nature, 343*, 266–269.

Davis, K. L., Kahn, R. S., Ko, G., & Davidson, M. (1991). Dopamine in schizophrenia: A review and reconceptualization. *American Journal of Psychiatry, 148*, 1474–1486.

Dunbar, R. M. (1988). *Primate social systems*. London: Croom Helm.

Gardner, R. J. (1982). Mechanisms in major depressive disorder: An evolutionary model. *Archives of General Psychiatry, 39*, 1436–1441.

Gawin, F. H., & Ellenwood, E. H. (1988). Cocaine and other stimulants. *New England Journal of Medicine, 318*, 1173–1182.

Gilbert, P. (1989). *Human nature and suffering*. London: Lawrence Erlbaum.

Gilbert, P. (1992). *Depression: the evolution of powerlessness*. New York: Guilford.

Glantz, K. Pearce, J. (1989). *Exiles from Eden: Psychotherapy from an evolutionary perspective*. New York: Norton.

Hamilton, W. D. (1963). The evolution of altruistic behavior. *The American Naturalist, 97*, 354–356.

———. (1964). The genetical evolution of social behaviour, parts 1 and 2. *Journal of Theoretical Biology, 7*, 1–52.

Harbour, S. N., Barchas, P. R. & Barchas, J. D. (1981). A primate analogue of amphetamine induced behavior in humans. *Biological Psychiatry, 16*, 181–196.

———. (1984). The regulatory effect of social rank on behavior after amphetamine administration. In P. Barchas (ed.), *Social hierarchies: Essays toward a sociophysiological perspective*. Westport, CT: Greenwood Press.

Hartmann, L. Presidential address: Reflections on humane values and biopsychosocial integration. *Psychiatry, 149*, 1135–1141.

Henry, J. P. (1982). The relation of social to biological processes in disease. *Social Science and Medicine, 16*, 369–380.

Henry, J. P., & Stephens, P. M. (1977). *Stress, health, and the social environment: A sociobiologic approach to medicine*. New York: Springer-Verlag.

Hill, D. (1968). Depression: Disease, reaction, or posture? *American Journal of Psychiatry, 125*, 445–456.

Janssen, P. A. J., Niemegeers, C. J. E., Awouters, F., Schellekens, K. H. L., Megeris, A. A. H. P., & Meert, T. P. (1988). Pharmacology of risperidone (R64 766), a new antipsychotic with serotonin-S2 and dopamine-D2 antagonistic properties. *Journal of Pharmacology and Experimental Therapeutics, 244*, 685–693.

Keenleyside, M. H. A. (1979). *Diversity and adaptation in fish behaviour*. Berlin: Springer-Verlag.

Kemper, T. D. (1984). Power, status, and emotions: A sociobiological contribution to a psychological domain. In K. Scherer & P. Ekman (Eds.), *Approaches to emotion*. Hillsdale, NJ: Lawrence Erlbaum.

Kosten, T. R. (1993). Pharmacotherapies for cocaine abuse: Neurobiological abnormalities reversed with drug intervention. *Psychiatric Times, 10*, 25.

Krebs, J. R., & Davies, N. B. (1981). *An Introduction to behavioral ecology*. Oxford: Blackwell.

Lewis, A. J. (1934). Melancholia: A clinical study of depressive states. *Journal of Mental Science, 80*, 277–378.

MacLean, P. D. (1990). *The triune brain in evolution*. New York: Plenum Press.

Mazure, A., Booth, A., & Dabbs, J. M. (1992). Testosterone and chess competition. *Social Psychology Quarterly, 55*, 70–77.

Meltzer, H. (1992). The importance of serotonin-dopamine interactions in the action of clozapine. *British Journal of Psychiatry, 160* (Suppl. 17), 22–29.

Mithen, S. (1996). *Prehistory of the mind.* London: Thames & Hudson.

Nesse, R. M. (1990). Evolutionary explanations of emotions. *Human Nature, 1,* 261–289.

Nestler, E., & Hyman, S. (1993). *The molecular foundations of psychiatry.* Washington, DC: APA.

Owens, M. J., & Risch, S. C. (1995). Atypical antipsychotics. In A. F. Schatberg & C. Nemeroff (Eds.), *Psychopharmacology.* Washington, DC: American Psychiatric Press.

Paglia, C. (1994). *Vamps and tramps.* New York: Vintage.

Powles, W. E. (1992). *Human development and homeostasis: The science of psychiatry.* Madison, CT: International Universities Press.

Price, J. (1967). Hypothesis: The dominance hierarchy and the evolution of mental illness. *Lancet, 2,* 243–246.

Price, J., Sloman, L., Gardner, R., Gilbert, P., & Rhode, P. (1994). The social competition hypothesis of depression. *British Journal of Psychiatry, 164,* 309–315.

Raleigh, M., McGuire, M., Brammer, G., & Yuwiler, A. (1984). Social status and whole blood serotonin in vervets. *Archives of General Psychiatry, 41,* 405–410.

Raleigh, M., McGuire, M., Brammer, G., Pollack, D. B., & Yuwiler, A. (1991). Serotonergic mechanisms promote dominance acquisition in adult male vervet monkeys. *Brain Research,* 181–190.

Ray, J. C., & Sapolsky, R. M. (1992). Styles of male social behavior and their endocrine correlates among high-ranking wild baboons. *American Journal of Primatology, 28,* 231–250.

Richards, R. (1987). *Darwin and the emergence of evolutionary theories of mind and behavior.* Chicago: University of Chicago Press.

Ruse, M. (1986). *Taking Darwin seriously.* Oxford: Basil Blackwell.

Sapolsky, R. M. (1989). Hypercortisolism among socially subordinate wild baboons originates at the CNS level. *Archives of General Psychiatry, 46,* 1047–1051.

———. (1990a). Adrenocortical function, social rank, and personality among wild baboons. *Biological Psychiatry, 28,* 862–878.

———. (1990b). Stress in the wild. *Scientific American, 262*(1), 116–123.

———. (1993). Endocrine alfresco: Psychoendocrine studies of wild baboons. *Recent Progress in Hormone Research, 48,* 437–468.

———. (1994). Individual differences in the stress response. *Seminars in the Neurosciences, 6,* 261–269.

Sloman, L. (1976). The role of neurosis in phylogenetic adaptation, with particular reference to early man. *American Journal of Psychiatry, 133,* 543–547.

Trivers, R. L. (1971). The evolution of reciprocal altruism. *Quarterly Review of Biology, 46,* 35–57.

Wilson, D. R. (1993). Evolutionary epidemiology: Darwinian theory in the service of medicine and psychiatry. *Acta Biotheoretica, 41,* 205–218.

Wyatt, R. J., Karoum, F., Suddath, R., & Hitri, A. (1988). The role of dopamine in cocaine use and abuse. *Psychiatric Annals, 18,* 531–534.

22

Darwinian Analysis of the Emotion of Pride/Shame

Glenn Weisfeld

Darwin observed that expressions of pride/shame are homologous to those for dominance and submission, as they came to be called, in other species. The idea that human pride and shame evolved from dominance behavior has been suggested before by William McDougall (1923) and others. Some recent data lend further support to this notion, as will be described.

For the most part, mainstream psychology has been disinterested in identifying the evolutionary origins of particular emotions. In the case of pride and shame, this has resulted in a proliferation of terms for identifying social motives in which pride and shame are involved. These include prosocial behavior, moral development, approval motivation, guilt, shame, evaluative reinforcement, and many others. It is unlikely that all of these terms represent fundamentally different behaviors; they probably all refer to the same basic behavioral system, which is concerned with social success or failure occurring in various situations. For example, the term "shame" seems to involve public humiliation for violating a social norm, whereas "guilt" often refers to a private misdeed (Buss, 1980). Because all these terms seem to involve the affects and emotional expressions of pride and shame, I think that these terms—pride and shame—should be substituted for the gamut of others. Further, other psychological terms that refer to social success or failure—achievement motivation, prestige striving, success striving, social comparison, the power motive, competitive behavior, self-esteem, and others—also seem to involve pride and shame and their typical emotional expressions and similarly can be subsumed under the same emotional modality. I do not deny that some consistent distinctions are made in the use of these terms, but semantic differences do not necessarily reflect fundamental behavioral distinctions. Surely all of these terms do not represent independent evolved behavioral tendencies.

In my view, the crucial test of the distinctiveness of a given emotion is not the situation or stimuli that elicit it, or the overt behavior that the affect prompts, because these are numerous and highly variable in humans (Weisfeld, 1997). Likewise, emotional expressions and visceral changes that accompany some emotions are not always highly specific. I believe that affects are the single best indicator of a basic emotion. As Panksepp (1994, p. 396) wrote, "[T]here seem to be as many internally experienced affective states as there are basic motivational and emotional systems of the brain." Moreover, by identifying emotions primarily by their affects, the sort of proliferation of terms for motives that characterizes pride and shame can be avoided. Not only does this serve the principle of parsimony, but also it makes comparison with other species easier, since ethologists postulate a fairly limited number of discrete motives for animals.

In failing to recognize that pride and shame underlie all of these categories of social behavior, psychology has also ignored the fundamental nature of this emotion, or motive. When this motive is referred to in textbooks, it is usually described as a "learned motive" rather than being recognized as specieswide and basic. The very idea of a learned motive may be oxymoronic. We indeed learn when to exhibit a particular motivated behavior, for example, when to fear speeding automobiles, and how to execute the motivated behavior, for example, how to earn money for food. But we can hardly learn to experience an affect. In the case of pride and shame, we indeed learn how to gain the approval of others—the criteria, or values, by which we are judged vary greatly across cultures—but the affect of pride and shame is invariant. Analogously, the means by which people satisfy the hunger motive vary greatly around the world, but hunger itself is basic and universal.

Aside from cultural variability in the means of gaining social status, pride and shame may actually be ignored because of their omnipresence. This emotion permeates virtually every human situation. When we plan any action, the fulfillment of any emotion, we usually also consider the consequences for our reputation. We may deceive ourselves into thinking that we are thereby behaving "rationally," but in fact we are being governed by anticipated pride or shame. In addition, we may be reluctant to admit to having a capacity for pride and shame because we are embarrassed by this human frailty. In conversation, we may say that we feel happy or sad when we mean proud or ashamed.

What, then, is the evidence that pride and shame constitute a basic, evolved human emotion? Darwin, in *The Expression of the Emotions* (1872), provided several research strategies for identifying an evolved behavior. I would like to describe how research utilizing these strategies offers evidence for the evolved basis of pride and shame.

One such test is for specieswide prevalence. The emotion of pride and shame indeed seems to be experienced universally (Edelmann, 1990; Sueda & Wiseman, 1992). In every culture, social standing is of concern to normal individuals from childhood on. Even in cultures in which social status is not formalized, everyone wishes to earn the respect of others.

Second, pride and shame have a typical onset that suggests a specific maturational, that is, evolved, basis. Whether it is termed rivalry, dominance competition, approval motivation, or achievement motivation, it and its expressions develop around the ages of two to three (Mascolo & Fischer, 1995; Weisfeld, 1997).

Incidentally, the notion that the cognitive capacity for self-recognition underlies the development of pride and shame has been refuted. Schneider-Rosen and Cicchetti (1991) showed that there was no relation between toddlers' touching a red spot on their noses reflected in a mirror (a sign of self-recognition) and their averting their gaze when they failed an easy task. Some general cognitive capacities are certainly necessary for the appreciation of cultural values and the ability to make comparisons, but a capacity for self-recognition does not appear to be essential. The capacity for pride and shame does not seem to depend upon specific cognitive capacities; sociopaths and neurological patients sometimes lack this affect but show no major cognitive deficits. Also, emotions present at birth cannot depend on higher cognitive processes, and numerous emotions that develop well after birth, such as sexual, maternal, romantic, and dominance motivation, seem to be induced by hormonal changes, not cognitive ones.

Third, as Darwin described, pride and shame have distinct, stereotypic expressions. Proud, successful people carry themselves expansively and conspicuously. Their gaze is direct (especially while speaking) and their manner is relaxed (figure 22.1). An ashamed or unsuccessful individual exhibits an antithetical demeanor (figure 22.2). Also, an ashamed or embarrassed person sometimes blushes or manifests a "nervous, silly smile," and inarticulate speech (Edelmann & Hampson, 1979, 1981a). In his commentaries on Darwin (1872), Ekman reveals that he found a similar pattern of gaze and head aversion in the expression of shame both in Western cultures and in pre-contact New Guineans. He regards the expressions of shame and guilt as identical.

The form of these expressions may help explain why this emotion has been neglected even by Darwinians. Whereas most other emotional expressions are facial, it is the bodily expressions of pride and shame that are most distinctive. Also, the facial expression of blushing may have been neglected partly because of its difficulty of elicitation and perhaps also because it requires the use of color graphics. Often, the facial expressions of happiness and sadness have been studied instead of the more precise expressions of pride and shame. However, joy and sadness are nonspecific expressions that reflect virtually any pleasant or unpleasant experience and so do not reflect specific evolved motives.

There are also stereotypic expressions of approval and disapproval—for example, attention, smiling, anger, ridicule, disgust (figure 22.3), and contempt (figure 22.4). This fact, as Pugh (1977, pp. 285, 352) argued, further supports

Figure 22.1 Pride

Figure 22.2 Shame

Figure 22.3 Disgust

Figure 22.4 Contempt

Table 22.1
Correlation of Erectness of Posture with Various Attributes

Grade	Class	N	Athletic ability	Attractiveness	Intelligence
9	1979	25	.46 (.008)	.48 (.007)	.16
9	1980	20	.40 (.08)	.55 (.02)	.48 (.04)
10	1979	12	.03	.04	.43
11	1980	32	.47 (.08)	.63 (.001)	.25
12	1977	16	.27	.67 (.02)	.29
Mean			.33	.47	.32

Table 22.2
Correlation of Grade School Toughness with High School Attractiveness and Athletic Ability

Class	Grade when toughness measured	High school grade	N	Attractiveness	Athletic ability
1979	3	10	4	.75	1.00 (.05)
	3	9	7	.52	.71 (.05)
1980	1	11	8	.67 (.05)	.76 (.05)
	2	11	16	.65 (.02)	.75 (.01)
	5	11	18	.80 (.01)	.75 (.01)

the idea of an evolved basis for pride and shame because nonverbal signals and responsiveness to them tend to evolve together.

Are the expressions of pride and shame valid and reliable? Several researchers have shown that success or failure in various situations is reflected by these nonverbal expressions of pride or shame. Longitudinal research at Wayne State University (Weisfeld, Muczenski, Weisfeld, & Omark 1987) has indicated that postural erectness reflects long-term dominance rank, which was highly stable from early boyhood through adolescence (tables 22.1, 22.2). Posture is also changed by short-term success or failure, such as performance on a university examination (figure 22.5, table 22.3) (Weisfeld & Beresford, 1982).

The expressions of pride and shame are identifiable by subjects. For example, subjects can indeed recognize the nonverbal display of embarrassment (Edelmann & Hampson, 1981b). In one study, the timing of gaze aversion by itself allowed subjects to discriminate embarrassed from nonembarrassed smiles (Asendorpf, 1990). Further, these expressions can influence behavior. Levin and Arluke (1982) found that a person who asked for volunteers received more compliance if he exhibited embarrassment. Ginsburg (1980) showed that bowing or crouching often cut off another child's attack. Zivin (1977) was able to predict on the basis of facial expression which child would win a dominance encounter. Direct gaze can cause the target person to move away (Ellsworth, Carlsmith, & Henson, 1972), especially a subordinate (Fromme & Beam, 1974). Rosa and

Figure 22.5
Examination Score as Related to Change in Erectness of Posture from Just before to Just after Receiving Examination Grade (College students. $N = 45$)

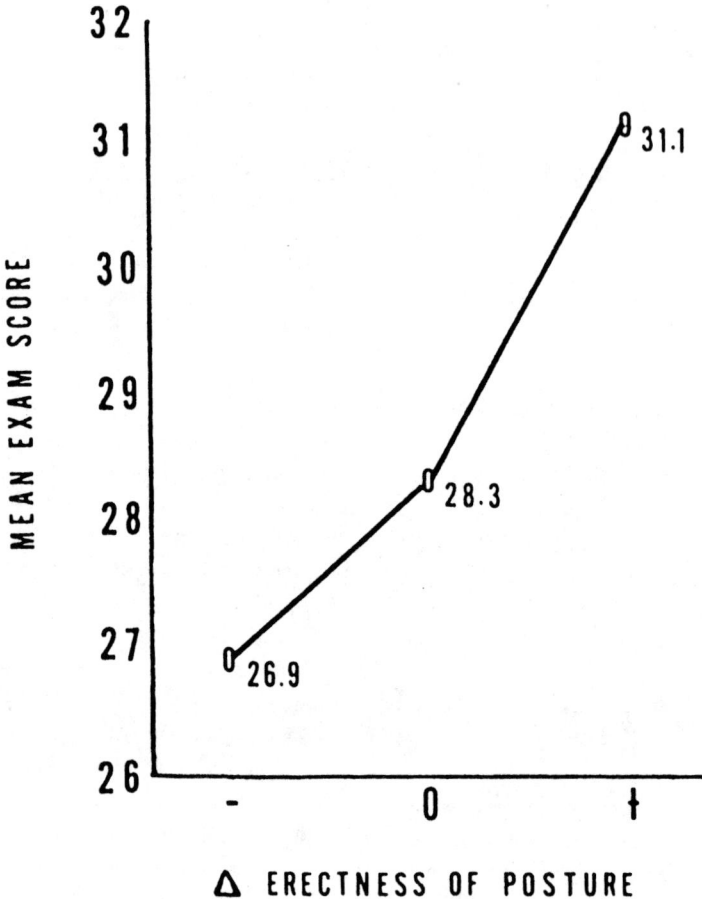

Δ ERECTNESS OF POSTURE

Mazur (1979) showed that the subject who averted his gaze first subsequently did less speaking. In a naturalistic experiment, we have shown that directing an angry or contemptuous facial expression at subjects caused a decrease in their postural erectness (Kahn & Weisfeld, 1993).

A fourth research strategy involves demonstrating specific physiological mediators of the behavior in question. Pride and shame seem to involve a particular neural structure, the orbitofrontal cortex (figure 22.6) (e.g., Devinsky, Hafler, & Victor, 1982). Patients with bilateral lesions to this area, such as the famous Phineas Gage, typically lose their former motivation to perform their jobs conscientiously or even to observe the basic social graces; Damasio (1994, p. 51)

Table 22.3
Correlation of Erectness of Posture after
Receiving an Examination Grade with
One's Examination Grade (College
Students)

Classroom	N	Correlation and level of significance
1	14	.79 (.001)
2	12	.65 (.05)
3	9	.62 (.05)
4	20	.63 (.01)
5	20	.63 (.01)
6	11	.33
7	8	.51
8	14	.79 (.001)
9	12	.65 (.05)
10	9	.62 (.05)
	129	.62 (.001)

concluded that Gage "lacked the feeling of embarrassment." What seems lost is the capacity for pride and shame. If damage is limited to this brain area, the other affects remain intact, such as hunger, sex, humor, and fear, so this is not a general loss of affect. Incidentally, the posterior orbitofrontal cortex is limbic in microstructural cytoarchitecture and in its connections with olfactory, hypo-

Figure 22.6
The Orbitofrontal Cortex

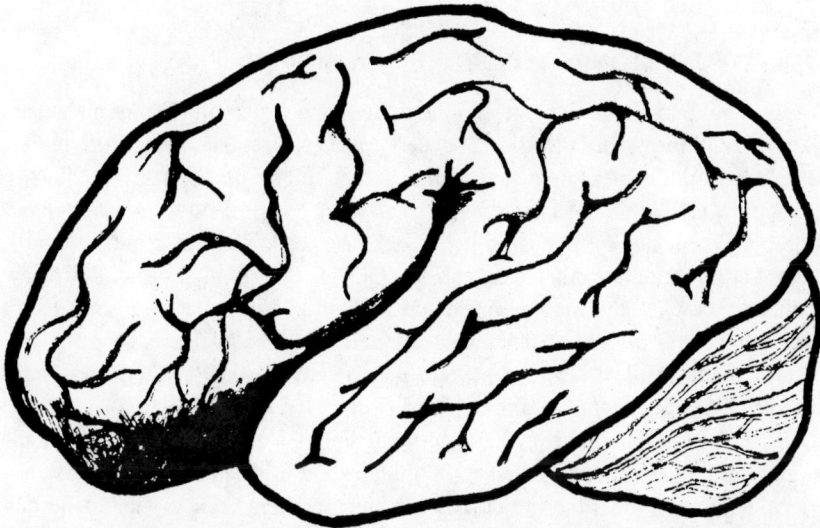

thalamic, autonomic, and other limbic structures (Fuster, 1980, pp. 39, 89, 93–94); this explains its direct role in emotion. In fact, this is the only affectively sensitive part of the frontal lobes; monkeys self-stimulate it electrically (Passingham, 1993). The orbitofrontal cortex is quite distinct functionally, ontogenetically, and phylogenetically from the dorsolateral frontal cortex, which is one of the last parts of the neocortex to develop and evolve (Fuster, 1980, p. 143), and, fittingly, is involved in complex sequences of behavior, categorization, and planning in the service of various motives.

A fifth source of evidence for the basic, evolved nature of pride and shame is phylogenetic. As already mentioned, Darwin noted the homology between the expressions of pride and shame in humans, on the one hand, and of dominance and submission in other species (figures 22.7, 22.8). There are many other such parallels.

Also, as stated before, people with orbitofrontal lesions show diminished motivation to succeed and to obey social standards. Phineas Gage was described as manifesting but little deference for his fellows (Damasio, 1994, p. 8). Similarly, homologous lesions in simians cause changes in aggressiveness and a fall in dominance rank (Fuster, 1980, p. 69) or inappropriate reactions to the ranks of their cagemates. Dominant monkeys have numerous serotonin-2 receptors in the orbitofrontal cortex and related structures (Damasio, 1994, p. 76). Recent research on serotonin levels confirms that human social status is homologous to dominance rank in simians. Men who are group leaders tend to have high serotonin levels, as do male vervet monkeys that have been submitted to (Masters & McGuire, 1994). Also, male rhesus monkeys that have risen in rank experience a surge of testosterone, as do men who have won various tests of skill (Mazur, 1983).

SOCIAL FUNCTIONING OF PRIDE/SHAME

Additional parallels exist between the dominance system in simians and social relations in humans. In both cases, success raises aspirations, and failure lowers them. Also, challenges are usually directed at others of similar or unknown ability. Further, dominant simians and successful people tend to welcome challenges, seek out others, persist at tasks, exercise leadership, discipline subordinates, and command others' attention. Despite these many parallels, few psychologists who have studied pride and shame (e.g., Tangney & Fisher, 1995) have utilized the dominance-hierarchy model proposed by Darwin and elaborated by McDougall (1923) and several modern ethologists (e.g., Mazur, 1983; Barkow, 1989; Weisfeld & Linkey, 1985; Pugh, 1977).

This model also helps to make functional sense of pride and shame. An additional parallel between dominance in simians and pride/shame in humans concerns the distribution of prerogatives among members of a group. In human groups, one gains in status by performing valued deeds, whereas dominance

Figure 22.7
Dog Approaching Another Dog with Hostile Intentions. By Mr. Riviere.

relations in animals are governed mainly by fighting ability. But both are psychological systems of regulated, equitable competition for social status. In both systems, a dominant or successful individual assumes prerogatives of rank. Gaining access to these prerogatives, in fact, constitutes the adaptive value of striving for dominance. A subordinate or unsuccessful individual, on the other hand, gives up prerogatives. Thus in a socially stable, or homeostatic, group, success is proportional to rewards.

Figure 22.8
The Same in a Humble and Affectionate Frame of Mind. By Mr. Riviere.

Further similarities concern the consequences of the violation of equity. A subordinate may usurp a dominant animal's prerogative, or a dominant animal may continue to attack a loser that has submitted. Such violations of social norms must be deterred, so the victim often reacts with rage—a vigorous attack. Similarly, equity in human relationships is enforced by anger—violations of social norms trigger anger. Interestingly, some monkeys, apes, and humans with orbitofrontal lesions exhibit a reduced capacity for anger when mistreated (Damasio, 1994, pp. 45, 57, 74).

Trivers (1971) described the punitive role of moralistic anger in reciprocal altruism and how this and other emotions enforce equitable exchanges. It may be possible to understand the evolution of these emotions underlying reciprocal altruism in terms of the dominance hierarchy. Trivers explained that guilt can preempt a punitive attack by appeasing the moralistic aggressor. But guilt may be the same affect as shame, which probably evolved from submissive behavior. In both guilt and shame, the subject relinquishes prerogatives in the manner of a submissive animal. Trivers also discussed the emotion of gratitude. The human recipient of a favor feels gratitude. But perhaps gratitude is affectively identical to shame, or guilt, and likewise evolved from submission. This is the same unpleasant feeling of lowered status, and it leads to relinquishment of resources. Also, the emotion of pride, not discussed by Trivers, may play a role in reciprocal altruism. Pride, being a pleasant feeling, may provide a psychological incentive for performing altruistic deeds. It also carries an expectation of reward, so that the altruist expects compensation. Subjects who feel that they have accomplished something tend to accept rewards readily. Pride complements gratitude in motivating the transfer of resources from recipient to altruist, and pride probably evolved from dominance. Animals seem to seek dominant status, which therefore must be pleasant, they claim the attendant prerogatives, and their demeanor resembles that of proud humans.

Thus human social behavior, in relationships of reciprocal altruism and other sorts, is supported by an emotional system that probably evolved from the capacity for dominance hierarchization. Human cooperation is not simply a bartering of goods and services; social status, or dominance, is affected by almost every transaction. That is, we are highly sensitive to considerations of equity because they can affect future material outcomes. This explains why people sometimes decline gratuitous material resources in order to maintain their status. When we receive something for nothing, we actually pay by suffering shame and a loss of status, which eventuates in the loss of some future prerogatives. Consequently, subjects often decline a favor if they will have no opportunity to repay it, apparently because of the burden of gratitude, or lowered status (Greenberg & Shapiro, 1971). Similarly, receiving charity is degrading. We may also decline a false or unstable rise in status, apparently because of fear of having to pay some price later. Thus, if we receive undeserved or excessive praise or attention, we may blush with embarrassment or decline a proffered reward. Lerner (1970), a social psychologist, referred to a "need to believe in a just world," but this is more than a belief; it is a motivational system to bring about actions that enforce justice.

In humans, reciprocal altruism is complicated by the capacity for language. We use words to express our intention to restore equity when circumstances permit. Threats, thanks, apologies, and boasting symbolize the behaviors that are prompted by anger, gratitude, guilt, and pride. But these words can be viewed as an extension of the dominance system. This is shown by the fact that nonverbal displays and compensatory actions can substitute for verbal expressions of emotion. Preschoolers who broke a rigged toy either tried to repair it—make compensation immediately—or else averted their gaze guiltily, but seldom did both (Barrett, 1995, p. 46). Similarly, embarrassed adults whose blushing was observed by the experimenter were less ingratiating than subjects who did not believe that their blushing had been observed (Landel & Leary, 1992).

EVOLVED SOCIAL VALUES?

Another possible parallel between dominance behavior in animals and in humans concerns the criteria of social success. In young children, dominance seems to depend on fighting ability or "toughness." In addition, some adult criteria may be universal and evolved. Izard (1977) suggested that nakedness may be inherently shameful to people because it would lead to our seeking privacy for sexual relations. Darwin thought that other people's paying attention to our appearance and especially our faces was inherently embarrassing.

Perhaps physical attractiveness is valued everywhere. We have gathered questionnaire data on adolescents in three countries: England, the United States, and China (Dong, Weisfeld, Boardway, & Shen, 1996; Boardway & Weisfeld, 1994). As expected, the criteria of social success varied greatly among the three countries. In China, adolescents whose schoolmates regarded them as leaders and as socially dominant, and who exhibited nonverbal expressions of dominance (erect posture, direct gaze, relaxation, and commanding attention), tended to be good students. In the United States, dominant boys were good athletes. In England, both these criteria operated (table 22.4). However, in all three countries, high-ranking adolescent boys were physically attractive and were predicted to be economically successful. It might make adaptive sense for such males to be valued by females; similarly, attractive females tend to be desirable. Likewise, it might be beneficial to seek out and defer to certain same-sex companions—for example, males who exhibited skill at hunting or warfare (Barkow, 1989, p. 188), or females competent at food gathering or infant care.

Table 22.4
Correlations between Measures and Expressions of Social Status with Possible Determinants of Social Status

	U.S.		China		England	
Determinants	Meas.	Express	Meas.	Express	Meas.	Express
Athletic	.51*	.48*	.32	.28	.58**	.52**
Intelligence	.24	.36	.80**	.61**	.57**	.53**
Attractiveness	.50*+	.48*+	.52**	.66**	.48**	.40**
Earn potential	.61*	.48*	.69**	.71**	.48**	.51**

* Significance at $p<.01$
**Significance at $p<.001$
\+ Ratings by girls only

Note: Correlations for U.S. and Chinese samples are for ratings *of* boys only for comparability with the present study. Except where noted, ratings *by* girls and boys were collapsed for the U.S. and Chinese samples. Social status measures (Meas) and expressions (Express) are composites of those items best representing the construct in a particular sample.

Thus some of the criteria of social success may have an evolutionary basis. Certain human traits, such as attractiveness, physical grace, courage, and other sorts of competence, may be inherently admirable or praiseworthy (Pugh, 1977, p. 267). One's fitness may have increased by associating with and imitating such individuals and by rewarding them with praise, deference, and resources. Furthermore, honesty and generosity in others, which would make them faithful reciprocators, may be inherently valued (Pugh, 1977, p. 377).

Other candidates for evolved social values concern our own attributes rather than those of others. These values may make us proud whenever we fulfill them, whether or not others praise us. For example, practicing kin altruism may make us proud. Also, Pugh (1977, p. 231) suggested that being smiled at by a potential mate may be inherently uplifting. Sexual failure or jealousy may be deflating; male vervet monkeys that observed another male copulating experienced a fall in serotonin (Masters & McGuire, 1994; Pugh, 1977 p. 328) also proposed that solving a complex problem may trigger pride, as well as a feeling of intellectual satisfaction. Further, the psychological tendencies to react with anger when wronged and with guilt when wronging another may be cases of evolved values (Pugh, 1977, pp. 273–274); the reciprocity norm is universal. Thus pride and shame may be induced by certain evolutionary salient stimuli just as other affects, such as fear, interest, humor, sexual arousal, and hunger, can be induced by prepotent stimuli.

CONCLUSION

It would seem to be important to study all of the human emotions, including pride and shame, just as Darwin attempted to survey all of the human emotional expressions (Weisfeld, 1997). If we can develop an inventory, or ethogram, of the basic human emotions, we will have a general model of human behavior. This would tell us what the varieties of human pleasure and displeasure are, that is, the dimensions of human happiness. We could use it to evaluate comprehensively the feasibility and desirability of social policies. We could strive to develop societal values that would maximize our collective happiness, in the manner of the pre-Darwinian utilitarian philosophers. Adherence to such cultural values occurs mainly through the emotion of pride and shame. Thus this emotion provides a nexus between cultural and biological influences on behavior—another reason why recognizing this emotion is important. These values would operate mainly through approval and disapproval, by rewarding people psychically (through pride and shame) and materially for behavior that promoted the general welfare. However, because pride/shame is a relative, or zero-sum, system, there is no possibility of raising a society's average self-esteem to a point above average. Thus pride/shame is a means to a social end, but not an end in itself.

Pride/shame warrants intensive study because it is crucial to many important social processes, including mate choice, role allocation, development of skills, and reciprocal altruism. These are not purely "cultural" phenomena, but rather are grounded in pride/shame and other emotions. Moreover, by recognizing this emotion, we avoid the dubious distinction between rational (or cultural) and emotional (or biological) behavior: acting rationally usually means simply taking into account the social consequences of a given course of action, which we do largely by anticipating pride or shame.

REFERENCES

Asendorpf, J. (1990). The expression of shyness and embarrassment. In W. R. Crozier (Ed.), *Shyness and embarrassment: Perspectives from social psychology* (pp. 87–118). Cambridge: Cambridge University Press.

Barkow, J. H. (1989). *Darwin, sex, and status: Biological approaches to mind and culture*. Toronto: University of Toronto Press.

Barrett, K. C. (1995). A functionalist approach to shame and guilt. In J. P. Tangney & K. W. Fischer (Eds.), *Self-conscious emotions: The psychology of shame, guilt, embarrassment, and pride* (pp. 25–63). New York: Guilford.

Boardway, R. H., & Weisfeld, G. (1994, August). *Social dominance among English adolescent boys*. Poster presented at the International Society for Human Ethology Congress, Toronto, Canada.

Buss, A. (1980). *Self-consciousness and social anxiety*. San Francisco: W. H. Freeman.

Damasio, A. R. (1994). *Descartes' error: Emotion, reason, and the human brain*. New York: Grosset/Putnam.

Darwin, C. (1872). *The expression of the emotions in man and animals.* With commentaries by Paul Ekman. Reprint. New York: Oxford University Press, 1998.

Devinsky, O., Hafler, D. A., & Victor, J. (1982). Embarrassment as the aura of a complex partial seizure. *Neurology, 32*, 1284–1285.

Dong, Q., Weisfeld, G., Boardway, R., & Shen, J. (1996). Correlates of social status among Chinese adolescents. *Journal of Cross-Cultural Psychology, 27*, 476–493.

Edelmann, R. J. (1990). Embarrassment and blushing: A component-process model, some initial descriptive and cross-cultural data. In W. R. Crozier (Ed.), *Shyness and embarrassment: Perspectives from social psychology* (pp. 205–229). Cambridge: Cambridge University Press.

Edelmann, R. J., & Hampson, S. E. (1979). Changes in non-verbal behaviour during embarrassment. *British Journal of Social and Clinical Psychology, 18*, 385–390.

———. (1981a). Embarrassment in dyadic interaction. *Social Behavior and Personality, 9*, 171–177.

———. (1981b). The recognition of embarrassment. *Personality and Social Psychology Bulletin, 7*, 109–116.

Ellsworth, P. C., Carlsmith, J. M., & Henson, A. (1972). The stare as a stimulus to flight in human subjects: A series of field experiments. *Journal of Personality and Social Psychology, 21*, 302–311.

Fromme, D. K., & Beam, D. C. (1974). Dominance and sex differences in nonverbal responses to differential eye contact. *Journal of Research in Personality, 8*, 76–87.

Fuster, J. M. (1980). *The prefrontal cortex: Anatomy, physiology, and neuropsychology of the frontal lobe.* New York: Raven Press.

Ginsburg, H. J. (1980). Playground as laboratory: Naturalistic studies of appeasement, altruism, and the omega child. In D. R. Omark, F. F. Strayer, & D. G. Freedman (Eds.), *Dominance relations: An ethological view of human conflict and social interaction* (pp. 341–357). New York: Garland Press.

Greenberg, M. S., & Shapiro, S. P. (1971). Indebtedness: An adverse aspect of asking for and receiving help. *Sociometry, 34*, 290–301.

Izard, C. E. (1977). *Human emotions.* New York: Plenum Press.

Kahn, E. S., & Weisfeld, G. (1993, August). *Facial expressions that influence subjects' postural erectness.* Paper presented at the convention of the Human Behavior and Evolution Society, Binghamton, NY.

Landel, J., & Leary, M. R. (1992, March). *Social blushing as a face-saving display.* Paper presented at the meeting of the Southeastern Psychological Association, Knoxville, TN.

Lerner, M. J. (1970). The desire for justice and reactions to victims. In J. Macaulay & L. Berkowitz (Eds.), *Altruism and helping behavior* (pp. 205–229). New York: Academic Press.

Levin, J., & Arluke, A. (1982). Embarrassment and helping behavior. *Psychological Reports, 51*, 999–1002.

Mascolo, M. F., & Fischer, K. W. (1995). Developmental transformations in appraisals for pride, shame, and guilt. In J. P. Tangney & K. W. Fischer (Eds.), *Self-conscious emotions: The psychology of shame, guilt, embarrassment, and pride* (pp. 64–113). New York: Guilford.

Masters, R. D., & McGuire, M. T. (Eds.). (1994). *Neurotransmitter revolution: Serotonin, social behavior and the law.* Carbondale: Southern Illinois University Press.

Mazur, A. (1983). Hormones, aggression, and dominance in humans. In B. B. Svare (Ed.), *Hormones and aggressive behavior* (pp. 563–576). New York: Plenum Press.

McDougall, W. (1923). *Outline of psychology.* New York: Scribner's.

Panksepp, J. (1994). Evolution constructed the potential for subjective experience within the neurodynamics of the neomammalian brain. In P. Ekman & R. J. Davidson (Eds.), *The nature of emotion: Fundamental questions* (pp. 396–399). Oxford: Oxford University Press.

Passingham, R. (1993). *The frontal lobes and voluntary action.* Oxford: Oxford University Press.

Pugh, G. E. (1977). *The biological origin of human values.* New York: Basic Books.

Rosa, E., & Mazur, A. (1979). Incipient status in small groups. *Social Forces, 58,* 18–37.

Schneider-Rosen, K., & Cicchetti, D. (1991). Early self-knowledge and emotional development: Visual self-recognition and affective reactions to mirror self-images in maltreated and non-maltreated toddlers. *Developmental Psychology, 27,* 471–478.

Sueda, K., & Wiseman, R. L. (1992). Embarrassment remediation in Japan and the United States. *International Journal of Intercultural Relations, 16,* 159–173.

Tangney, J. P., & Fischer, K. W. (Eds.). (1995). *Self-conscious emotions: The psychology of shame, guilt, embarrassment, and pride.* New York: Guilford.

Trivers, R. L. (1971). The evolution of reciprocal altruism. *Quarterly Review of Biology, 46,* 35–57.

Weisfeld, G. E. (1980). Social dominance and human motivation. In D. R. Omark, F. F. Strayer, & D. G. Freedman (Eds.), *Dominance relations: An ethological view of human conflict and social interaction* (pp. 273–286). New York: Garland.

———. (1997). Discrete emotions theory with specific reference to pride and shame. In N. L. Segal, G. E. Weisfeld, & C. C. Weisfeld (Eds.), *Uniting psychology and biology: Integrative perspectives on human development* (pp. 419–443). Washington, DC: American Psychological Association.

Weisfeld, G. E., & Beresford, J. M. (1982). Erectness of posture as an indicator of dominance or success in humans. *Motivation and Emotion, 6,* 113–131.

Weisfeld, G. E., & Linkey, H. E. (1985). Dominance displays as indicators of a social success motive. In S. Ellyson & J. Dovidio (Eds.), *Power, dominance, and nonverbal behavior* (pp. 109–128). New York: Springer-Verlag.

Weisfeld, G. E., Muczenski, D. M., Weisfeld, C. C., & Omark, D. R. (1987). Stability of boys' social success among peers over an eleven-year period. In J. A. Meacham (Ed.), *Interpersonal relations: Family, peers, friends* (pp. 58–80). New York: Karger.

Zivin, G. (1977). Facial gestures predict preschoolers' encounter outcomes. *Social Science Information, 16,* 715–730.

Index

"i" indicates an illustration; "t" indicates a table.

About the Editors and Contributors

ROBIN ALLOTT has for many years pursued a special interest in the relation of language, gesture, and sound symbolism (as in his 1973 work, *The Physical Foundation of Language*). More recently, following retirement from government service as undersecretary in the United Kingdom Department of Industry, he has concentrated on the relation of language, vision, and motor organization in the central nervous system (developed more fully in his *The Motor Theory of Language Origin*, 1989).

LUCIO FERREIRA ALVES graduated in pharmacy and worked for several years in the areas of organic chemistry (phytochemistry) and chemical ecology. His current research interest is Darwinism and Marxism, with an emphasis on study of the collapse of Marxist regimes from a sociobiological perspective. Among his many works are *Química de Lepidopteros, Sociobiology versus Reducionismo Biológic, Chemical Ecology and the Social behavior of Animals, Bioethics and Environmental Education, The Rise of Science in an Endangered Planet*, and *Sociobiology, Marxism and Religion*.

MICHAEL BRADIE is Professor of Philosophy at Bowling Green State University in Bowling Green, Ohio. He has published numerous articles on the philosophy of science and epistemology. His most recent publications include several journal articles and a book on evolution and ethics, *The Secret Chain: Evolution and Ethics* (1995). He is currently working on a series of papers exploring the role of models and metaphors in scientific representation.

M. L. BUTOVSKAYA is Professor at the Institute of Cultural Anthropology, Russian State University for Humanities, Moscow, and Leading Research Scientist at the Russian Academy of Sciences Institute of Ethnology and Anthropology, Moscow. Her main interests are human ethology and primate social behavior and human evolution.

JOHN CONSTABLE read English at Magdalene College, Cambridge, receiving his Ph.D. in 1993. He has edited *The Selected Letters of L. A. Richards* (1992) and *Critical Essays on William Empson* (1993). He is currently Associate Professor in the Faculty of Integrated Human Studies, Kyoto University, where he is engaged in empirical research into the epidemiology of cultural forms. His current project concerns the universality of metrical structures, particularly verse.

PETER A. CORNING serves as Director of the Institute for Study of Complex Systems and works as a consultant. He has authored more than 100 papers and other scholarly communications and is president-elect of the International Society for Systems Sciences and Treasurer of the International Society for Bioeconomics. Dr. Corning is also a contributing member of the International Society for Human Ethology, the Human Behavior and Evolution Society, and the European Sociobiological Society and was recently the recipient of a research fellowship of the Collegium Budapest institute for advanced study.

CHARLES ELWORTHY teaches at the Otto Suhr Institute, the political science department of the Free University of Berlin, and is director of the European Academy in Schloss Wartin. His work centers on the application of evolutionary psychology to the social sciences, and he is the author of *Homo Biologicus: An Evolutionary Model for the Human Sciences* (1993). His current research interests include the application of evolutionary psychology to the new institutional economics, in particular the way cognitive programs facilitate the resolution of collective action problems.

HARALD A. EULER is a Professor of Psychology at the Universitat Gesamthochschule Kassel. He has published on behavior modification, aggression, and topics in evolutionary psychology, which has been his main research interest for several years. He is currently working on psychological adaptations of intergenerational family relationships, psychological aspects of sperm competition, stuttering therapy, and relationships between children and horses, and is co-editing a handbook of emotions.

A. G. KOZINTSEV is Leading Research Scientist at the Museum of Anthropology and Ethnography in St. Petersburg, Russia. He earned a Doctor of Sciences (History) from the Institute of Ethnography at Leningrad University and was a postdoctoral fellow in the Academy of Sciences (USSR). Dr. Kozintsev is the author of more than 140 publications concerning physical anthropology, primate and human ethology, the population history of Eurasia, and the evolution of laughter.

ADA LAMPERT is a Senior Lecturer in the Department of Behavioral Sciences at the Ruppin Institute, Israel. Her research combines the psychology of emotions, sex differences, love, and the family with genetic, hormonal, and evolutionary thinking. Praeger published her book *The Evolution of Love* in 1997.

PETER MEYER is teaching sociology at the Institute of Socioeconomics, University of Augsburg, Germany. His main fields of research are biosociology, conflict theory, and human sociobiology. Author of *Evolution und Gewalt* (1981), he has also published a number of works in English, including "The Problem of Certainty in Human Communication: An Evolutionary View" in *Origins of Semiosis, Sign Evolution in Nature and Culture* (1994).

A. J. NABULSI works in Hamburg as an associate researcher. His main scientific interest is the biology of the Jordanian population in past and present times. He has published a number of studies, especially on Jordan, in the field of population genetics, and as an anthropologist, he joined a number of excavations in Jordan, Germany, Gaza (Palestine), and Bahrain. Since 1995 he has headed the Khirbet Es-Samra Byzantine Cemetery Excavation Project in Jordan.

JOHN S. PRICE served as a member of the Scientific Staff of the MRC Psychiatric Genetics Research Unit London, where he developed his interest in the application of evolutionary biology to psychiatry. He developed a psychiatric service for the National Health Service and served as a consultant psychiatrist to the NHS in the newly created city of Milton Keynes. Since retiring from the NHS in 1991, he has held academic posts at the University of Otago in New Zealand and the University of Tasmania in Australia. His book *Evolutionary Psychiatry: A New Beginning*, co-authored with Anthony Stevens, was published in 1996.

J. PHILIPPE RUSHTON is Professor of Psychology at the University of Western Ontario. He holds two doctorates from the University of London and has published nearly 200 articles in scientific journals as well as six books, including a best-selling introductory psychology textbook. His latest book, *Race, Evolution, and Behavior*, was published in 1995. Rushton has been elected a Fellow of the John Simon Guggenheim Foundation, the American Association for the Advancement of Science, and the American, British, and Canadian Psychological Associations.

NORMA J. SCHELL received her Ph.D. in clinical psychology from the University of Detroit Mercy in 1993. She is now a school psychologist and a practicing clinical psychologist in Michigan.

POUWEL SLURINK is a member of the Department of Philosophy, University of Nijmegen, the Netherlands. His lifelong interest in field biology and nature photography form the background to his naturalistic approach to philosophy. He has written a series of papers on evolutionary epistemology, evolutionary ethics, and the philosophy of mind and a Dutch book on evolution and morality. His current focus is on the origin of culture, and he has integrated his papers on this subject in an (English-written) volume.

DAVID SMILLIE served as a developmental psychologist for forty years, serving as a Professor at New College, University of South Florida, during the last twenty-five of those years. Since 1993 he has been a Visiting Professor in the Department of Zoology at Duke University where he has been pursuing a study of social evolution. He has published articles on both fields in a variety of journals and books.

SEAN STANTON is currently Director of the Florida Psychopharmacology Research Center in Orlando. He holds a B.A. from the University of Cincinnati and is formerly Research Associate at the University and Ohio Department of Mental Health.

JOHAN M.G. VAN DER DENNEN studied behavioral sciences (psychology, ethology, psychobiology, sociobiology, evolutionary ecology) at the University of Groningen, and is a researcher at the Section Political Science of the Department of Legal Theory, University of Groningen, the Netherlands. He has published more than 100 works on all aspects of human and animal aggression, sexual violence, neuro- and psychopathology of human violence, political violence, theories of war causation, macroquantitative research on contemporary wars, ethnocentrism, and the politics of peace and war in preindustrial societies. *The Origin of War: The Evolution of a Male-Coalitional Reproductive Strategy*, was published in 1995. He is secretary of the European Sociobiological Society (ESS).

TATU VANHANEN is a political scientist who has studied the problem of democratization and attempted to apply evolutionary principles to the study of politics. His current comparative study concerns ethnic conflicts and ethnic nepotism. He served as a docent of political science at the University of Helsinki from 1972 to 1995. His most recent books include *Politics of Ethnic Nepotism: India as an Example* (1991), *On the Evolutionary Roots of Politics* (1992), *Prospects of Democracy: A Study of 172 Countries* (1997), and *Prospects for Democracy in Asia* (1998).

MICHAEL J. C. WALLER has taught at the postgraduate level at the University of Staffordshire and has worked in both the public and private sectors of personnel management, organizational development, and internal consultancy in the United Kingdom. Among his research interests is the difficulty of reconciling Dawkins' notion that all organisms are ruthlessly honed survival machines for their genes with the growing body of evidence that prolonged periods of unresolved stress have psychosomatic effects, with no apparent survival value and all too obviously deleterious consequences.

CAROL C. WEISFELD is Professor of Psychology at the University of Detroit Mercy, where her research interests focus on determinants of marital satisfaction in different cultural settings and ethological analyses of male-female interactions. She is co-editor, with Nancy Segal and Glenn Weisfeld, of *Uniting Psychology and Biology* (1997).

GLENN WEISFELD is an Associate Professor of Psychology at Wayne State University. In 1997, with Nancy L. Segal and Carol C. Weisfeld, he published *Uniting Psychology and Biology: Integrative Perspectives on Human Development*. Dr. Weisfeld has also published several theoretical papers on human adolescence and *Evolutionary Principles of Human Adolescence* (1999). He formerly edited the *Human Ethology Bulletin*, and is currently engaged in cross-cultural research on correlates of marital satisfaction.

BARBARA WEITZEL received her M.A. in ethnology and anthropology from the University of Göttingen, Germany, and is currently a Member of the Parliament of the state of Hesse.

DENNIS WERNER is Professor Titular at the Universidade Federal de Santa Catarina, Florianópolis, Brazil. He is author of *Amazon Journey: An Anthropologist's Year among Brazil's Mekranoti Indians, Introducão às Culturas Humanas: Comida, Sexo, Magia e Outros Assuntos Antropológicos*, and *O Pensamento de Animais e Intelectuais: Evolucão e Epistemologia*, as well as dozens of scholarly articles in U.S. and Brazilian journals.

DANIEL R. WILSON is Professor of Psychiatry at the University of Cincinnati and Medical Director of the Lewis Center of Ohio Department of Mental Health. He has authored more than one hundred scholarly communications, mostly in evolutionary medicine and psychopharmacology. He is a member of the APA Council on International Affairs and has lectured throughout the Americas, Europe, India, and Asia. Dr. Wilson was educated at Yale, Iowa, Harvard, and Cambridge.

SANDRA WILSON is a management consultant and part-time Research Associate at the Ohio Department of Mental Health. She holds a B.A. from the University of Massachusetts and was Visiting Member of Clare Hall, Cambridge.

ISBN 0-275-96436-1

90000>

9 780275 964368

HARDCOVER BAR CODE